THE HISTORICAL DEVELOPMENT OF DIPTERA

THE HISTORICAL DEVELOPMENT OF DIPTERA

by

Boris Rohdendorf

Translated from the Russian
by
J. E. Moore and I. Thiele

Edited by
Brian Hocking   Harold Oldroyd
George E. Ball

The University of Alberta Press

1974

First English language edition published by
The University of Alberta Press
Edmonton, Alberta, Canada
1974
Original Russian language edition published
by "Nauka" as volume 100, Transactions of
the Institute of Paleontology, Academy of Sciences
of the USSR, Moscow, 1964.

ISBN 0-88864-003-X

Printed in Canada by Printing Services
of The University of Alberta

This book is dedicated to the illustrious memory of
Vladimir Nikolayevich Beklemischev, the greatest
Soviet zoologist, whose works and advice were
of great assistance to the author.

CONTENTS

EDITORS' NOTE

This English version of Rohdendorf's book on the Historical Development of the Diptera is based on a literal translation from the Russian by Dr. J. E. Moore and Dr. I. Thiele. We have sought to retain the Russian flavour of the work while editing for terseness and contemporary terminology of English speaking dipterists. It has been necessary to reproduce the illustrations direct from the published Russian edition; handwork has been restricted to reinforcing lines and tones which failed to reproduce.

We are happy to acknowledge the courtesy and cooperation of the author, the advice of the many dipterists and students of Russian we have consulted, the technical assistance of the staff of the University of Alberta Press, typing of the final manuscript by Mrs. Susan Hamilton and Mrs. Frances Rowe, and the technical assistance of Mrs. Margaret Abraham in editing copy and in final layout of proof prior to printing.

Brian Hocking
Department of Entomology
University of Alberta
Edmonton, Canada T6G 2E3

Harold Oldroyd
British Museum (Natural History)
Cromwell Road
London, S. W. 7

George E. Ball
Department of Entomology
University of Alberta
Edmonton, Alberta T6G 2E3

INTRODUCTION

In this study the historical development of the dipterous insects is examined as
a process dependent on modifications of the conditions of life. The resultant outline
of the historical development of the order is the first appraisal of the history of this
vast group of insects. Based on the investigation of rich fossil remains and the study
of all available information on the contemporary fauna, this review had as its goal
the illumination of the paths of historical development; a chief consideration was
the functional significance of the observed morphological characteristics. This
allowed a consideration of the development of different groups and an attempt
at an analysis of the general problem of modifications of organisms in time as a
process of development.

It is necessary to point out as an important condition of the success of such
an investigation the obligation to study the features of the group in the contemporary
fauna. Inasmuch as a system of organisms is the result of the phylogenetic development of the group, its formation — the abundance or scarcity of taxa of different
ranks, their proximity or separation — reflects the course of historical development.
The evaluation of the character of a system therefore is exceptionally significant:
this historical evidence is valuable for clarification of the features of the phylogenetic process, particularly considering the paucity of direct historical documents,
fossil remnants. In the very elementary determination of the character of a system,
carried out by considering the corresponding features of the separate groups, I
thought a study of the numbers of species in different taxa necessary.

Although the functional value of the morphological characteristics is the chief
goal of the study it is frequently a difficult and sometimes an impracticable task.
However, in the structure of the wing for instance, (Rohdendorf, 1951, 1958-59)
a whole series of features which at first appear to be functionally inexplicable
can be reliably explained with a more intent and thorough examination. We are
still far from a full knowledge of the main characteristics of the organization of
all groups of Diptera, but already light may be thrown upon the chief distinguishing
features of the principal superfamilies and families of the order with an adequate
degree of authenticity and completeness.

The first part of this study gives the data on the relation of the Diptera to
other orders and an account of the causes of their separation and the special
features of the organization and structure of contemporary forms. The consideration of characteristics of the different groups of Diptera is developed according to a definite plan. First, the principles which have had prime attention for
revealing the peculiarities of life and development of present day organisms are
explained. The features of a system are established on the basis of a study of all
the systematic investigations known to me, which in a series of cases are critically
reviewed and, as a rule, are modified by an evaluation of the sum of the peculiarities of the given group. After a survey of the modern organization we analyze the
characteristics of insect organization directing attention to the functional and
ethological significance of given structural features. The account of these data
on the basis of the characteristics of the different groups of Diptera is completed

by an explanation of the trends of historical development of the group.

The second part is devoted to a survey of the existing data on the dipterous insects of the geological past. This part consists of several sections which cast light upon the Mesozoic and Tertiary faunas. The outlines of the Mesozoic Diptera based on the preeminence of materials recovered in the territory of our country have a special significance. The most important first section is devoted to a description of the vast Upper Triassic fauna, unknown until recently, first discovered in central Asia. The sections of the second part comprise the most complete account of the Triassic and Jurassic faunas and are based on the diverse materials investigated by the author.

The third part is devoted to a consideration of the phylogenetic relationships of the different groups of the order Diptera and opens with a critical review of the existing literature. The concrete phylogenesis of the individual groups of Diptera under consideration is illustrated by schemes and ethological characteristics of the phylogenetic changes which they have passed through.

The fourth and final part generalizes on the theoretical results of the whole investigation. The first chapter of the section considers the methodological foundations of investigation — the value of the evidences of historical development being obtained by different branches of biological science and the importance of the study of a system of organisms as the basis of a solution of the problem. The second chapter of the section examines the historical development or organisms as a dialectical process, analyzing examples of concrete inherent conflicts. In this chapter the inconsistencies in the development of organisms are assessed.

Since this is the first work of its kind numerous faults have not been positively eliminated. In further works of the Laboratory of Arthropoda of the Paleontological Institute of the Academy of Sciences of the U.S.S.R. for other groups of arthropods a similar kind of study will be carried out, of course on a higher level.

I thank my colleagues at the Paleontological Institute of the Academy of Sciences, U.S.S.R. for numerous suggestions and critical observations which assisted me in the compilation of the work. I thank particularly, for reading and criticizing the manuscript, the esteemed scientist Professor Dimitri Michailovich Fedotov, and Doctor of Biological Science Viktor Nikolaevich Shimanskom: their advice undoubtedly improved many parts of the work and assisted in the avoidance of many errors. I am also very grateful to the artists Vitali Ivanovich Dorofeev and Tatyan Leonidov Savranska for the faultlessly made illustrations. The greater part of the work in the preparation of the manuscript for press was done by a laboratory worker in the Laboratory of Arthropoda of the Paleontological Institute, Irina Dmitrievna Sukacheva.

Table 1

Abbreviations of the taxon names of Diptera used in the diagrams and schemes

Roman numerals denote infraorders

I — Nymphomyiomorpha
II — Deuterophlebiomorpha
III — Blephariceromorpha
IV — Tipulomorpha
V — Bibionomorpha
VI — Asilomorpha
VII — Musidoromorpha
VIII — Phoromorpha

IX — Termitoxeniomorpha
X — Myiomorpha
XI — Braulomorpha
XII — Nycteribiomorpha
XIII — Streblomorpha
XIV — Dictyodipteromorpha
XV — Diplopolyneuromorpha

Two Latin letters denote superfamilies

An — Anthomyiidea
As — Asilidea
Bb — Bibionidea
Bl — Bolitophilidea
Bm — Bombyliidea
Br — Borboridea
Cc — Cecidomyiidea
Ch — Chironomidea
Cl — Culicidea
Cn — Conopidea
Cr — Chloropidea
Di — Dictyodipteridea
Dr — Drosophilidea
Dx — Dixidea
Dy — Dyspolyneuridea
Em — Empididea
Eo — Eopolyneuridea
Ep — Eoptychopteridea
Fn — Fungivoridea
Ga — Gastrophilidea
Gl — Glossinidea
He — Heleomyzidea
Hi — Hippoboscidea
Hy — Hyperpolyneuridea
Me — Mesophantasmatidea

Mu — Muscidea
Oe — Oestridea
Or — Orphnephilidea
Pc — Pachyneuridea
Ph — Psychodidea
Pl — Platypezidea
Pm — Phragmoligoneuridea
Pr — Protoligoneuridea
Ps — Psilidea
Py — Pleciodictyidea
Rp — Rhyphidea
Rt — Rhaetomyiidea
Sc — Scatopsidea
Sh — Sarcophagidea
Sm — Somatiidea
Sp — Sapromyzidea
Sr — Syrphidea
St — Stratiomyiidea
Ta — Tanyderophryneidea
Tb — Tabanidea
Tc — Tachinidea
Tp — Tipulidea
Tr — Trypetidea
Ty — Tipulodictyidea

Three Latin letters denote families

Aca — Acanthomeridae
Acd — Acridomyiidae
Acr — Acroceridae
Agz — Agromyzidae

All — Allactoneuridae
Anh — Anthomyzidae
Ant — Anthomyiidae
Api — Apioceridae

Arh — Archisargidae
Arn — Architendipedidae
Art   — Architipulidae
Arz — Archizelmiridae
Asi   — Asilidae
Bib   — Bibionidae
Ble   — Blephariceridae
Bol   — Bolitophilidae
Bom — Bombyliidae
Bor   — Borboridae
Bra   — Braulidae
Cal   — Calliphoridae
Cec   — Cecidomyiidae
Cel   — Celyphidae
Cer   — Ceratopogonidae
Cha   — Chaoboridae
Chi   — Chironomidae
Chl   — Chloropidae
Coe   — Coenomyiidae
Col   — Coelopidae
Con — Conopidae
Cor   — Cordyluridae
Cra   — Cramptonomyiidae
Crp   — Ceroplatidae
Cry   — Cryptochaetidae
Cul   — Culicidae
Cut   — Cuterebridae
Cyl   — Cylindrotomidae
Cyp — Cypselomatidae
Cyr   — Cyrtosiidae
Dex — Dexiidae
Dia   — Diadocidiidae
Dic   — Dictyodipteridae
Dio   — Diopseidae
Dip   — Diplopolyneuridae
Dit   — Ditomyiidae
Dix   — Dixidae
Dol   — Dolichopodidae
Dro   — Drosophilidae
Dxm — Dixamimidae
Dys   — Dyspolyneuridae
Egi   — Eginiidae
Emp — Empididae
Eol   — Eolimnobiidae
Eom — Eomyiidae
Eop   — Eopleciidae
Eor   — Eopolyneuridae
Eos   — Eostratiomyiidae
Eot   — Eoptychopteridae

Eph   — Ephydridae
Eyr   — Eurychoromyiidae
Fan   — Fanniidae
Fun   — Fungivoridae
Fut   — Fungivoritidae
Gas   — Gastrophilidae
Glo   — Glossinidae
Hel   — Heleomyzidae
Hes   — Hesperinidae
Het   — Heteropezidae
Hil   — Hilarimorphidae
Hip   — Hippoboscidae
Hpr   — Hyperoscelididae
Hyp   — Hypodermatidae
Hyy   — Hyperpolyneuridae
Les   — Lestremiidae
Leu   — Leucostomatidae
Lim   — Limoniidae
Lon   — Lonchaeidae
Lpt   — Lonchopteridae
Lyg   — Lygistorrhinidae
Mac — Macroceridae
Man — Manotidae
Meg — Megamerinidae
Mes   — Mesophantasmatidae
Mic   — Micropezidae
Mil   — Milichiidae
Msd — Musidoromimidae
Mus — Muscidae
Myc — Mycetobiidae
Myd — Mydaidae
Nem — Nemopalpidae
Nms — Nemestrinidae
Not   — Nothybidae
Oes   — Oestridae
Olb   — Olbiogastridae
Oli   — Oligophryneidae
Orp   — Orphnephilidae
Oti   — Otitidae
Pac   — Pachyneuridae
Pal   — Palaeophoridae
Pap   — Palaeopleciidae
Pas   — Palaeostratiomyiidae
Pax   — Paraxymyiidae
Pen   — Penthetriidae
Per   — Perissommatidae
Pha   — Phasiidae
Phl   — Phlebotomidae
Phr   — Phragmoligoneuridae

Pio — Piophilidae
Pip — Pipunculidae
Pla — Platypezidae
Plc — Pleciodictyidae
Plf — Pleciofungivoridae
Plm — Pleciomimidae
Plt — Platystomatidae
Pro — Protendipedidae
Prt — Protobrachycerontidae
Pry — Protocyrtidae
Psi — Psilidae
Psy — Psychodidae
Ptb — Protolbiogastridae
Ptl — Protoligoneuridae
Ptm — Protomphralidae
Ptp — Protopleciidae
Pts — Protoscatopsidae
Pty — Ptychopteridae
Pyr — Pyrgotidae
Rac — Rachiceridae
Rha — Rhagionidae
Rhe — Rhaetomyiidae
Rhg — Rhagionempididae
Rhi — Rhinophoridae
Rho — Rhopalomeridae
Rhy — Rhyphidae
Sap — Sapromyzidae
Sar — Sarcophagidae
Sca — Scatopsidae
Scd — Sciomyzidae

Sce — Scenopinidae
Sci — Sciaridae
Sco — Sciadoceridae
Sep — Sepsididae
Sim — Simuliidae
Sin — Sinemediidae
Sol — Solvidae
Som — Somatiidae
Sta — Stackelbergomyiidae
Stb — Streblidae
Str — Stratiomyiidae
Sty — Stylogastridae
Syn — Synneurontidae
Syr — Syrphidae
Sys — Systropodidae
Tab — Tabanidae
Tac — Tachinidae
Tan — Tanyderidae
Tch — Tachiniscidae
The — Therevidae
Tip — Tipulidae
Tnp — Tanyderophryneidae
Tpd — Tipulodictyidae
Tpl — Tipulopleciidae
Tri — Trichoceridae
Try — Trypetidae
Usi — Usiidae
Vil — Villeneuviellidae
Xyl — Xylophagidae

PART I

CHARACTERISTICS OF THE DIPTERA

Derivation of the order Diptera

The interrelations of Diptera with other groups

The order Diptera is a member of the vast group of neopterous insects with complete metamorphosis, which make up the cohort Oligoneoptera. This major systematic taxon combines a number of superorders which possess certain special features of the structure of the jugal area of the wings, and which was first characterized by Martynov (1923), who called it a 'subdivision'.

The principal characteristic of the Oligoneoptera is the perfecting of development through the elaboration of complete metamorphosis. This feature has enabled insects to take possession of many concealed habitats in which their larvae could live, and thus enormously increased their distribution and ecological range. It is still not known when the Oligoneoptera arose. The earliest representatives of this cohort that are known to us are scorpion flies of the family Metropatridae, which were already present in the lowest strata of the Middle Carboniferous, but the greatest development and predominance of the group did not come about until the Permian. By the beginning of the Mesozoic they had become the most numerous and diversified of winged insects.

Oligoneoptera fall into four sharply distinct superorders: the coleopteroid, neuropteroid, mecopteroid and hymenopteroid complexes. These differ from each other in the development of special methods of feeding (Hymenopteroidea and some Mecopteroidea and Coleopteroidea), in transformation and development (Mecopteroidea and Hymenopteroidea), in evolution of special integumentary features (Coleopteroidea), and in the perfecting of the instincts of social life, and care of offspring (Hymenopteroidea).

In the fauna of the present day, Mecopteroidea are the most numerous and diversified superorder, with nearly 300,000 species, divided among six orders: Mecoptera, Trichoptera, Zeugloptera, Lepidoptera, Diptera, and Siphonaptera. Collectively, Mecopteroidea form a distinctive group, with characteristic methods of feeding, transformation, and flight. The larvae are usually wormlike, and sometimes legless. A particularly characteristic feature of Mecopteroidea is the specialization of the mouthparts of the adult insects into a proboscis of varied structure, adapted to the sucking of liquid food, and sometimes even to piercing the surface of the nutritive substrate. Larvae, too, have modified mouthparts, and an associated development is the perfecting of the salivary glands, the secretions of which often possess chemical properties which permit external digestion of the food before it is sucked into the gut (Diptera and some others).

Flight is very important in the biology of these insects, and the number of wingless and flightless Mecopteroidea is limited; in fact, apart from fleas (Siphonaptera) there are only a few brachypterous and apterous forms in the other orders. In Mecopteroidea the wings seldom have a protective function, since the insects can usually fly well, or have powerful running legs.

Mesozoic Paratrichoptera – specialized scorpion flies

An examination of the interrelations and characteristics of the whole superorder Mecopteroidea is beyond the scope of the present work which is concerned with the order Diptera. However, we are interested in the nearest relatives of the Diptera, the order Mecoptera (scorpion flies), and in particular in its suborder the Mesozoic Paratrichoptera which, from their wing venation, seem to have been closest to the dipterous forms.

Before we analyze the morphological resemblances between Paratrichoptera and Diptera, we must realize how little we know about these fossil insects apart from their wings. The only impression of an entire insect so far known is a single specimen from the Jurassic of Karatau (*Pseudopolycentropus latipennis* Martynov) (fig. 1). Such a lack of data is extremely disappointing, because the distinctions between the living orders of Mecopteroidea are only partly based on wing characters. There are far more substantial differences in their development and feeding, especially in the mouthparts of the adult, and in larval structures. Lack of knowledge of the metamorphosis and the mouthparts of the Paratrichoptera means that any conclusions about the evolution of these insects, and their differences from Diptera, can be only tentative.

The evolution of the Diptera was characterized by the perfection of feeding methods and metamorphosis, in parallel with the improvement of flying abilities. Feeding was improved by the development of specialized mouthparts, in the form of a characteristic suctorial proboscis, able to pierce a substratum in search of

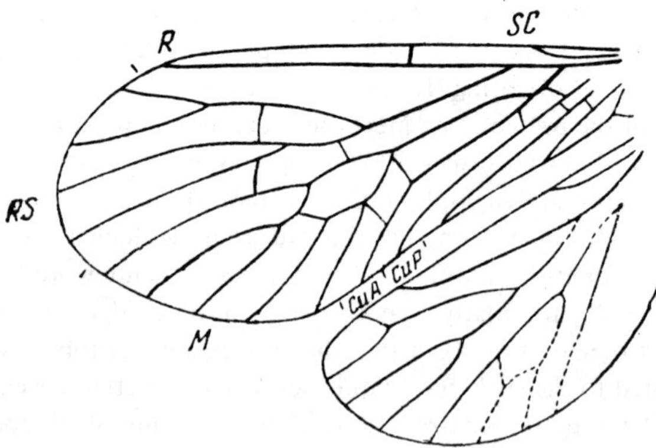

Fig. 1. *Pseudopolycentropus latipennis* Martynov (Mecoptera: Polycentropidae). Wings. Length of forewing 8 mm. Middle Jurassic of Karatau. (After Martynov, 1927.) Standard abbreviations.

food, and accompanied by the perfecting of glandular structures which allowed
some degree of external digestion, so that the nutrient could be partially liquefied
and chemically decomposed before being sucked up. This process is seen not only
in the imagines of living orders, but also in their larvae, and so extends to all
stages of development of the insects.

No less characteristic of the evolution of Diptera is the perfection of
development and metamorphosis, leading in the first place to a larva with a
wormlike body form and reduction of the thoracic legs. This was a very
advantageous evolution, which enabled the larva to live, move, and feed in
nourishing substrata which could be very moist. The development of an active,
apodous, wormlike larva is very characteristic of the Diptera, and has been one of
their principal evolutionary assets, setting them apart from their mecopteroid
ancestors.

Finally, a very important aspect of the evolution of Diptera has been the
concentration of flight functions into the mesothorax, relying entirely upon the
forewings. The tendency towards a two-winged condition is by no means confined
to the Diptera (Rohdendorf, 1943, 1949) and many other groups of
Mecopteroidea show it in various forms. Thus many Lepidoptera have the
hindwings sharply reduced in size (e.g. Sphingidae, Syntomididae, etc.) leading up
to their complete disappearance in the genus *Macrochila*, and in some Trichoptera,
e.g. Bajkalinini. Similar striking resemblances to Diptera occurred among extinct ·
Mesozoic Mecoptera of the suborder Paratrichoptera, the hindwings of which
were sometimes greatly reduced (Martynov, 1937). Thirteen genera, belonging to
seven families, have been described, differing from Mecoptera and Trichoptera in
their reduced venation and approaching the Diptera in the important character
of the reduced hindwing.

Until recently, the only hint of a trend towards a two-winged condition in
Paratrichoptera was given by *Pseudopolycentropus latipennis* Martynov (fig. 1), in
which the hindwings were approximately the same length as the forewings, and
only insignificantly narrower; such a difference is very common in Lepidoptera,
and of course is a long way from the complete reduction that we see in Diptera. A
far greater development of the dipterous condition was reached in the family
Pseudodipteridae (*Pseudodiptera gallica* Laurentiaux, fig. 2), in which the
hindwings were reduced to narrow blades not more than a quarter as long as the
forewings. The general appearance of this fly immediately suggests some
two-winged fly of the superfamily Tabanidea.

The development of a two-winged condition, and reduction of the venation of
the forewings, are both features which suggest a strong degree of probability that
the Paratrichoptera and the Diptera evolved along parallel lines. In both cases
flight was improved by the tendency to concentrate flying functions into a single
pair of wings. Costalization of the wing – i.e. a pronounced strengthening of the
fore margin – which is so clearly seen in many Paratrichoptera, particularly in
Pseudodiptera, as well as the relatively large mesothorax, both point to
intensifying the speed and power of the wing stroke.

Such a similarity in the structure of the wing in Paratrichoptera and Diptera

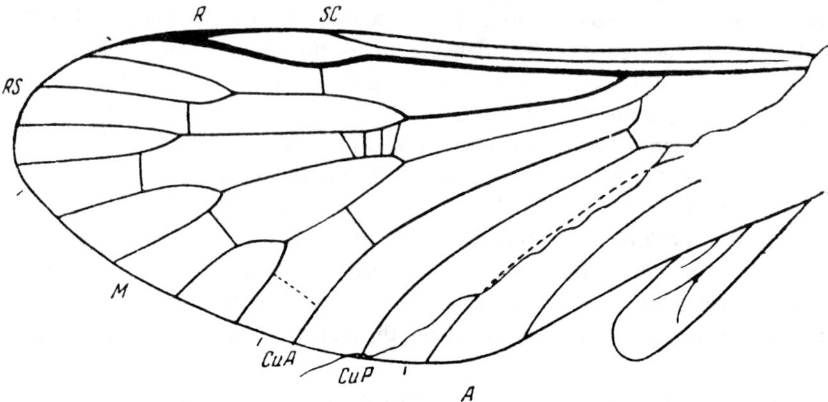

Fig. 2. *Pseudodiptera gallica* Laurentiaux (Mecoptera: Pseudodipteridae). Wings. Length of forewing 10 mm. Triassic of France. (Original.) Standard abbreviations.

gives rise to a legitimate speculation whether these two orders might not be allied; could they be ancestor and descendant? To decide this question we need to return to a comparison of the special features of the two orders, and palaeontological evidence is confined to the structure of the wings. The wing venation of Diptera has only one special feature that is readily observed in fossils: the way in which the two cubital veins (Cu_1, Cu_2) run closely together, with Cu_2 attenuated, so that the two together form a characteristic fold which separates the anal lobe of the wing from the rest. This approximation of the two cubital veins was long ago noted by palaeoentomologists, among whom Martynov, for example, wrote (1937:36) of it as the predominant distinctive characteristic of dipterous wings, enabling them to be distinguished from those of other Mecopteroidea.

These closely approximated cubital veins in Diptera have a functional importance as a longitudinal fold at the base of the wide anal lobe of the wing membrane. It is known that the wings of insects are subdivided into different sections, and that the mechanical operation of the wing must be affected in some way, but the details are still little understood. The anal lobe of Diptera is such a subdivision, which possesses great flexibility, and the cubital fold is an axis about which this anal lobe can vibrate. One must assume that this feature is beneficial, since it has been evolved by Diptera, and distinguishes them from related Mecopteroidea, especially from Paratrichoptera.

The important mechanical significance of the cubital fold is demonstrated if examined under polarized light, when Cu_2 is found to have different optical properties from Cu_1. In contrast to nearly all other veins, Cu_2 is anisotropic, and thus must have a different molecular structure from the normal fibrillar cells of the other veins.

Those wings of Paratrichoptera that are very similar in structure to Diptera, e.g. *Pseudodiptera gallica* Laurentiaux (fig. 2), or *Ptychopteropsis mirabilis* Martynov (fig. 3), lack a cubital fold, both cubital veins being equally strong and widely separated from each other. On the other hand, the mechanical specialization of these wings is quite complete, and as well advanced as it is in many Diptera. Thus, the degree of costalization in *Pseudopdiptera* is much greater than it is in many Tipulidea (Diptera), while in *Ptychopteropsis* the radial veins are much further reduced than they are, for instance, in Tanyderidae, Psychodidae or Culicidae. It is particularly impressive to note how specialized are these forms of Paratrichoptera in comparison with Upper Triassic Diptera, such as species of the families Hyperpolyneuridae, Diplopolyneuridae, Dictyodipteridae, which already had a clearly formed cubital fold, but which still retained an even more archaic wing venation, with little costalization, multiple branches of the radial and medial systems, and even remnants of the archedictyon.

The cubital fold persists in the vast majority of Diptera, and disappears only in the most specialized wings, when the posterior branch of the median vein has become the axis of vibration of the posterior area of the wing. This is so, for example, in the superfamily Muscidea, which possess wings of the high-lift type, with reduced venation (Rohdendorf, 1951, 1958-9).

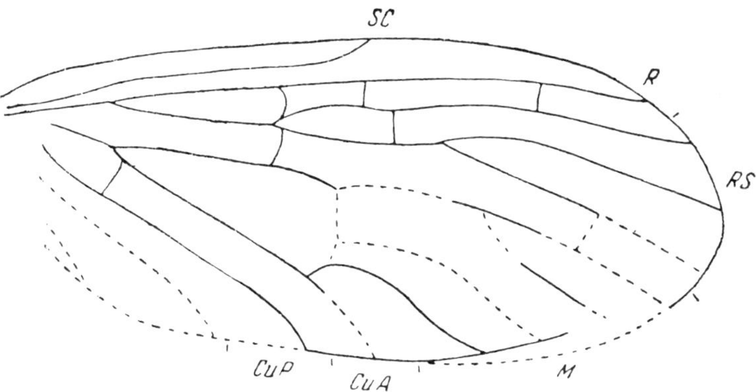

Fig. 3. *Ptychopteropsis mirabilis* Martynov (Mecoptera: Ptychopteropsidae). Wings. Length 11 mm. Lower Jurassic of central Asia. (After Martynov, 1937.) Standard abbreviations.

To conclude this review of venational features, with special emphasis on the cubital veins, it remains to mention some Mecopteroidea that have the two wings alike, but which also show the well-known approximation of the cubital veins in both wings. First in this group is the genus *Permotipula*, from the Upper Permian of Australia, a representative of the peculiar family Permotipulidae of the order Paratrichoptera. The author of this genus, Tillyard (1929, 1937), describing the venation, draws attention to peculiarities in the radial, medial, and cubital fields, and assumes that this form may be ancestral to the Diptera. It is impossible to agree with this because, although the advanced specialization of the venation actually suggests certain features of the Diptera, including the similar (but by no means homologous) approximation of the two cubital veins, yet on the other hand the presence of two pairs of almost homonomous wings separates these Permian insects completely from the Diptera, and indicates that they belong to the Paratrichoptera, or even to the Mecoptera. On the basis of these two forms, Tillyard established a new order of Mecopteroidea, calling it the Protodiptera, and regarding it as the ancestral group of all the Diptera. Of course this action was premature, and ill-founded. Everything that is known indicates clearly that there is no direct connection between Diptera and Permotipulidae, which seem to be a peculiar group of scorpion flies, having acquired many of the special features of dipterous wing venation a long time ago, far back in Permian times, but yet remaining four-winged insects.

Another mecopteroid which had the cubital veins approximated was *Permotipula patricia*, which Tillyard first described from an impression of a single wing — apparently a hindwing — from the Australian Upper Permian. As shown by the work of O.M. Martynova (1948), this wing was misidentified by Tillyard, and really belongs to a representative of a completely different, specialized genus of scorpion flies, *Robinjohnia* O. Martynova. This wing, which is constricted in the main part, carries four branches of the radial, and four branches of the medial veins, but has the cubital veins parallel and approximated, or more correctly, merging. The wing is 5 mm long. Thus this species has the appearance of belonging to the Diptera, but is easily distinguished by the way in which the two cubital veins come together, as well as by the large size of the wing. These features sharply distinguish *Robinjohnia* from the Diptera, especially from the Triassic Dictyopdipteridae, and preclude the possibility of its being directly ancestral.

The main features which separated the Diptera from other orders

By studying the differences between one mecopteroid order and another it is possible to draw conclusions about their main lines of evolution, and hence about the divergences that segregated them. Even though we limit our studies to the special characteristics of the one order Diptera, and its distinctions from the related Mecoptera and Paratrichoptera, we can see that the data on fossil insects are very different from those available on living ones. From the study of living flies we can build up a relatively complete picture of the organization of Diptera, whereas the fossil remains allow us only to come to conclusions about the structure of their wings, and to some extent about their powers of locomotion. Hence the characteristics, as well as the evolution, of different orders of insects can only be worked out on the basis of a close study of both fossil and living representatives.

Diptera differ from the related mecopteroids on the one hand by the wormlike form of their larvae, and on the other by the development of the two-winged condition, which is associated with extremely rapid wing movements. These features in turn indicate, on the one hand, a larval habitat in a mass of food material within which movement was both restricted and difficult; and on the other hand, an increased activity of the winged phase, and an adult life in which flight played an increasingly important part.

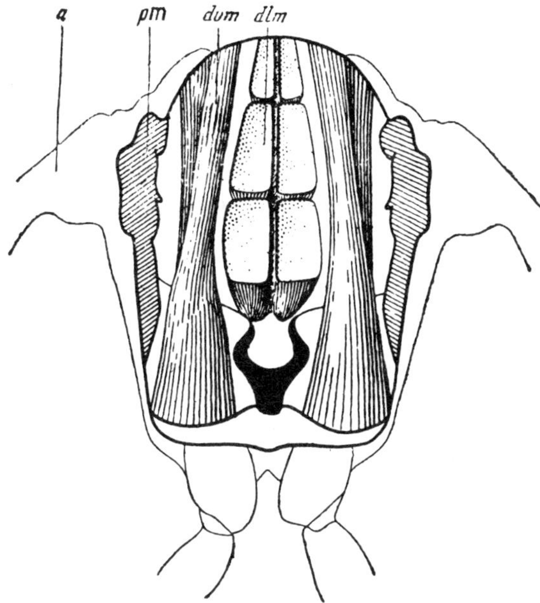

Fig. 4. *Pales* sp. (Tipulidae). Female, cross-section through the mesothorax just in front of the bases of the wings, and through the middle coxae. Only the principal muscles are shown. (Original, diagrammatic.) Abbreviations: *a* − base of wing; *dlm* − longitudinal muscles; *dvm* − dorsoventral muscles; *pm* − pleural muscles.

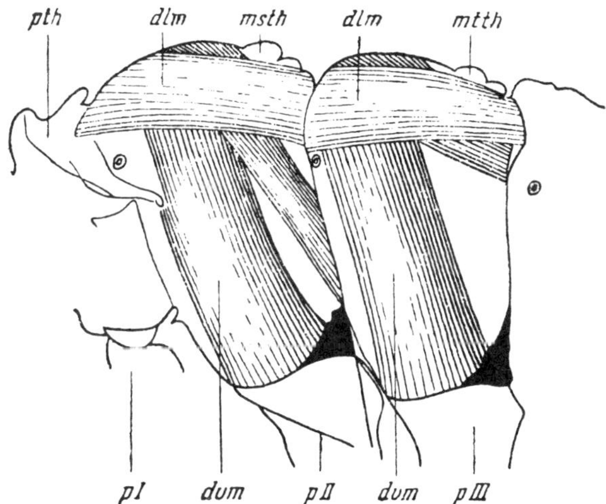

Fig. 5. *Panorpa communis* (L.) (Mecoptera). Male, longitudinal section of mesothorax to show longitudinal and dorsoventral muscles. (Original, diagrammatic.) Abbreviations as in fig. 4. plus *pth* − prothorax; *msth* − mesothorax; pI, pII, pIII − coxae of anterior, middle and posterior legs.

These features are undoubtedly closely linked with each other, and point to a conflict in the evolutionary progress of the early Diptera. As a larval habitat they populated semi-liquid media, such as detritus or decaying plant material, which were at once quite local and possibly also temporary. This set a premium on the evolution of improved organs of locomotion and in particular of the flight mechanisms. Another conflict perhaps arose, that of supplying nutriment for the adult flies. The limited supply of larval food, and at first inadequate powers of larval digestion, may have compelled the adult insect to seek food for itself, and thus again stimulated improved powers of locomotion (Rohdendorf, 1949, 1950a).

Diptera of the present day fauna

A complete study of any group of animals must begin with their structure, because an understanding of their evolution can be obtained only by relating their changes of form and function to the changing conditions of their environment. For this kind of analysis we must turn to those groups which can give us the most information, in short those which are most completely known, and whose biology can be most thoroughly investigated. All this points to groups, and faunas, which are best preserved and most readily available for a complete investigation of their conditions of existence; for most organisms this obviously means the contemporary fauna and flora living on earth at the present time.

Only by studies of the animals and plants of the present day can we relate the differences that exist among living organisms to the environment. If the group of organisms being studied is extinct we may have to rely on evidence from related contemporary groups, but this is never completely satisfactory. The group of animals that I have chosen – the Diptera – is richly represented in the fauna of the present day and this fact allows us to study variation within the order. That is why we start with the living flies rather than with the fossil ones.

On the basis of abundance and variety of different forms, the order Diptera is one of the most prominent groups of animals of the present day, and is often a dominant one. Only three other orders of insects surpass the Diptera in number of species. Leaving aside the beetles, which with over 300,000 species exceed all the other insects put together. It should be noted that the relative 'insignificance' of Diptera in number of species is to a large extent a reflection of the small extent to which they have been studied. Hymenoptera and Lepidoptera are much better known than Diptera. For this reason the estimates of numbers of species in the different orders published by, for example, Shvanvich (1949) – Lepidoptera 140,000, Hymenoptera, 90,000, Diptera, 85,000 – have only temporary value[1] and are a better index of the relative degree of study of the three orders than of their real diversity.

1. It is necessary to point out that these numbers seem to be much too high. Thus according to data given by Zerny and Beier (1936) the number of species of Lepidoptera was then about 102,000 and there is no reason to think that during the 13 years from then to 1949 this number increased by almost 40,000. On the other hand, data for Diptera, too, are undoubtedly too high. I have no information about Hymenoptera.

Table 2

Numbers of known Diptera of the contemporary fauna

Note: Bold face totals denote infraorders; italics, superfamilies; ordinary numerals, families. Hendel's data, as shown by their detailed consideration, refer to the years 1934-35. Questionable figures are enclosed in parentheses. Under "other data" the abbreviation H.48 refers to Hennig, 1948, and H.50 to Hennig, 1950; the abbreviation R.46 refers to Rohdendorf, 1946. Figures without abbreviations denote original data.

Groups of Diptera. *Infraorder, superfamily, family.*	*Hendel* *1937*	*Other data*	*Additions* *to 1958*	*Total*
Nymphomyiomorpha				**1**
Nymphomyiidae	1	—	—	1
Deuterophlebiomorpha				**6**
Deuterophlebiidae	2	—	4	6
Blephariceromorpha				**161**
Blephariceridae	(50)	128 (H.50)	33	161
Tipulomorpha				**16,402**
Pachyneuridea				*5*
Pachyneuridae	5	—	—	5
Tipulidea				*8,297*
Trichoceridae	(40)	60 (H.48)	5	65
Cylindrotomidae	(1,000)	25 (H.50)	2	27
Limoniidae	(1,000)	(5,000)(H.48)	1,557	5,572
Tipulidae	(1,400)	2,100 (H.48)	432	2,532
Tanyderidae	30	—	6	36
Ptychopteridae	40	—	25	65
Psychodidea				*921*
Psychodidae		180	269	449
Nemopalpidae	305	5	7	12
Phlebotomidae		120	340	460
Culicidea				*1,967*
Chaoboridae	(50)	50 (H.50)	10	60
Culicidae	1,450	1,452 (H.50)	455	1,907

Dixidea					*107*
Dixidae	(100)	92	(H.50)	15	107
Chironomidea					*5,036*
Chironomidae	2,000	–		1,135	3,135
Ceratopogonidae	400	500	(H.50)	645	1,145
Simuliidae	300	500	(H.50)	256	756
Orphnephilidea					*68*
Orphnephilidae	30	40	(H.50)	28	68
Rhaetomyiidea					*1*
Perissommatidae	–	–		1	1
Bibionomorpha					**7,378**
Bolitophilidea					*32*
Bolitophilidae	20	–		12	32
Fungivoridea					*3,214*
Allactoneuridae	–	1		1	1
Ditomyiidae	20	–		7	27
Ceroplatidae	200	–		58	258
Diadocidiidae	10	–		1	11
Macroceridae	50	–		16	66
Manotidae	15	–		1	16
Lygistorrhinidae	5	–		2	7
Mycetobiidae	(20)	7	(H.48)	1	8
Fungivoridae	1,400	–		848	2,248
Sciaridae	(100)	500	(R.46)	72	572
Cecidomyiidea					*3,403*
Lestremiidae ⎫		200	(H.48)	–	200
Cecidomyiidae ⎬	3,000	3,000	(H.48)	173	3,173
Heteropezidae ⎭	30	–		–	30
Scatopsidea					*139*
Hyperoscelididae ⎫	–	6		–	10
Synneurontidae ⎬	100	1		–	1
Scatopsidae		73		55	128
Bibionidea					*492*
Penthetriidae ⎫	52			50	102
Hesperinidae ⎬	2	380	(H.48)	–	2
Bibionidae ⎭	250			138	388

Rhyphidea				*98*
Rhyphidae	} 70	50	16	66
Olbiogastridae		18	13	31
Cramptonomyiidae	—	—	—	1
Asilomorpha				**22,532**
Tabanidea				*4,262*
Rhagionidae	400	350 (H.48)	71	421
Coenomyiidae	7	—	4	11
Tabanidae	(2,200)	3,100 (H.48)	450	3,550
Nemestrinidae	200	(150)(H.48)	50	250
Acanthomeridae	(10)	(29)(H.48)	1	30
Stratiomyiidea				*2,226*
Solvidae		61	9	70
Xylophagidae	} 93	20	—	20
Rachiceridae		9	6	15
Stratiomyiidae	1,300	1,500	328	1,838
Acroceridae	200	250	33	283
Asilidea				*5,956*
Therevidae	460	670 (H.48)	10	680
Apioceridae	20	(25)(H.48)	92	112
Asilidae	4,000	4,200 (H.48)	625	4,825
Mydaidae	200	—	54	254
Scenopinidae	50	75 (H.48)	10	85
Bombyliidea				*3,330*
Bombyliidae				2,950
Usiidae	} 2,000	} 2,900 (H.48)	} 430	70
Cyrtosiidae				250
Systropodidae				60
Empididea				*6,758*
Empididae	2,600	2,800 (H.48)	210	3,010
Hilarimorphidae	5	—	—	5
Dolichopodidae	(2,000)	3,600 (H.48)	143	3,743
Musidoromorpha				**24**
Lonchopteridae	20	—	4	24
Phoromorpha				**1,589**
Phoridae	775		84	1,560
Aenigmatiidae	10	(1,500)(H.48)	8	18
Thaumatoxenidae	5		6	11

Termitoxeniomorpha				33
Termitoxeniidae	10	30 (H.48)	3	33
Myiomorpha				32,569
Platypezidea				*142*
Sciadoceridae	–	2	–	2
Platypezidae	100	120 (H.48)	20	140
Syrphidea				*4,900*
Syrphidae	3,000	(2,500)(H.48)	1,370	4,370
Pipunculidae	300	(350)(H.48)	230	530
Conopidea				*662*
Conopidae		–	157	632
Stylogastridae	} 500	–	5	30
Somatiidea				*2*
Somatiidae	2	–	–	2
Trypetidea				*4,192*
Pyrgotidae	50	160 (H.48)	50	210
Tachiniscidae	2	–	–	2
Platystomatidae		–		1,000
Otitidae	} 1,200	– }	330	530
Trypetidae	1,400	–	1,050	2,450
Psilidea				*937*
Cypselosomatidae	–	–	–	5
Micropezidae	325	545 (H.48)	70	615
Psilidae	–	118 (H.48)	22	156
Megamerinidae	–	15 (H.48)	1	16
Nothybidae	–	7	–	7
Diopseidae	–	130 (H.48)	8	138
Heleomyzidea				*999*
Rhopalomeridae	–	25 (H.48)	–	25
Coelopidae	–	45 (H.48)	3	48
Sepsididae	–	159 (H.48)	1	160
Sciomyzidae	120	117 (H.48)	100	220
Anthomyzidae	–	290 (H.48)	16	306
Heleomyzidae	–	220 (H.48)	20	240
Sapromyzidea				*1,741*
Piophilidae	–	77 (H.48)	3	80
Lonchaeidae	–	250 (H.48)	16	266
Sapromyzidae		1,265 (H.48)	95	1,360

Celyphidae	20	34	—	34
Eurychoromyiidae	—	1	—	1
Borboridea				*1,598*
Agromyzidae	—	612 (H.48)	388	1,000
Milichiidae	—	240 (H.48)	20	260
Borboridae	—	250 (H.48)	83	333
Cryptochaetidae	—	5	—	5
Drosophilidea				*2,390*
Ephydridae	—	1,020	30	1,050
Drosophilidae	—	776 (H.48)	564	1,340
Chloropidea				*1,325*
Chloropidae	—	1,200 (H.48)	125	1,325
Gastrophilidea				*26*
Gastrophilidae	—	25 (H.48)	1	26
Anthomyiidea				*2,049*
Cordyluridae		500 (H.48)	36	536
Acridomyiidae	—	1	2	3
Anthomyiidae	—	—	—	1,500
Eginiidae	—	—	—	10
Muscidea				*3,000*
Fanniidae	—	—		
Muscidae				3,000
Sarcophagidea				*3,251*
Calliphoridae				882
Sarcophagidae				2,262
Rhinophoridae				106
Stackelbergomyiidae				1
Oestridea				*121*
Oestridae				58
Hypodermatidae				33
Cuterebridae				20
Villeneuviellidae				10
Tachinidea				*5,080*
Phasiidae				
Leucostomatidae				
Dexiidae				5,080
Tachinidae				

Glossinidea				*31*
Glossinidae	–	30 (H.48)	1	31
Hippoboscidea				*123*
Hippoboscidae	–	100 (H.48)	23	123
Braulomorpha				3
Braulidae	–	3 (H.48)	–	3
Streblomorpha				85
Streblidae	40	70 (H.48)	15	85
Nycteribiomorpha				**150**
Nycteribiidae	50	90 (H.48)	60	150

Total species of Diptera	80,933

The total number of species in the order exceeds 80,000. This figure surpasses the data of Hendel in 1935 by approximately 30,000: such a large increase in the number of species during a 22 year period depends not only on the description of new forms, but also on the refinement and revision of incorrect information on individual families.

There are curious variations in the known species composition for the individual infraorders:

Table 3

	Numbers of species		Factor of increase	Increase in number of species, abso-lute numbers
	Hendel			
	1935	*1958*		
Nymphomyiomorpha	1	1	–	–
Deuterophlebiomorpha	2	6	3.00	4
Blephariceromorpha	50	161	3.22	111
Tipulomorpha	7,151	16,402	2.28	9,251
Bibionomorpha	5,344	7,377	1.38	2,033
Asilomorpha	15,745	22,532	1.40	6,787
Musidoromorpha	20	24	1.20	4
Phoromorpha	792	1,591	2.00	799
Termitoxeniomorpha	10	33	3.30	23
Myiomorpha	20,365	32,551	1.60	12,186
Braulomorpha	2	3	1.50	1
Streblomorpha	40	85	2.20	45
Nycteribiomorpha	50	150	3.00	100

The number of present day species

Better than comparisons with other orders are statistics of different groups within the order Diptera, though even to these the same limitations apply: comparative figures show only the relative sizes of the groups *as known at this moment*. Absolute numbers of known species change, partly by the addition of new, hitherto undescribed forms, and partly by reduction as a result of better study, and the consequent merging of earlier 'species' into synonymy. The latter is insignificant compared with the vast number of new species described. Data on the number of described species of Diptera are presented by many authors, for example by Hendel (1937) and Hennig (1948-52). These figures were supplemented by me on the basis of a survey of *The Zoological Record*, for the years 1935-58, and by a study of some other individual monographs.

An examination of these data allows us to draw a series of conclusions. The chief weakness of basing a comparative study on these figures is that of the incomplete and approximate figures given by Hendel and Hennig for Tipulidae on the one hand, and for the muscoid flies on the other. *The Zoological Record*, too, is weak on these groups. Hence the total numbers can be accepted only as a best approximation possible on the contemporary evidence (fig. 6).

Outstanding increases in numbers of known species have occurred in the infraorders Tipulomorpha,[2] Myiomorpha, Asilomorpha, and parts of Bibionomorpha and Phoromorpha, and it is these five taxa that make up the main mass of recent Diptera. Increase of knowledge of these infraorders has been comparatively uniform; thus the species composition of the two largest infraorders — Myiomorpha and Asilomorpha — increased by about one third, while the known extent of the Phoromorpha almost doubled. The smallest increase occurred in Bibionomorpha. Infraorders made up of relict or parasitic species increased irregularly, sometimes being doubled or trebled, but the actual numbers of species involved are small, and scarcely call for discussion.

It is important to note that in spite of the considerable increase in our knowledge of the recent fauna of Diptera, the relative size of the various infraorders remains unchanged, and is evidently characteristic of the fauna of the present epoch. This is further confirmation that comparative statistics of the relative size and diversity of the different groups of Diptera living today are valid evidence of their probable evolution, in particular as indicating various stages of decline into relict groups.

These data on the relative numbers of species, and the rate of increase of knowledge of different groups of Diptera, are interesting but have only an indirect application to our main problem. One contribution that they make is to direct attention to groups that have been insufficiently studied taxonomically, including some that are of economic importance such as mosquitoes (Culicidae) and non-biting midges (Chironomidae), and others that are of no practical importance, such as Limoniidae, Empididae, and the majority of species of Myiomorpha.

---

2. The great increase in the number of known species of tipulids noted above depends not so much on the description of new forms as on the fact that Hendels' original figure was too low.

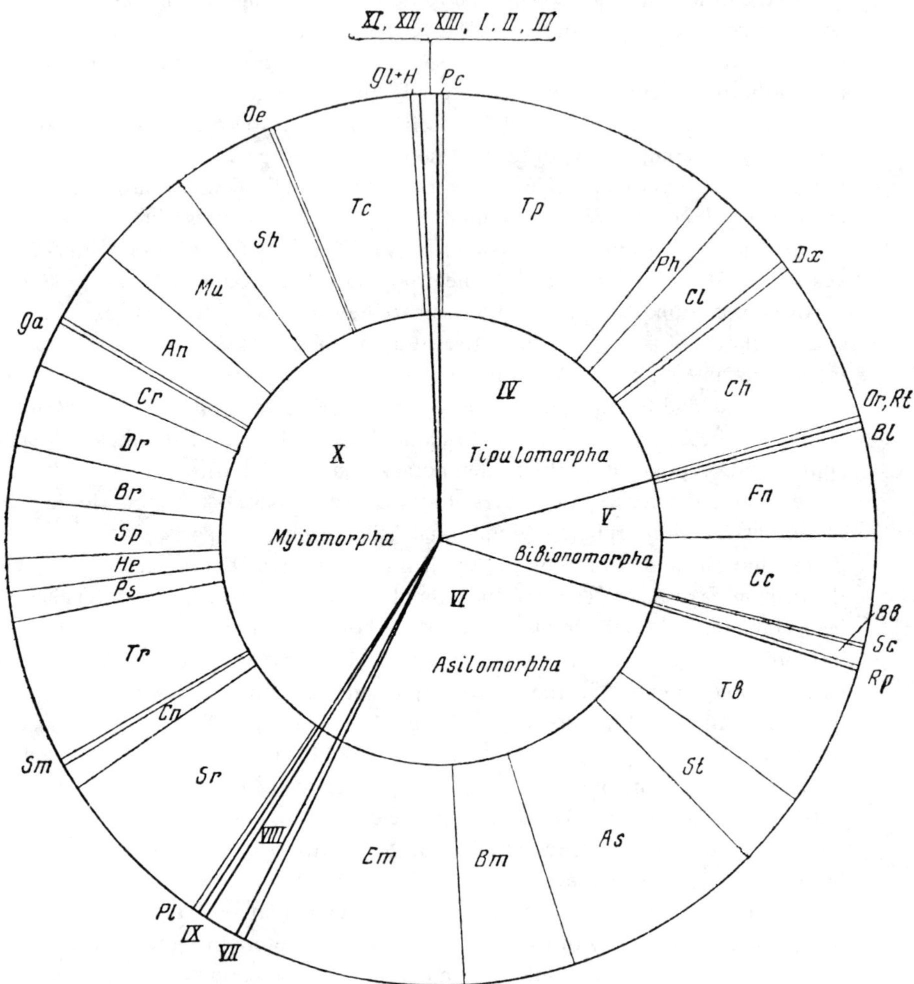

Fig. 6. Diagram of the composition of the order Diptera in the contemporary fauna. Infraorders and superfamilies are shown in a series of sectors of a circle, the size of each sector corresponding to the number of species in a given taxon (1° of arc is equivalent to 2,225 species). In the centre of the circle are set forth the infraorders, and on the circumference the superfamilies. (Original.) For abbreviations, see p. xiii.

Division of the order Diptera into subordinate taxa

Before turning to an appraisal of the modern diversity of the separate groups of the order Diptera, it is necessary to examine the classification in general. Schemes for the systematic subdivision of the order Diptera have been drawn up by a succession of authors, beginning with Latreille, continuing through the nineteenth century with Brauer, Schiner and Coquillet, and in more recent times by Hendel.

A critical examination of all existing systematic schemes proposed during the 150 years of study of the Diptera shows that most of them were based upon either a single feature or a group of structural characteristics. Each author proposing and defending his own system has usually not attempted to evaluate the categories that he was establishing in any ecological or ethological sense. All subdivisions of the various classifications are characterized by their authors on purely morphological grounds, without any functional appraisal whatever. Classifications differ from one another merely in their different rearrangements and recombinations of families and other taxa. Moreover, certain groupings are more 'popular' than others, and are accepted by nearly every author, while others are 'unstable', in the sense that they are shifted about, and altered in size from one author to another.

To put it briefly, attempts to establish a realistic classification of the Diptera have so far been predominantly subjective, extremely inexact evaluations of the interrelations of different groupings. Perhaps it is unnecessary to add that this is true for the state of classification of any group of organisms, and is by no means peculiar to Diptera. There is no doubt that the reason for this situation is the absence of a proper theory of classification of organisms, or of an exact understanding of the nature of subordinate taxa, and hence the lack of any reliable criterion for distinctions between taxa of different rank. The answer to this difficult question lies in the purely philosophical problem of the relationship between the whole and its parts, an exercise in dialectical materialism that lies beyond the scope of the present, comparatively narrow investigation. I must therefore try to solve these problems as far as they relate to Diptera, solely within the limits of the group concerned, and what is more important, not to stray too far into the development of general theories. Obviously my conclusion must to some extent relate Diptera to other orders of insects.

A sound basis for the classification of a given group, including a system of subordinate categories or taxa, must be the establishment of a unit of classification or biotype that is associated with definite conditions of existence, which are reflected in similarities of structure, function and ontogenetical development. The biotype must have a single origin, and may contain either subordinate taxa or discrete populations of individuals, or both. These considerations affect a taxonomic investigation in every respect, and not only morphologically. The determination of every taxon, be it species, genus, family or any other, must involve not only an understanding of its morphology, but necessarily also of its particular conditions of existence, evolution and phylogeny. In practice this leads to an appraisal, to an understanding of the functional significance of these morphological peculiarities that are the 'diagnostic features' of each particular group. This is the only way to understand what has been happening, and so to account for the divergencies that have led to this line of evolution.

An accurate understanding of the phenomenon of subordination in taxonomy, of the relations between higher and lower organisms as a special case of the

general and the particular, turned out to be especially difficult. Without discussing
in detail the whole complex question, I see a preliminary solution to it if we can
establish the reality of the groups in question by a comparative study of the
distinguishing features of the supposed subordinate groups, and by trying to trace
their evolution. By this means we may discover whether subordinate taxa exist or
not, and this is clearly a better method than the subjective appraisal of 'systematic
features' that has hitherto predominated in taxonomy. Once a given taxonomic
group has been shown to have a real existence, it then has to be related, on the
one hand to some higher taxon of which it forms a part, and on the other to
lower, subordinate taxa that are parts of it. Without going into further detail, I
consider that the determination of the reality of a taxon is the means by
which we must approach the solution of all taxonomic problems, and
this is what I shall try to do in the following account of the systematics of
Diptera.

Finally, it is necessary to say something about the value of the many systematic
schemes for the order Diptera that already exist in the literature. Almost without
exception, all classifications of Diptera have been constructed by their authors on
the basis of certain selected features. Nevertheless many of the groups that have
been established are undoubtedly genuine, and have a real existence, for example
many families and superfamilies. Of course there is nothing surprising in the fact
that wrong methods may sometimes lead to correct results: the history of science
is full of examples of success by trial and error. Keenness of observation enables
taxonomists to note the reality of many systematic groups of Diptera even though
their metaphysical approach may have been unsound. Their success has depended
upon the famous 'spiritus systematicus', or 'instinct', in reality a finely developed
keenness of observation, and the ability to evaluate objectively resemblances and
differences. From this it follows that the task of the present-day investigator is
not so great, nor so infinitely difficult, as it would have been if all pre-existing
systems had been shown to be valueless. We are able to start from the established
metaphysical systems of classification, and subject them to re-valuation or
revision in the light of our understanding of correct theory and methodology.

The order Diptera is subdivided into various suborders, and up to the present
there have been two points of view. The first system divided the order into
Nematocera and Brachycera, on the basis of the state of evolution of the
antennae; this system has existed since the times of Latreille and Meigen, i.e. from
the beginning of the nineteenth century. The other division of the order was
based on details of metamorphosis, in particular the existence of a puparium,
formed from the integument of the last larval instar, which is not discarded and
which forms a kind of false cocoon. Brauer (1869, 1883) made this the basis of
his two suborders, Orthorrhapha and Cyclorrhapha, according to the form of the
slit which appeared in the puparium at the time of the emergence of the winged
fly.

These two subdivisions of the Diptera do not agree with each other. Not only
does the group Brachycera include both orthorrhaphous and cyclorrhaphous
Diptera, but also among Nematocera there are some which have produced

pseudococoons, i.e. puparia (notably some Cecidomyiidae and Scatopsidae).

Examining the significance of these features, which served for a time as diagnostic criteria for the distinction of suborders, one may draw the following conclusions. Long antennae with many homonomous small segments are a characteristic feature of most mecopteroid orders. Simultaneous reduction in the number of segments and in the length of the antennae hardly ever occurs in mecopteroid orders other than Diptera, and although heteronomy does occur in Lepidoptera it is not accompanied by shortening of the antennae, nor by reduction in the number of small segments. Obviously the 'long antenna' is the original, ancestral condition in mecopteroid ancestors of Diptera, a fact that is almost universally accepted, and which must be emphasized for a clear understanding of the process of development of the 'short antenna'.

Short antennae are widely distributed among a high proportion of present-day Diptera, and characterize the youngest groups of the order. Decrease in size of the antenna is closely linked with the visual function: longer antennae undoubtedly reduce the field of vision, shading the eyes to some degree or another. The development of short, specialized antennae, which are located between the eyes, and which do not project far, undoubtedly improves visual ability. Furthermore, a decrease in length of the antennae improves the functioning of sensory organs related to feeding, while specialization of the antennae, and differentiation of parts such as has occurred in brachycerous Diptera (a scape, a pedicel with Johnston's Organ, a large bulky terminal section, and a fine arista) adapt these short, small structures into compact, complexly-built organs, which undoubtedly serve various, different functions. Finally, specialization of the antennae is often combined with an increase in the relative dimensions of the head, usually as a result of the enlargement of the compound eyes. It may be said that there are no Diptera with a large head and at the same time with long, undifferentiated antennae: on the other hand, small-headed Diptera with short antennae do exist, of secondary origin, and associated with the processes of aphagia, about which we shall have more to say later on.

All that has been said so far may be summarized by stating that the process of differentiation of the antennae, with reduction in the number of segments, is in line with the general development of the cephalic sense organs of insects. It indicates an increase both in the physical dimensions and in the importance of the activity of the cephalic section of the body, which is indirectly reflected in the enlargement and development of the supra-oesopageal ganglion, or 'brain' of the insect. We can view the development from 'long antennal state' to 'short antennal state' as a very essential process, and as a visual indicator of the increasing perfection of the principal, indeed the most vitally important component of the insect organization — the nervous system.

A comparison of the diagnostic features of Brauer's two suborders of Diptera, Orthorrhapha and Cyclorrhapha, clearly shows that the primary 'orthorrhaphy' is characterized by the absence of any puparium, the last larval skin being cast, as is usual among insects, and the pupa exposed; whereas the secondary 'cyclorrhaphy' involves transformation from larvae, through pupa to adult, without casting the

last larval skin, which constitutes a special pseudococoon, or 'puparium', an
envelope for the pupa. This results from a development of the process of
metamorphosis with the production of a beneficial sheathing structure, a strong
envelope in which lies the exposed, immobile pupa. Furthermore, it must be
added that from now onwards the production of a puparium or pseudococoon is
often closely linked with improvement of larval feeding. The larva passes through
the well-known process of 'acephalization', or reduction of the cephalic skeleton
and mouthparts, together with complication of the enzymes of the saliva, and the
development of extrintestinal digestion, with development of an amphi- or even a
metapneustic respiratory system, and with thickening of the cuticle of the body.

The phenomenon of 'cyclorrhaphy' (which is better described as the formation
of a puparium), reflects the important perfection of metamorphosis that is going
on as an adaptation to living in the feeding medium of the apodous larva. Hence it
is abundantly clear that the production of a puparium is an important step
forward and is the most significant aspect of the unique process of 'cyclorrhaphy'.

It must be pointed out that the group Cyclorrhapha does not include all
Diptera that have a pseudococoon in their metamorphosis, but only those in
which the insect emerges from its puparium by a circular slit, lifting off a kind of
lid; hence the name Cyclorrhapha. There are a few exceptions in Phoromorpha
and Musidoromorpha (= Lonchopteroidea), which are included in Cyclorrhapha
even though their puparia do not open with a circular slit. On the other hand
some Diptera which possess highly developed puparia (Stratomyiidae), and which
may even have acephalized larva (Coenomyiidae, Hyperoscelididae) are never
included among Cyclorrhapha because they are so very obviously closely related
to various 'orthorrhaphous' groups.

Having considered those features that are regarded as being diagnostic of the
suborders of Diptera, and having established the great importance of these
features in the active life of the flies concerned, it is now necessary to test the
reality of the groupings Nematocera/Brachycera on the one hand and
Orthorrhapha/Cyclorrhapha on the other. It has already been pointed out that
these divisions into suborders are rather inconsistent with each other. Only Nematocera
on the one hand, and Cyclorrhapha in the narrow sense on the other, can be
formally defined. 'Brachycera' and 'Orthorrhapha' overlap with these, because
Brachycera includes both orthorrhaphous and cyclorrhaphous forms, while
Orthorrhapha may be either nematocerous or brachycerous. These two categories
are heterogeneous; if we rely upon antennal structure the taxon 'Brachycera' is
unreal, while if we make use of peculiarities of metamorphosis, then
'Orthorrhapha' is a mixed group.

What other structures can we use to supplement these? Examination of
different systems of organs draws attention first of all to the special features of
the central nervous system, in particular to its greater or lesser degrees of
concentration, and of the interconnections and fusions of the individual ganglia of
thorax and abdomen. Such concentration is symptomatic of a general increase in
complexity of organization, and improvement of nervous control of the organism,
and has occurred to a varying extent in different groups of Diptera.

We know enough to be able to say that Nematocera are characterized by a low degree of concentration of the central nervous system, and by the presence of many individual ganglia in the thorax and abdomen. Firstly there are the midges (Chironomidae), with three thoracic and six abdominal ganglia, Tipulidea with three thoracic and up to eight abdominal ganglia, and Mycetophilidea with three thoracic and five or more abdominal ganglia. In bibionids and the bloodsucking mosquitoes the number of thoracic ganglia is reduced to two, whereas abdominal ganglia are comparatively numerous. (Brandt, 1879; Hendel, 1937; Beklemishev, 1944). It should be noted that in all known Nematocera the larvae as well as the adult insects have the central nervous system dispersed.

'Brachycerous' Diptera vary considerably more than Nematocera in the degree of fusion of ganglia, and the thoracic section nearly always has one large and complex ganglion. Only rarely does the thorax have one other ganglion (Empididae and Phoridae). The number of abdominal ganglia in brachycerous forms is variable, and only rarely does it reach six (some Tabanidae), or three-five (the majority of the families Tabanidae, Asilidae, Empididae, and some Stratiomyiidae). Usually the number is considerably less, not more than two in Syrphidae, and only one in various Myiomorpha. Abdominal ganglia may be absent from some Muscidea and Tachinidea.

Another characteristic of the 'non-Nematocera', which is no less peculiar, is that the nervous system of the larvae may be considerably more concentrated than of the winged adults. A sharp concentration of the nervous system of the larvae is observed in Cyclorrhapha, and in some 'Brachycera', for example in Stratiomyiidae, and especially among those larvae which have reduced mouthparts and a puparium.

Considering the significance of the process of concentration of the central nervous system, and noting the similarity in this respect of all the Nematocera, it is possible to draw the following inference: although B.N. Shvanvich (1949, p. 628) assumes that "to link the concentration of the abdominal chain [of ganglia] with anything else is difficult as yet", it is a fact that concentration and fusion of separate ganglia must lead to an acceleration and greater integration of all the nervous processes, by reducing the number of connections and the complexity of junctions. This obvious deduction can be excellently illustrated by considering examples of greater or lesser concentration of ganglia.

Thus, if we compare the larvae of the aculeate Hymenoptera with those of the cyclorrhaphous Diptera we see a superficial similarity of appearance (wormlike body, absence of legs, reduction of sensory organs) combined with sharp differences in the central nervous system, which is highly concentrated in flies, and completely dispersed in Hymenoptera. This difference in ganglionic concentration becomes intelligible if we compare the life-histories of the two groups of larvae. The larvae of Hymenoptera are inert, relatively immobile insects, living in artificially constructed cells. The apodous and acephalous larvae of Diptera, on the other hand, are generally highly active, and move freely in a mass of nutritive substratum, or at least constantly perform active movements within the tissues of another organism, either plant or animal. Concentration of the nervous system in Dipterous larvae clearly reflects an active way of life, and shows how the whole organization of the insect has become adapted to its needs.

When the nervous system has become concentrated in insects that are only feebly mobile, or even altogether static — as for example in the female coccid — this can be explained as a by-product of the specialization of the whole organism. Moreover we must not forget that the act of feeding, the operation of a piercing proboscis which makes frequent and complicated movements into the tissues of plants, is both a complex and a necessary part of the activity of coccids, and therefore requires a correspondingly complex nervous system to operate it.

What has been said leads to the inevitable conclusion that concentration of the central nervous system is highly important, and a very significant step in evolution. Nematocera as a group are characterized by having little apparent concentration, and are thereby distinguished from all other Diptera, more especially from the Cyclorrhapha. 'Brachycera-Orthorrhapha' are again a heterogeneous group, since they include forms with highly concentrated nervous systems, as well as others with nerve chains that are comparatively diffuse.

The morphology of the digestive, secretory, vascular, and respiratory systems is still very insufficiently known for most Diptera, and therefore has only a limited use in testing the validity of the suborders of Diptera that have been proposed. One example is the number and type of Malpighian tubules. There are usually four, but a few Diptera have five (Ptychopteridae, Psychodidae, Culicidae); some have the four arranged in two pairs, and occasionally they are reduced to three or even to two (some Phoridae and Nymphomyiidae). All that we can deduce from this list is that a section of nematocerous Diptera have the larger number of Malpighian tubules; this is undoubtedly a primitive feature, relating them more closely to the Mecopteroidea, which have six. Apart from this, the number of Malpighian tubules does not fit in with either of the two primary divisions into suborders: i.e. Nematocera/Brachycera or Orthorrhapha/Cyclorrhapha. Nor does the structure of the vascular system provide a basis of classification into these suborders.

The respiratory system of Diptera provides rather more data. The most primitive feature of the system is the presence of transverse commissures between the longitudinal trunks, and these are seen in the larvae of many Nematocera (up to 10 commissures in Culicidae, eight in Heteropezidae, and some others). In the larvae of Cyclorrhapha the number of transverse commissures is very reduced, down to two or even one; an exception is the larvae of Phoridae, which have eight. Moreover, cyclorrhaphous Diptera are characterized by a spacious widening of the tracheal system, with numerous and complicated air-sacs in the abdomen and thorax; though even this feature does not separate Cyclorrhapha completely from Orthorrhapha, since tracheal air-sacs are well developed in the thoracic section of Chironomidae and other Nematocera.

It should be pointed out that this kind of formal examination of each of the different systems in turn is purely arbitrary, and is permissible only as a temporary, convenient method of exposition. In practice the systems are very closely connected and mutually interdependent, and consideration of feeding and metamorphosis in different flies will show clearly how provisional such an analysis is. In fact such phenomena as the morphology of the various stages of development, the characteristics of larval organization, the processes of specialization and reduction, cannot possibly be understood except in conjunction with the transformations and special features of the chief systems of organs of feeding, respiration, nervous regulation, and integumentary (protective) organs.

Thus, although the trend of specialization in the metamorphosis of Diptera can be formally defined as the production of an acephalous larva and of a puparium, in fact it is seen to consist of a number of contradictory tendencies working together; dwelling in a mass of substrate, the attainment of greater mobility and protection, the acceleration of development, and the capacity to feed on a wider range of foodstuffs. Dwelling in a mass of substrate brought about the wormlike form, with reduction of the legs and other projecting appendages. Simultaneously there is a need to be able to move about freely through a substrate that is semiliquid, with solid particles, and this is reflected in the thickening and armament of the cuticle. Living in a mass of substrate also brings the larva into continuous contact with its source of nutriment, and so stimulates the evolution of extra-intestinal digestion, with the salivary fluids being discharged into the nutritive substrate, and in this way accelerates development, improving and facilitating feeding.

All these things are interconnected, and sometimes contradictory processes which have brought about the evolution of acephalous amphi- or even metapneustic larvae, which change into pupae without discarding the tough last larval skin, which becomes the puparium. This complex process is one of the chief characteristics of the evolution of the Diptera, and is expressed in some degree or another in all the different groups of the order. It appears most clearly of course in Cyclorrhapha, but this does not mean that puparia and acephalous larvae have no place in any other groups. Similar developments are to be seen in Stratiomyiidae of the 'brachycerous Orthorrhapha', as well as in some Nematocera (the major group Cecidomyiidea, the relict family Hyperoscelididae, and the subfamily Leptoconopinae of the Ceratopgonidae). Furthermore a partial movement in this direction, with reduction of mouthparts but not the evolution of a puparium, has taken place in some Empididea and Tabanidea.

It will be realized that the various groups of Diptera which have evolved acephalous larvae and puparia (Cyclorrhapha, Stratiomyiidae, Cecidomyiidae, Hyperoscelididae and Ceratopogonidae) are not fundamentally similar, nor are they linked phylogenetically. These peculiarities of development evolved independently in very different sections of the order. Hence it is not unexpected that there should be inconsistencies in the subordinal classifications that have been proposed, since the concept of Cyclorrhapha is the only one that unites all the flies that are alike in this respect, and excludes all others. The concepts of Nematocera and Brachycera-Orthorrhapha both include similar diversities of larval forms, and so do not provide a simple and reliable classification into two suborders.

It remains to examine the phylogenetic relationships of the taxa which the author himself has in view for the division of the Diptera into suborders. The vast group of 'nematocerous' Diptera is undoubtedly the most ancient, and was the starting point from which the 'Brachycera' arose. This suborder comprises no fewer than 15 superfamilies, many of which reveal their interrelationships (see p. 288). However, by no means all the superfamilies can be combined in this way to form larger complexes: some superfamilies remain sharply isolated, not directly linked with any other. Thus the superfamilies Fungivoridea, Bibionidea, Scatopsidea, Bolitophilidea and Cecidomyiidea are undoubtedly in a group by themselves, with the superfamily Rhyphidea close by, and the whole agreeing well with the Oligoneura of former authors. The superfamilies Culicidea, Dixidea, Chironomidea and Orphnephilidea are another distinct association, with which the Tipulidea and Psychodidea are closely linked, together forming the Polyneura.

These two complexes, Oligoneura and Polyneura, have very little in common, and their apparent resemblances are not evident before Triassic times. Any resemblance seems to be due to convergence, and not to any proved relationship. Three present-day infraorders, Blephariceromorpha, Deuterophlebiomorpha and Nymphomyiomorpha occupy a still more isolated position in the system, being sharply separated, relict groups. The differences between them are even clearer and more profound than those between Oligoneura and Polyneura, while their relationships with each other are completely obscure.

On the whole, therefore, the suborder 'Nematocera' does not possess any real unity but consists of an assembly of groups of superfamilies not directly related to one another. This artificial concept includes the most ancient Diptera alive today, and is characterized by a whole range of primitive, ancient features of organization.

Looking at the group Brachycera as a whole, it is remarkably homogeneous phylogenetically. The majority of its superfamilies form two huge associations: Cyclorrhapha with not fewer than 19 superfamilies, and 'Brachycera-Orthorrhapha' with four superfamilies: Tabanidea, Stratiomyiidea, Asilidea and Empididea. Apart from these clear groups there are several superfamilies the interrelations of which are almost completely obscure, and which show no clear links with either of the two major groups just mentioned. In the first place we have the Musidoridea and Streblidea, which give some slight indications of relationships with some extinct representatives of the 'Brachycera-Orthorrhapha', or even with the Nematocera, though at the same time the presence of a puparium, reduction of larval mouthparts, and extra-intestinal digestion show that these forms are rightly placed in the Cyclorrhapha. The taxonomic position of the parasitic Braulidae and Nycteribiidae, and even more their relationships, are quite uncertain. Thus in the same way as the 'Nematocera', the 'Brachycera' also prove to be an artificial grouping combining both real associations of phylogenetically related superfamilies and quite unrelated extraneous groups which have become like them by convergence.

If we consider the systematic classification of the Diptera into suborders as it is practised at the present time, we can draw certain quite definite conclusions. The suborders Nematocera, Brachycera-Orthorrhapha and Cyclorrhapha as they are widely accepted by almost all taxonomists, are not natural groups. The widespread acceptance of these groupings bears witness to their importance as some kind of step in the evolution of Diptera, but these groupings place too much stress on single features of the sense organs and nervous system and of the nature of metamorphosis. It is obvious that 'suborders' characterized in this way cannot be considered as true systematic categories. A complete review of these 'suborders', and particularly of their internal phylogenetic relationships, clearly shows that it is necessary to remove the isolated, unrelated groups until each suborder has become a sound taxon. Such a grouping of Diptera into major taxa, which have a real existence, can be carried out without any particular difficulty, but this still leaves the question of what rank these taxa should be given. The resolution of this question is part of the big problem of co-subordinate taxa. Hitherto each individual case has been determined subjectively, on its own merits, attributing to it a definite rank that was more or less arbitrary.

As I have noted earlier, a detailed examination of this problem is outside the scope of the present work, and I must confine myself to a few comments only.

Thus first and foremost it is necessary to note that the relations between taxa of different rank cannot be resolved independently in different groups of organisms. The relations between, for example, species and genus, genus and family, family and order must be similar for all groups of animals, or at least among related groups of animals. Such a bold generalization seems to result directly from the existence of a general classification of animals, and this in turn is a consequence of having them all classified on the same system of orders, families, genera, and species. It follows that any arbitrary change in rank of any of these systematic categories is admissable only after a thorough appraisal of the nature and features of a given category. Any kind of expansion or contraction of the concept of a taxon — whether a genus, a family, or any other category — must not be made arbitrarily without a thorough review of its place in the system. There is no room for pragmatic arguments about 'convenience', and 'practical advantage' in favour of such changes. This is particularly a problem in groups of organisms where taxa are apt to be highly diverse, and to include a great range of variation, and where there is resulting pressure to narrow the concepts of certain categories within the group. Many examples of this are to be seen among insects, as in every class of abundant animals.

Without going any further into the general problem of co-subordinate taxa in organisms, we must now return to the main theme of this book, and try to evaluate the rank of superfamily. It has already been suggested that superfamilies should not be grouped into higher taxa, but should be arranged directly within the order Diptera. It is natural to assume that between superfamilies and the order Diptera there should be an intermediate category of suborder, but we consider this assumption to be premature, because of the relatively small difference in taxonomic status between such subordinal groupings and the superfamilies themselves. The previous artificial 'suborders' Nematocera,

Brachycera and Cyclorrhapha have far greater claim to acceptance than any proposed grouping of superfamilies.

In an attempt to determine more precisely what is the nature of 'suborder' it seems to me that we should look upon it as one rank below the order, and its relations with the order to resemble those between a family and a subfamily, or a genus and a subgenus. Such a relationship implies a relative similarity, a close relationship with the higher taxon, with the possibility of comparing and contrasting the features of the two groups. More precisely, the characteristics of a suborder should rightly be close to, and comparable with the features of the order, should concern the same structures of the organism, and should differ only in that the range of variation of these structures should be less in the suborder than in the order. Looked at in this way it is evident that groupings of superfamilies do not rank as suborders.

Two conclusions may be drawn: (1) that present-day Diptera comprise only two suborders; the most peculiar family Nymphomyiidae, which may be called the Archidiptera, and all the rest, which are the Eudiptera (Rohdendorf, 1961b); and (2) that almost all living Diptera can be arranged in 12 groups of superfamilies, and that, following the practice in some other groups of animals (e.g. Simpson, 1945), such groups may be called *infraorders*.

Finally it remains to deal briefly with the position of a number of groups which are represented in the contemporary fauna only by an insignificant number of species, and which are usually rare, or very limited in the geographical or ecological distribution. Most of these are relicts, remnants of groups that were formerly more widespread and diverse, and one can say with certainty that the variety and numbers of a group as they exist today cannot be used as a basis for evaluating its position and status in the taxonomic system. To determine their real status, and their relative position in the system, as well as the variation of these with time, we must rely on the system of co-subordinate categories. Their true systematic position is determined by studying resemblances and differences in relation both to ecological conditions at the time and to ancestry. From this it follows that the size of any group is irrelevant to its position in the system. Relict groups that are poor in modern representatives, and which differ sharply from any other in many features, must be given careful consideration, and not reduced in status just because they are rare. They must not be dismissed as mere 'transitional' or 'extreme forms', because this is to construct a purely metaphysical system which ignores both ancestry and variation with time.

Characteristics of the individual Groups of Diptera

Suborder Archidiptera

Rohdendorf, 1961b, p. 154; Rohdendorf 1962, p. 308.

Characteristics. – Head of pupa directed forward, prognathous. Ocelli very large, two in number. Prothorax large, well isolated. Wings of pupa parallel-margined, elongate.

Composition. – Among living Diptera only the infraorder Nymphomyiomorpha belongs to this suborder, and fossil representatives have been found only in the Upper Triassic. Unfortunately the two peculiar groups Dictyodipteromorpha and Diplopolyneuromorpha are known to us only from fragments of the wings, and this seriously hampers their comparison with living representatives of the suborder. (Rohdendorf, 1961a). (see p. 136-140).

Infraorder Nymphomyiomorpha, infraorder novus

Extent, evolution, systematics. – This little-known group of Diptera was first reported in 1935 on the basis of a single species, *Nymphomyia alba* Tokunaga, discovered in the mountains of Japan (Tokunaga, 1935; Hennig, 1948-52). Nothing is known of this unique form beyond the structure of the winged adult and of the pupa, and some very scanty ecological information. Fossil records of the evolution of Nymphomyiidae are completely lacking, and the unusual characteristics of this family can be compared only with the Dictyodipteridae, from the Upper Triassic of Central Asia (p. 138). The two families have similar wings, especially in the pupa, where the wings are extremely elongate and moderately pointed at the tip. This association of Nymphomyiidae with the extinct Dictyodipteridae is consistent with the great isolation of the Nymphomyiidae from all contemporary Diptera, which compels us to assume that they belong to a line that became divergent a long time ago.

Chief features. – It is noteworthy that Nymphomyiidae have been found near mountain streams, in which their still undiscovered larvae presumably live. Such conditions are characteristic ecological niches for many relict forms which found a refuge there, and by a narrow but perfect stenobiosis have succeeded in outliving most of their near relatives (Rohdendorf, 1958, 1959a).

Their development is unknown. The morphology of the pupa suggests that this stage is unusually mobile, since it has a wormlike form, with armoured abdominal segments and a free head. Feeding habits of Nymphomyiidae are unknown, but the adult has no mouthparts, and therefore the larva must accumulate reserves sufficient both for the activity of the winged insect and for the maturation of the gonads.

Respiratory organs are unusual, the adult having no abdominal spiracles, and depending entirely on the two pairs of thoracic spiracles. The intestine has no crop, the only two Malpighian tubules. The central nervous system is made up of sharply isolated ganglia, three thoracic and eight abdominal. The sense organs as a whole are peculiar: abruptly shortened antennae of the 'brachycerous' type, with enlarged compound eyes fused with each other on the lower surface of the head (fig. 7*A*); a liberal development of macrotrichia on the wings.

The organs of movement are irregularly developed. Legs are weak and short with short femora, but tarsi very long, with the five tarsomeres distinctly subdivided. The wings are very big, and evidently powerful, since the mesothorax is unusually elongate, as if to accommodate exceptional longitudinal muscles, which power the downstroke of the wings. A characteristic feature of the venation is a great reduction in the number of veins (fig. 7*C*), and even these are

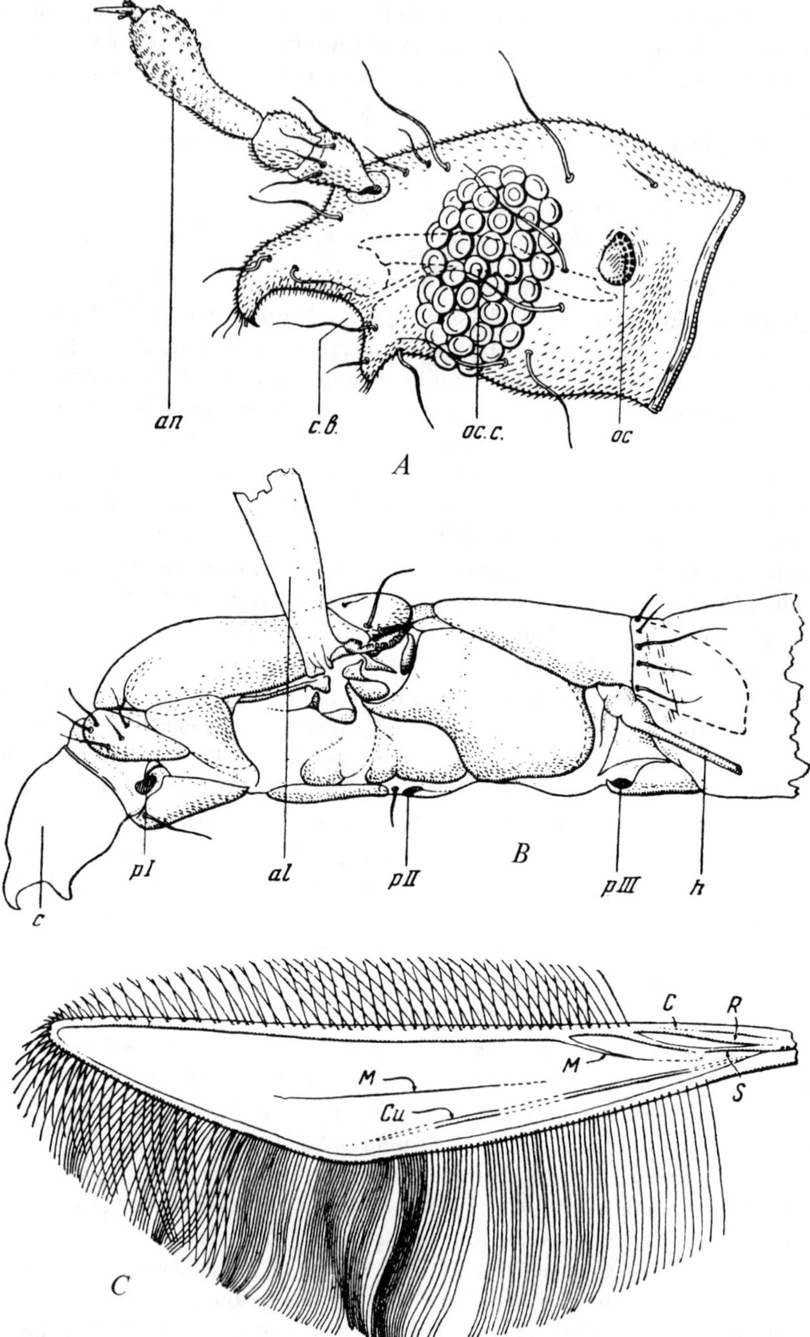

Fig. 7. *Nymphomyia alba* Tokunaga (Nymphomyiidae). *A.* Head from the side. *B.* Thoracic section from the side. *C.* Wing. Present-day fauna of Japan. (According to Tokunaga, 1935.) (for abbreviations, see p. 29).

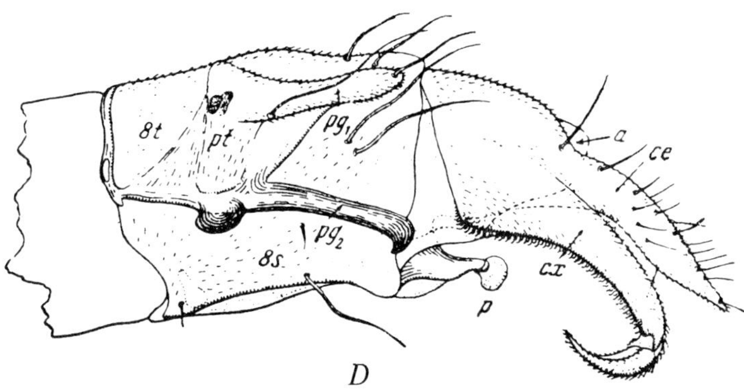

Fig. 7. (cont.) *Nymphomyia alba* Tokunaga. *D*. End of abdomen of male from the side. (According to Tokunaga, 1935.) Abbreviations: *a* – anal aperture; *al* – base of wing; *an* – antennae; *c* – head; *cb* – mouth indentation; *ce* – cercus; *cx* – coxite of ninth segment; *h* – haltere; *oc* – ocellus; *occ* – compound eye; *p* – aedeagus; *pg₁*, *pg₂* – paired protuberances of paratergites of eighth segment; *pt* – paratergite (lateral plate) of eighth segment; *pI, pII, pIII* – places of attachment of anterior, middle and posterior legs; *s* – radial sector; *8c, 8t* – eighth sternite and tergite; *C, R, M, Cu* – veins of wing.

evanescent in some of their sections. At the same time tiny hairs and bristles (micro- and macrotrichia) are very abundantly developed on the wings. The halteres are very large, and located relatively far from the wings. The protective features of most winged insects are absent: the wings at rest are not folded back, but raised upwards. The structure of the abdominal integument is also peculiar, the eighth sternite of the male bearing separated protuberances, the paratergites (fig. 7*A*)(Tokunaga, 1932, 1935, 1936).

Nothing is known about reproduction of Nymphomyiidae.

Factors which influenced the evolution of the Nymphomyiidae. – Lack of information about the development and living conditions of the larvae makes it impossible to say how far the biology of Nymphomyiidae is unique among Diptera, and so it is possible only to discuss their evolution in very general terms. A major peculiarity of these flies is that feeding seems to be confined to the larval stage, since the winged adults have no mouthparts. This extreme development of aphagia has been accompanied by the reduction of legs, by a paleopterous condition, and presumably by the shortening of the imaginal life. The evolution of such features can be related to the great difference between the environment in which larva and adult live; the larval habitat is rich in organic substances, whereas the adult lives in a mountainous terrain, on exposed rocks, in relatively barren surroundings. Aphagia under such conditions is not peculiar to Nymphomyiidae, but has occurred in the evolution of some other groups, notably mayflies and Chironomidae. Deficiency of food in the winged phase, and the necessity to guarantee the formation and deposition of sexual products is the most important problem facing many groups of Diptera, and one which has resulted in this particular solution No doubt the feeding peculiarities of Nymphomyiidae is a

recent development in their evolution, and does not throw light on the
deep-seated differences that exist between them and all other Diptera, as shown
by the pupal structure, locomotory apparatus and enormous ocelli.

The absence of any data on the organization, and hence the living conditions of
the larva, the chief feeding stage, is also a bar to any understanding of the
differences between these and all other Diptera. It is hardly profitable to draw
any preliminary conclusions about the habitat of the larva from the features of
the pupa (e.g. wormlike shape and probable mobility), to assume that the habitat
of the larva is particularly concealed because it lives in an abundance of
food material, or even to judge the activity of the winged insect from a superficial
examination of its morphological features, with its powerful flying mechanism
which resembles that of broadwinged insects. Comparisons with the extinct
Triassic Dictyodipteridae, which possess slightly costalized veining, show some
ways in which the venation of the Nymphomyiidae may have been derived, but
this conclusion is all that can be said about the evolutionary development of these
groups.

Conclusions. — The small amount of information that exists in the literature
about these peculiar Nymphomyiidae permits only very general and preliminary
conclusions to be drawn. Apparently the prevailing tendency was the organization
of the larval phase for feeding, with a parallel degeneration of the adult, which
became aphagic. This is a process of desimaginization (Rohdendorf, 1960). Lack
of data makes it impossible to throw any more light on these peculiar insects,
which are far removed from the rest of the order.

Suborder Eudiptera

Rohdendorf, 1961b, p. 154; Rohdendorf, 1962, p. 910.

Characteristics. — Head of pupa directed downwards or backwards,
opisthognathous; wings not parallel-sided, but gradually tapered. In the imago the
ocelli are only moderate in size, and generally three in number. Compound eyes
not united below the head. Prothorax of pupa more or less reduced, fused with
mesothorax, not isolated.

Composition. — All contemporary Diptera except Nymphomyiidae belong to
this suborder, which comprises 12 large systematic units, infraorders, made up of
many superfamilies. The most ancient infraorders of Diptera — Tipulomorpha and
Bibionormorpha — are known to have existed since Upper Triassic times, and in
the Jurassic another infraorder — Asilomorpha — appeared. The rest are known to
us only from Tertiary deposits.

Infraorder Deuterophlebiomorpha
Rohdendorf, 1961b, p. 158

Extent, evolution, systematics. — The five known species of the sole genus
Deuterophlebia Edwards are widespread in several regions of the mountainous
system of the northern hemisphere — Altai, Tian-Shan, the Himalaya, Japan and
the Rocky Mountains of North America. Described for the first time in 1922,

these insects have remained incompletely studied until recently, with the greater part of all known data concerning the larvae. Fossil remains are still unknown.

Deuterophlebiidae occupy a very isolated position in the systematics of the Diptera, showing some points of resemblance to Blephariceridae in larval life and structure, and some to Chironomidae in antennal structure of larvae and adult, and in the respiratory system of the pupa. The Deuterophlebiidae are unusual enough for us to assume that they broke away from the general line of Tipulomorpha and Blephariceromorpha a long time ago.

Chief features. – The characteristic way of life in a high mountain terrain, and the habitat of the larva on the rocks of fast mountain streams, determine the features of these obviously relict forms and agree closely with similar features in Blephariceridae which also live under similar conditions.

The larvae (fig. 8*A*) show little obvious concentration of the individual parts. They have a free head with enormous antennae, a three-segmented thorax without appendages, and an eight-segmented abdomen, which bears on lateral protuberances of the first seven segments strong circular suckers which allow the larva to live and move about in a rapidly-flowing stream of water. The complex mandibles of the larva suggest that it feeds on various algae and protozoa growing on the rocks under water, which the larva obtains by filter-feeding. The respiratory system is closed, and gas exchange is apparently carried out by means of peculiar cutaneous protuberances or gills on the last segment of the abdomen.

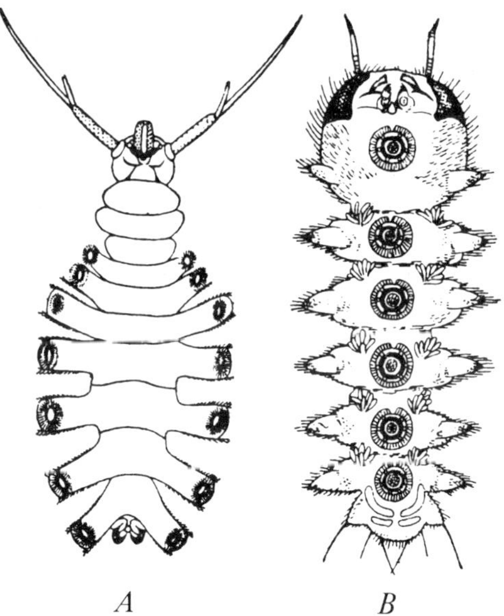

A   *B*

Fig. 8. Larvae of a deuterophlebiomorph and a blephariceromorph. *A. Deuterophlebia japonica* Kitami. Ventral view. *B. Blepharicera fasciata* Westwood. Ventral view. *(A.* after Kitami, from Hennig, 1948-52; *B.* after Lindner, 1930).

The pupa is oval, convex above and concave below, shorter than the larva, without abdominal protuberances, and with a pair of branching anterior thoracic respiratory structures. It is immovably attached to the substrate by three pairs of glandular suckers situated on segments 3-5 of the abdomen. Typical desimaginization is expressed by the imaginal aphagia and the reduced mouthparts of the winged insect. The head of the adult is small, concealed under the large anterior section of the very convex thorax, which projects in front. The eyes are enormous, and so are the antennae, which however comprise only six filiform segments. The wings are very large, and this together with the strong convexity of the mesothorax bear witness to considerable powers of flight. Central nervous system, excretory and productive systems are unknown. The larva and pupa are protected by a very thick cuticle; the winged insects are paleopterous (see p. 27), and the tarsi have strongly developed empodia, and are sexually dimorphic in the development of the claws. (Lindner, 1930a).

Factors which influenced the evolution of the Deuterophlebiidae. — The desimaginization and aphagia of the winged insect reflect the chief factor which has influenced the evolution of the Deuterophlebiidae. The feeding of the larvae in pools, combined with the severe conditions of existence which were encountered by the winged adult (altitude, deficiency of food, changes of temperature) contributed to the evolution of a short-lived, non-feeding adult insect. The decisive factor may have been the perfecting of feeding by the larva, combining the power of attachment to the substrate under fast-flowing water, by means of the abdominal pseudopodia, with the ability to collect the available food by means of the freely moving thorax and head. This made possible a high degree of stenobiosis, with the larva withdrawing into those parts of the stream with the swiftest current. Meanwhile the winged insects developed their powers of flight, their antennae and their prehensile legs against the hazards of aerial life, while abondoning feeding, and shortening the length of adult life.

Conclusions. — Deuterophlebiomorpha are clearly a relict group of flies, which have developed desimaginization, and perfected the larval phase, while becoming a well-protected stenobiont, living in extreme conditions in rapid mountain streams. Living a short time only, the winged insect is characterized by aphagia, with retrograde development of the digestive organs, together with great development of the flying apparatus and sense organs. Inadequacy of data about the feeding of the larvae, and the way of life and reproductive behaviour of the adults does not allow us to outline more accurately the evolution of this rare and peculiar group.

Infraorder Blephariceromorpha
Rohdendorf, 1961b, p. 158.

Extent, evolution, systematics. — The sole family Blephariceridae contains about 161 families, arranged in five subfamilies: Edwardsininae with one genus and approximately 20 species; Blepharicerinae with eight genera and over 70 species; Paltostomatinae with six genera and upwards of 40 species; Hapalotrichinae with three genera and about 10 species; and Apistomyiinae with

three genera and about 20 species. There is no doubt at all that at least twice as
many species exist, so that the figures quoted above have only relative
significance; evidence for this lies in the narrow distribution of the majority of the
known species, which are associated with mountain streams. Blephariceridae are
world-wide, but reach their greatest variety in the Oriental and Neotropical
Regions.

The individual subfamilies are characterized by a definite geographical
distribution; Edwardsininae inhabit Australia and the southern Neotropical
regions; Blepharicerinae the Holarctic and Oriental regions; Paltostomatinae the
Neotropical, Ethiopian (southern Africa), Australian and part of the Oriental
regions; Hapalotrichinae chiefly the Palaearctic regions (one species known from
Japan), and Apistomyiinae chiefly the Oriental regions, with one species known
from Australia and one from southern Europe. There are no fossil records and so
their relationships must be judged entirely from the living forms. They are peculiar insects, with most affinity to the Deuterophlebiomorpha and the Tipulomorpha, especially to the Chironomidea as a whole.

Examining the relationships of the Blephariceridae more precisely, it must be
recognized that they are profoundly different from all other living Diptera, and
any resemblances are convergent features arising out of their aquatic way of life in
the larval stage. This isolation, in conjunction with the presence of certain primitive features – notably the large, free mandibles in the adult, the thin legs, the
homonymous segments of the antennae, and the multisegmented abdomen –
compels us to assume that this group must have separated from the general ancestors of the Tipulomorpha long ago, probably not later than the Lower Jurassic.

Chief features. – The life-history and way of life of these peculiar Diptera
invites comparison with that other obviously relict group, the Deuterophlebiomorpha. The larvae of both groups live (sometimes together!) in the fast-flowing
water of mountain streams, crawling over, or tightly attached to underwater
rocks. These two groups, only distantly related to each other, developed similar
features through convergence as a result of living under similar conditions in these
fast mountain streams: particularly convergent features of larvae and pupae,
mainly of a protective nature (fig. 8*B*). A comparison of the features in these two
groups should therefore give us an insight into the evolutionary problems associated with this environment (Lindner, 1930a).

The larvae exhibit considerable integration of the body-segments, during which
the head has become united with the thorax and the first abdominal segment to
constitute a peculiar 'cephalothorax'. The posterior end of the abdomen forms
another intricate complex of smaller dimensions, and only the middle four or five
segments of the abdomen remain free. The abdominal segments usually have
lateral protuberances armed with protective or tactile spines, 'pseudopods'.
Sometimes the whole body is strongly contracted and the lateral protuberances
poorly developed, so that the larva acquires the shape of an elongate oval. The
body is strongly convex dorsally, with a thick cuticle sometimes provided with
spines; ventrally it is flat, and the cuticle is far thinner.

The larva is provided with very characteristic equipment for anchoring it firmly to the substrate against the pull of a strong current. Five large, unpaired circular suckers are located along the midline of the abdomen, and one on the 'cephalo-thorax'. Mouthparts are of a chewing type; the mandibles are simple, but the maxillae and labium are provided with thick bristles. The respiratory system is a closed one: gas exchange is accomplished in the larva by means of clusters of gills located in pairs on the ventral side of the abdominal segments, whereas in the pupa there are short, closed, anterior thoracic structures, 'horns', of complex structure. The pupa is depressed, highly sclerotised, almost 'plated', and reminiscent of the *pupa obtecta* of the Lepidoptera. The internal organization of the larva is unknown.

The winged insects are characterized by very large wings with peculiar veining, but without any obvious mechanical improvements such as additional phragma. The thorax is moderately enlarged, but convex, thus suggesting that it is the dorso-ventral muscles — which raise the wings — that are specially developed. The head is large, with moderately long antennae, but has enormous eyes that are often complex in structure. The mouthparts are quite primitive, with free, toothed mandibles; in some, Edwardsininae and others, the mouthparts may be reduced. Apparently aphagia is a recent evolution in these flies, because many of them are still active predators; they capture small insects (Komárek, 1914; Lindner, 1930a), and visit flowers.

The structure of the central nervous system, digestive, excretory, and reproductive systems, as well as details of reproduction, are all quite unknown.

Protective features are thus clearly evident in larva and pupa, but not obvious in the adult which remains paleopterous. The legs are of a clearly-defined, thin type.

Evolutionary problems and determining factors. — The chief problem encountered during the evolution of the Blephariceridae was undoubtedly that of providing protection for the larvae as they browsed on algae growing on the rocks and bottom surface of pools. As they browsed they were obliged to move slowly, and their protective responses were to develop a thick, armoured dorsal cuticle, and finally to withdraw into the fast, cold mountain streams which had fewer predators, even though they were also poorer in food supply. The algal growth was not a very efficient source of food, and the larva had to consume large quantities of it, hence its slow-moving habits. The simple mouthparts suggest that mechanical improvement in feeding-mechanisms was not an evolutionary advantage.

It was probably only late in the evolution of this group, after it had withdrawn into mountainous areas, that the winged adult began to be affected, and the primitive entomophagia and the acquired nectarophagia began to be replaced by adult aphagia during the process of desimaginization. It is probable that larval feeding improved at this time, but there is no proof of this.

It is very interesting to compare Blephariceridae with Deuterophlebiidae at this point (fig. 8*A*, 8*B*). Blepharicerid larvae have the head and thorax fused together, and there is little mobility of the sucker complex of the mouthparts. These features are associated with rather inefficient algal browsing by the larvae, and in turn are matched by a more active adult life, the adults being comparatively long-lived, and most species feeding actively on insects. Deuterophlebiid larvae have less concentration of head and thorax, greater feeding mobility, and hence probably enjoy more efficient feeding at the larval stage; deuterophlebiid adults have carried the regressive process further than Blephariceridae, and have ceased to feed at all.

In both groups, as in some Tipulomorpha (notably Psychodidea), the larvae have become adapted to feeding on the surface of the substrate, and have not evolved the wormlike larval form so common among Diptera. The adult form of Blephariceridae underwent relatively little change during evolution, and the determining factors were sexual activity and reproductive behaviour. These influenced the wing-structure and the most peculiar eyes, which occur in both sexes.

Conclusions. – The determining factors in the evolution of the Blephariceromorpha have been those associated with a habitat of extreme stenobiosis, arising from withdrawal into mountain streams. In general there was no improvement in the efficiency of larval feeding to induce aphagia in the adult, and so the adults mostly retained the ancestral ways of feeding. The structure of the adults of modern Blephariceridae is attributable to the requirements of feeding and reproductive behaviour under the difficult conditions of mountainous terrain: improved flight, delicate legs, bigger eyes.

Infraorder Tipulomorpha
Rohdendorf, 1961b, p. 154 (syn. Polyneura Brauer, 1880;
Tipuliformia – Culiciformia Hennig, 1948)

Extent, evolution, systematics. – This is a vast complex of eight superfamilies: Pachyneuridea, Tipulidea, Psychodidea, Culicidea, Dixidea, Chironomidea (syn. Tendipedidea), Thaumaleidea (syn. Orphnephilidea) and Rhaetomyiidea. There are 18 families, with more than 16,000 species, which hitherto have been grouped as 'Nematocera Polyneura'.

Representatives of this infraorder are among the oldest known Diptera, having been found even in the oldest fossils of the Upper Triassic of Central Asia. Among the living Tipulomorpha the superfamilies are very unequal in size. Tipulidea is the biggest superfamily, with about 8,300 species, followed by Chironomidea with about 5,100 and Culicidea with 2,000. The remaining superfamilies are much poorer in species: Psychodidea includes only about 900 species, and the others still fewer, with Dixidea about 100, Orphnephilidea 70 and Pachyneuridea only five species. Finally, the relict family Rhaetomyiidea is represented by two species. The most diverse superfamily of the contemporary fauna, Tipulidea, is also the most ancient, and was present even in Upper Jurassic and Triassic times as families very close to those of the present day. Next in order of diversity, the superfamily Chironomidea has also been discovered in the fauna of the Upper Triassic, but its representatives at that time were still quite archaic, and obviously much more primitive than those of the present day.

Separation of the superfamily Culicidea undoubtedly took place in the Cretaceous, since it is completely absent from the Jurassic, yet representatives of
present-day families were already present in the Baltic amber of the Upper
Eocene. The history of the Psychodidea is very poorly known, since evidence so
far discovered of its existence in the Jurassic is very unreliable, and the first
authentic Psychodidea do not appear until the Tertiary. The history of Dixidea is
curious in that they appear in the Upper Jurassic of Karatau in the form of a
peculiar extinct group, Dixamimidea. Those clearly relict superfamilies of the
present day, Orphnephilidea and Pachyneuridea, are not well known from fossils.

The systematic composition of the various superfamilies helps to explain why
some of the families appear to be relicts. Thus the biggest superfamily, Tipulidea,
has six surviving families of very unequal size. The largest of these are Limoniidae
(with about 5,600 species) and Tipulidae (2,500 species), whereas the other four
have many fewer: Trichoceridae and Ptychopteridae have 65 species each,
Tanyderidae 36 and Cylindrotomidae only 27. The last four have the appearance
of relict families. The second superfamily, Chironomidea, comprises three families
of which the largest is Chironomidae (syn. Tendipedidae), with no fewer than
3,100 species, and thus considerably exceeding the second, Ceratopogonidae (syn.
Heleidae), with 1,150 species, and the third, Simuliidae with only about 750. The
superfamily Culicidea contains only two families, and the vast majority of the
species (about 1,970) belong to the one family Culicidae. The superfamily Psychodidea includes three families, of which the Psychodidae and Phlebotomidae each
have several hundred species, and of which the Nemopalpidae are clearly a small
relict family. The relict nature of the superfamilies Rhaetomyiidea (only two
species of the family Perissommatidae) and Pachyneuridea (five species of the
family Pachyneuridae) is especially evident.

The extent and character of the different groups is shown in fig. 9.

Chief features. – The most characteristic feature of Tipulomorpha is possession
of an aquatic larva, as in Chironomidae, Culicidae, Dixidea, Orphnephilidea, most
Tipulidea and some Psychodidea. Only a minority of Tipulomorpha have terrestrial larvae: many Psychodidae develop in decaying substances; some Tipulidea,
notably many Tipulidae, live as larvae in soil or rotting wood, but the former at
least is a secondary development (Gilyarov, 1949); a few Chironomidea, especially
Ceratopogonidae, live as larvae in ant hills, plant residues and moss.

Larval food is very varied, and in addition to a widespread practice of feeding
on detritus, many groups have predatory members. Tipulomorphs rarely feed on
living plants nor live as larvae in their tissues. The so-called 'microphagia' of larvae
of Culicidae and Simuliidae – i.e. feeding on microorganisms which they collect
and filter by means of specially converted antennae and mouthparts – is a very
peculiar development.

Adult feeding habits, too, are very varied. Feeding on nectar and sucking
various plant juices are the 'norm' for Diptera (Tipulidea and others), but
predation and parasitism towards other insects also take place (Ceratopogonidae).
Especially characteristic of Tipulomorpha is the development in some groups of
blood-sucking on vertebrates – as in the majority of Culicidea (Culicidae), some

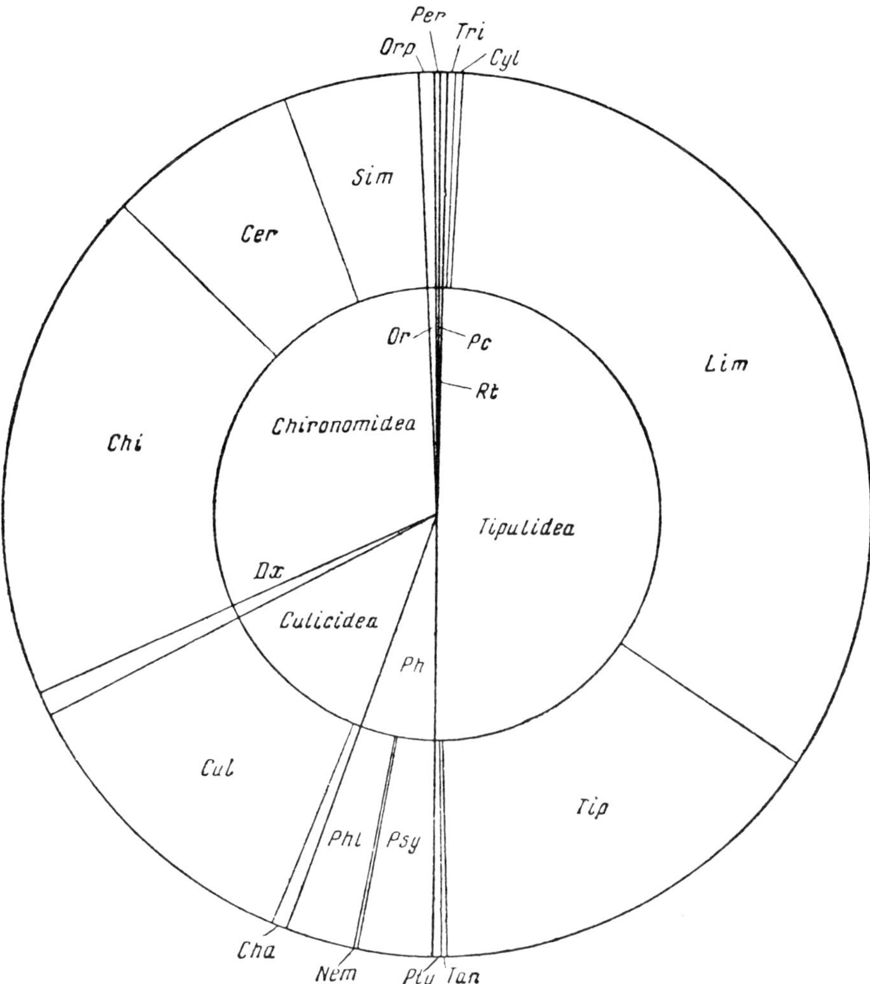

Fig. 9. Diagram of the composition of the infraorders of living Tipulomorpha. Superfamilies and families are indicated as the sectors of a circle, the sizes of which correspond to the number of included species (1° of arc equivalent to 46 species). In the centre of the circle are shown the superfamilies, and on the circumference the families. (Original.) For abbreviations, see p. xiii.

Psychodidea (Phlebotomidae), and Chironomidea (Ceratopogonidae) — or else complete aphagia — as in the majority of the Chironomidea (Chironomidae), Dixidea, and probably some Tipulidea, Culicidea (Chaoboridae) and Psychodidea.

The general geographical distribution of Tipulomorpha is very wide, and the chief superfamilies of the infraorder — Tipulidea, Culicidea and possibly Chironomidea — reach their greatest variety in the humid tropics. Tipulomorpha are ecologically diverse too; they occur mainly in damp vegetation, but some members of the infraorder may be found in quite arid places: larvae of some

Chironomidea and Culicidae inhabit brackish or even salt pools, while some
Psychodidea (Phlebotomidae, among the smallest of Diptera), find larval 'microhabitats' in the soil of arid and semiarid regions.

The morphology of the larvae of Tipulomorpha is very diverse, particularly in
the many modifications of the respiratory system which result from aquatic
habitats and have led to the evolution of a metapneustic or even an apneustic
tracheal system (Chironomidea). The morphology of the larval head is very varied,
and comprises both examples of unspecialized mouthparts of a biting type (some
Tipulidea – Trichoceridae), and of a range of peculiar adaptations for feeding on
microorganisms (Culicidae, Simuliidae). The larval mouthparts are only very
rarely reduced in Tipulomorpha. Only in one small group, Leptoconopinae of the
Ceratopogonidae, does the extreme reduction of the head capsule and its
appendages result in a peculiarly 'muscoid' type of larva, apparently adapted to
the external digestion of food that is so often a feature of highly evolved Diptera.
Another feature of many Tipulomorpha (Tipulidea, Chironomidea and Orphnephilidea, some Psychodidea Culicidea and Dixidea) is the mobility of the head
capsule, associated with hypognathism (together with the production of peculiar
protuberances on the first body-segment); this testifies to the active selection of
nutritive material, and has led to the development of powerful muscles which
move the head (Hennig, 1948-52).

Adult Tipulomorpha in general may be characterized by the almost complete
absence of oligomerization of the body; the vast majority of members of this
infraorder possess a long abdomen, which is always much longer than head plus
thorax, and which consists of many homonomous segments. A few exceptions are
the blackflies (Simuliidae) with short bodies, and perhaps the Psychodidae and
Orphnephilidae. The head is relatively small, and usually narrower than the
thorax. The sense organs of the head – eyes and antennae – are developed in a
peculiar way. The eyes are large, often occupying a great part of the head, and
sometimes there is a sharp sexual dimorphism, with holoptic males. Only rarely
are the eyes small, as in some Tipulidea. Especially characteristic is the structure
of the antennae which are almost always long, and provided with numerous
patches of hairs, and which are often sexually dimorphic, the males having huge,
fluffy antennae. Oligomerization of the antennae, with conversion into short
appendages with few segments, is quite rare, as in Simuliidae, Orphnephilidae and
some Limoniidae.

The mouthparts are very diverse. In some members a piercing type of proboscis
incorporates a complete set of mouthparts, mandibles as well as maxillae, whereas
in others the mouthparts are reduced to form a soft proboscis suitable for sucking
the juices of plants.

The thoracic structure is characterized by the great development of the prothorax, the dorsal section of which often projects forwards and is visible from
above. The legs of Tipulomorpha are mostly of the thin, aerial type (Rohdendorf,
1951), and only a few forms possess running legs, notably some Psychodidea, and
Simuliidae of the Chironomidea. The structure of the wings is more varied. The
primitive lifting type, scaly, wide and costalized, is seen in only one group of the

Psychodidea, and the others are developed into various high-thrust types. The chief characteristic of the wings of Tipulomorpha is their elongation, the few exceptions — Psychodidae, Simuliidae, Perissommatidae, some Ceratopogonidae — being purely secondary, and in no way contradicting the general principle. The elongation of the wings indicates an important feature of the wings of the ancestral tipulomorphs, namely the position of the radial veins, branches of the radial sector, which are directed towards the apex of the wing, and not curved forward as they are in Bibionomorpha, as a result of the shortening of the wing. The phenomenon of costalization is not a characteristic of Tipulomorpha, and comes about very differently in different groups.

The abdomen of Tipulomorpha is usually cylindrical, and elongate, and hardly ever undergoes reduction, except in a few groups such as Simuliidae, Orphenephilidae, and Ceratopogonidae.

Factors which influenced the evolution of the Tipulomorpha. — The major problem in the evolution of Tipulomorpha was the ancestral colonization of an aquatic medium by a terrestrial, soil-living larva moving into freshwater pools. This proved to be a very significant step in the evolution of Diptera, since moving the larva away from living in a mass of soil meant that the adult had to become more mobile in order to find pools which were few, scattered and transient. Although the water was an 'extensive medium' it introduced new problems, particularly that of being able to breathe under water. Legless larvae were not mobile enough for an aquatic habitat, and a deficiency of food developed. The solution of these problems is the history of the Tipulomorpha, the decisive factors being an improvement of the respiratory system of the larvae, with development of first amphi- and then metapneustic types, and then the development of predatory habits. In this instance the winged adult phase did not in fact become more mobile, because the margins of pools, swamps, etc., provided an abundance of oviposition sites. Instead, it was insufficiency of food in the larval stage that led the adults to preserve and even to improve their adult feeding. These problems, and their solutions, differ sharply from those experienced by other infraorders, and particularly from the Bibionomorpha which were evolving during the same period (Rohdendorf, 1951, 1958, 1959, 1961a).

Let us now look at the superfamilies of the Tipulomorpha individually, and briefly consider the various factors that have influenced their evolution.

Superfamily Tipulidea

The largest superfamily of the infraorder, comprises six families of very different size and distinctiveness (fig. 10, 11, 12). It will be seen at once that two of the six families, Tipulidae and Limoniidae, contain an overwhelming majority of the species in the superfamily, over 8,000 in all. The rest are certainly relict forms, and comprise only 200 species. Tanyderidae are clearly a relict family, with only about 40 species divided among 11 genera, and found chiefly in the southern hemisphere. Cylindrotomidae are cosmopolitan, with about 30 species in nine genera. Finally, Ptychopteridae, with 65 species in three genera, and Trichoceridae with 65 species in five genera are families of about the same size, the former with two subfamilies almost entirely in the northern hemisphere, and the latter found in all

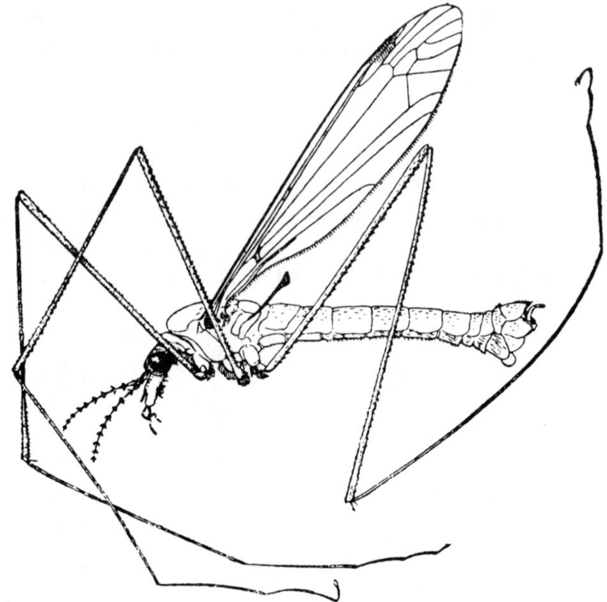

Fig. 10. *Tipula paludosa* Meigen (Tipulidae). Male, general view. Length of body 20 mm. (After Lindner, 1928.)

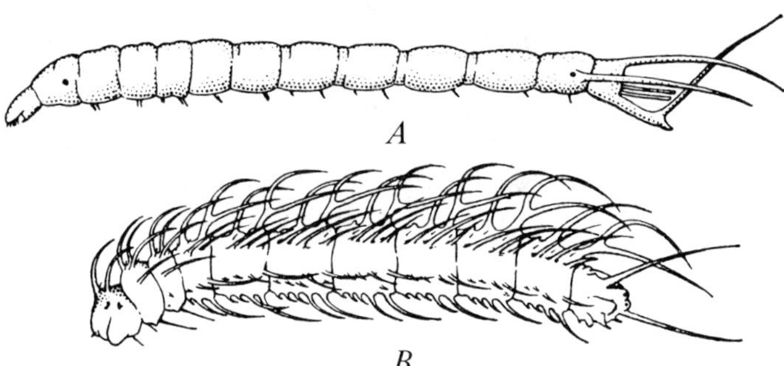

Fig. 11. Tipulidea: larvae. *A. Protoplasa fitchii* Osten Sacken (Tanyderidae). General view from the side. *B. Phalacrocera replicata* (L.) (Cylindrotomidae). General view from the side. (After Hennig, 1948, enlarged.)

continents except the Ethiopian region. The relict character of these families has been accepted by Curran (1934), Hendel (1937), and Hennig (1948-52, 1953, 1954).

On the whole this superfamily is well characterized by the development of elongate wings, of low thrust; a complex system of numerous pleural muscles (fig. 13); thin legs of the bristly subtype; and by polyphagous larvae which live on various rotting plant substances, on living plant tissues, or even as active predators. Larvae of Tipulidea (fig. 11) are characterized by the powerful head, having a solid head capsule, (epicranium), sometimes with deep grooves on the posterior margin (Limoniidae, and part of Tipulidae), capability of retraction well into the thorax by the contraction of strong muscles, and finally, by bearing very powerful and large mandibles which are undoubtedly protective organs. All these features indicate an improvement in the feeding of the larvae, which became able to feed on a wide variety of substances. They originated from the aquatic ancestors of the Tipulomorpha, and initially they did not leave the water. Only later on did a continued improvement in feeding ability allow the larva to widen the range of its ecological distribution, and to pass into terrestrial habitats. The evolution of the ability to feed in the tissues of living plants allowed the larvae to become soil-dwellers (Tipulidae, Cylindrotomidae, some Limoniidae).

The evolution of the individual families of the Tipulidea can at present only be deduced from a thorough study of the contemporary fauna of the group, which is especially rich and widely distributed in the tropics. The data to be found in the literature about the two principal families Tipulidae and Limoniidae are almost entirely concerned with taxonomy, and throw very little light on the habitat of these groups. The evolution of these families is a problem of great importance (see Gilyarov, 1949), but the families are so big that the problem will need to be treated as a study in itself. The evolution of the other, relict families, suffers from lack of data, but the peculiarities of these families are obviously responsible for their becoming relicts, and so in a general way we can deduce some of the factors which have been responsible both for their peculiarities and for their prolonged survival.

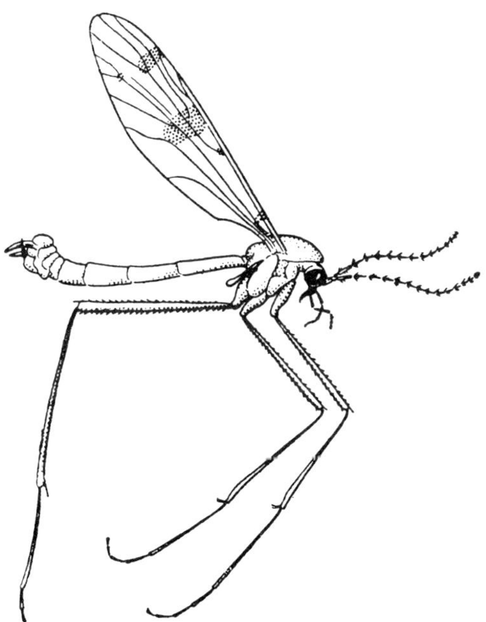

Fig. 12. *Ptychoptera contaminata* (L.) (Ptychopteridae). Male, general view. Length of body 9 mm. (After Lindner, 1928.)

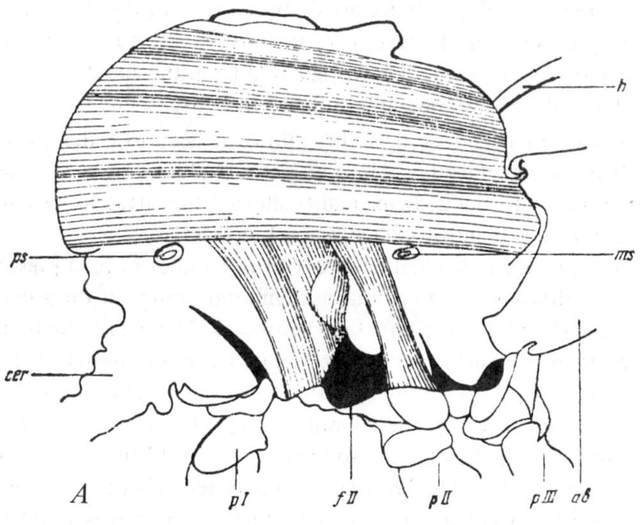

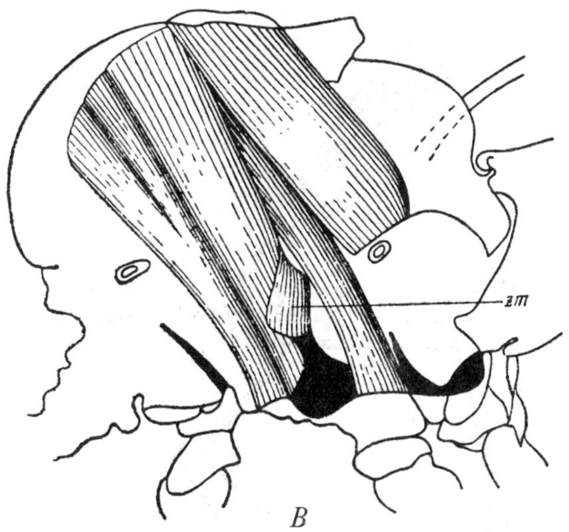

Fig. 13. *Limonia quadrimaculata* (L.) (Limoniidae). Male, muscular apparatus of right half of a thoracic section. Length 4.9 mm. *A*. Longitudinal muscles. *B*. Dorsoventrals. (Original, diagrammatic). Abbreviations: *ab* – base of abdomen; *cer* – cervical section; *FII* – furcasternite of mesothorax; *h* – haltere; *ms* – posterior thoracic spiracle; *pi, pii, piii* – anterior, middle and posterior legs; *2m* – intermediate muscle between furcasternite and pleural rib; *8, 10, 14, 15, 17, 19, 25, 27* – corresponding pleural muscles.

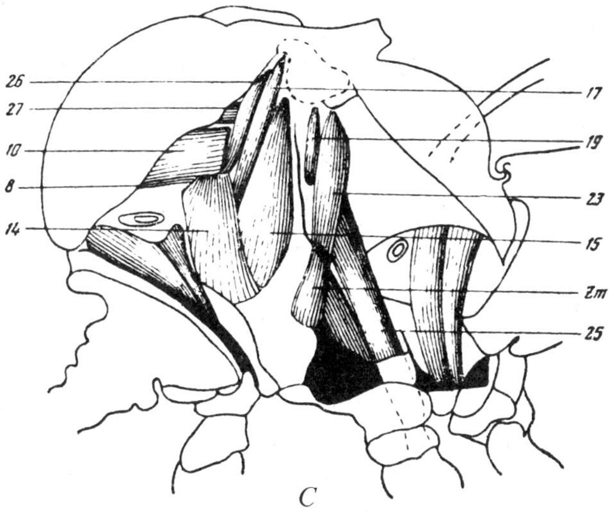

Fig. 13 (cont.) *Limonia quadrimaculata* (L.) *C.* Pleural muscles. (Original, diagrammatic.) (for abbreviations, see p. 42).

The larvae of the Ptychopteridae are unusual in combining a free, isolated and quite primitive head capsule with extremely elongate posterior spiracles at the end of a telescopic tube, features which point to a significant expansion of the ecological range. The larva is now able to inhabit deeper layers of shallow pools, and has greater freedom of movement away from the surface film (Hendel, 1928). Another feature is that both larvae and, even more, pupae of Ptychopteridae have much thicker cuticle, sometimes provided with a dense spiny armament; the protective nature of such an armament is obvious. This present-day group of Tipulidea is far removed from other families, being a remnant of some Jurassic groups absent in the Cenozoic (see p. 292).

The other family of this group, Cylindrotomidae, is poor in species, but we know enough about it to indicate something of its evolution, which is partly similar to that just considered. In particular there was a change in the larval cuticle, which acquired a heavy protective covering of long, thin protuberances which cover the whole body (fig. 11*B*). At the same time the Cylindrotomidae are interesting in that their larvae have evolved away from an aquatic medium, living in mossy outgrowths where the conditions may be quite different from those of a fully aquatic medium. This family is comparatively near to the Limoniidae and presumably, like the latter, is a derivative of the Mesozoic Architipulidae.

A particularly important problem is the evolution of the well-known winter gnats of the family Trichoceridae. These insects are characterized on the one hand by a series of ancient features (larva with an isolated head of simple structure, and adult with elongate tipulid-type wings, with a primitive, basal articulation)

(Rohdendorf, 1951, p. 49, fig. 22); and on the other hand by being able to exist in caves, burrows, at great altitudes, and in low temperatures, where they remain active even in winter. Furthermore these little gnats have become partly synanthropic insects, having populated basements and cellars in towns, and occurring in large numbers in tunnels and in mines. In fact, the activities of man have favoured these flies by greatly increasing the number of their possible habitats.[3]

The ability of Trichoceridae to colonize such 'extreme' habitats as these points to the way in which they have managed to survive into the present day, and makes their relict status apparent. It is still impossible, however, to say just how these insects are able to tolerate conditions that are so abnormal for other insects. It must be assumed that the peculiarity of Trichoceridae lies somewhere in their metabolism, but this is a problem for the future.

Tanyderidae are an obviously relict group of Tipulidea about which we still know very little, and so it is pointless to speculate about their possible evolution. Tanyderidae are remarkable for combining many archaic features in the adult with various specialized features in the larvae (fig. 114), which have not yet been sufficiently investigated (Hennig 1948, 1952).

The second superfamily of Tipulomorpha in point of size is the Chironomidea, consisting of three very dissimilar families (fig. 14). The most numerous family, Chironomidae, includes 3,135 species, about two-thirds of the total number of species in the superfamily, and is divided into about eight subfamilies and a large number of genera. The family is cosmopolitan, ranging up to high latitudes, and is one of the most important constituents of the entomobiocenoses of humid terrains.

Next in order of size come the Ceratopogonidae, containing about 1,150 species, and divided into three subfamilies and about 30 genera, with world-wide distribution. The third family, Simuliidae, has upwards of 750 species, and is made up quite differently from the other two, being a relatively monolithic assembly of only 15 to 20 closely related genera.

The superfamily Chironomidea as a whole is characterized by the strongly costalized wings of the thrust type, sometimes very broad. The muscular apparatus is very powerful, with well-developed dorsoventral and longitudinal muscles, and with a consequent strengthening of the episternites and posterior phragma of the mesothorax (fig. 15). The legs are of the thin type, often prehensile or even tactile, but never belonging to the bristly subtype. A particularly important feature of the larvae is that they are generally cut off from atmospheric air, and respire by means of tracheal gills. These two chief features of the Chironomidea — breaking away from the surface film of the water on the one hand, with respiration by means of tracheal gills; and on the other hand the perfection of the flying apparatus, with a powerful muscular system to give quick strokes to a strongly costalized wing — are most important to an understanding of the influences that have shaped the evolution of this superfamily of Tipulomorpha.

3. The adoption of synanthropic habits by Trichoceridae is an example of evolution in action. The majority of synanthropic insects were unimportant members of the fauna before the advent of man, probably mostly scattered relict species that lived in concealed habitats as a kind of refuge. The development of human activity brought these insects out from their natural retreats and permitted them to spread more widely. Such undoubtedly are the cockroaches; various Coleoptera that feed on detritus, in particular the Ptinidae; and some Diptera, including the relict window flies, Scenopinidae, the Rhyphidae (Anisopodidae), and possibly the Psychodidae. Similar considerations probably also apply to other groups of synanthropic animals. (See the work of V. N. Beklemishev, 1948, 1951, 1954, 1957.)

The capacity of breathing directly in an aquatic medium, without contact with the atmosphere, has allowed the descendants of the ancient Tipulomorpha to extend their ecological range within a given piece of water, and in particular to colonize rapidly flowing water and even deep water. The primitive metapneustic or amphipneustic larvae of the original Tipulomorpha had been tied to the surface film, or to the shallow littoral, and had not been able to live in rapidly flowing water nor on the bottom because of the difficulty of obtaining sufficient air. No less important are those biocenotic changes that the original forms of Chironomidea underwent when they passed from the surface layers to the bottom; especially the advantages in the way of protection that were enjoyed by the ancestors of the Chironomidea when they were able to move away from the surface, where they had been exposed to many predators, including dragonflies, beetles, fish and birds.

The stronger wings and more powerful flight of the adults can be seen as an outcome of their acquisition of a benthic habit as larvae. The new larval habitats were local and deficient in food material, and stimulated the adults to disperse more widely, and to feed more actively.

The phylogeny of the different families of the Chironomidea is poorly known, and it is possible only to indicate some of the most general characteristics of the principal groups. The families of the Chironomidea are very different, as exemplified by the feeding, respiration, and by the morphology, feeding habits and locomotion of the adults.

The midges, Chironomidae (Tendipedidae) are most numerous in species, and are characterized by having very mobile larvae, living a benthic life in various aquatic habitats, from shallow water near the snowline to tropical lakes and to the seacoast. Oxygen content varies from saturation in mountain streams to mere traces in stagnant pools and ponds. Larvae of Chironomidae have been found at the remarkable depth of 325 m in Teletsk Lake (Lipina, 1949), where apparently Chironomidae are the only insects able to populate such deep zones.

Adult Chironomidae are characterized by aphagia, accompanied by a marked reduction of the mouthparts, and by strong sexual dimorphism, whereby the males have enormous fluffy antennae and large eyes, with the females having inconspicuous antennae and smaller eyes. The wings of midges are long, quite narrow and only moderately costalized compared with other families of the group; they belong to the narrow subtype of the thrust type (Rohdendorf, 1951). The middle and hind legs are very unusual, bearing peculiar spines or combs at the tips of the tibiae, and obviously being developed as supporting limbs. The anterior pair differ markedly from the others, and apparently do not take any part in supporting the body, but act as tactile organs[4] in both sexes, but especially in the males. These flies almost always hold the fore-legs freely suspended, extended forwards. Thus Chironomidae clearly show desimaginization.

The vast family of midges comprises a series of quite ancient secondary groupings, subfamilies and tribes, but a consideration of the phylogeny of the whole family is beyond the scope of this book. It can only be noted that different

---

4. The name of the genus *Tendipes* (now suppressed) well expressed this condition, from the Latin *tendere*, to extend, and *pes*, a leg.

groups vary in the way of life of the larvae — predatory, phytophagous and polyphagous — in the structure of the wings, and in body size.

The main evolutionary trends were determined by the perfecting of larval
feeding and respiration, accompanied by the development of aphagia in the
adults. In this connection it is important to note what has happened in the well-
known Pacific Ocean genus *Pontomyia* Edwards (fig. 14*D*. 14*E*)(Hendel, 1937), a
marine chironomid. Here the females have undergone extreme reduction, having
lost both wings and legs, and reduced the head and thorax, and become wormlike
animals which do not come out of the water. Males of *Pontomyia* have highly
reduced and modified wings, and have lost the power of flight, but run over the
surface film of the sea.

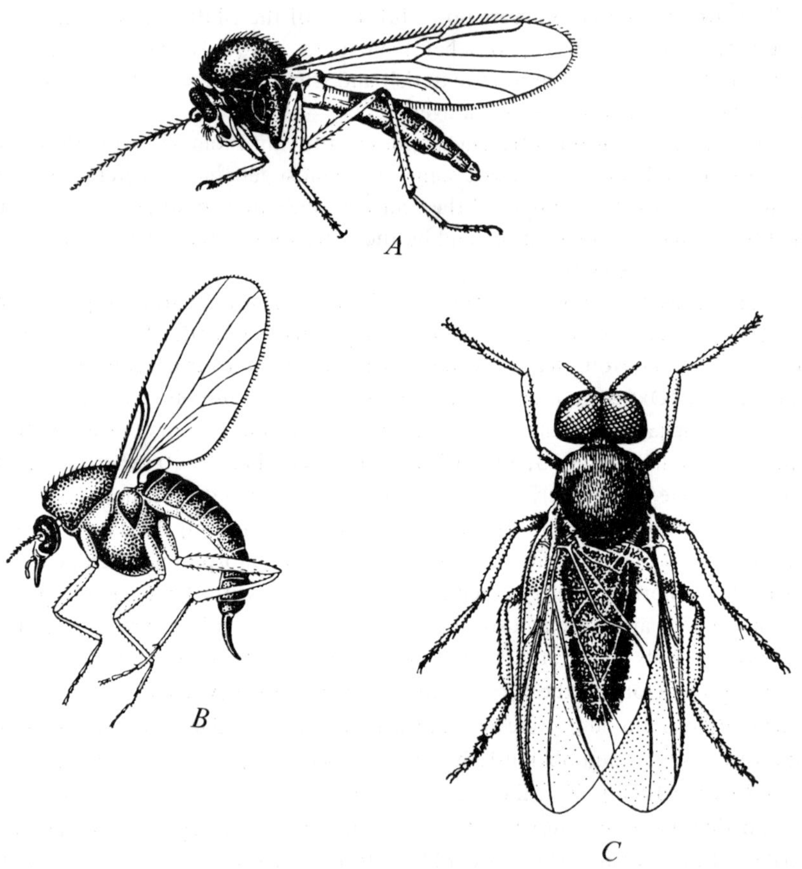

Fig. 14. Chironomidea. *A. Palpomyia flavipes* Meigen (Ceratopogonidae). Female, general view. Length
3 mm. *B. Leptoconops bezzii* Noé (Ceratopogonidae). Female, general view. Length 1.2 mm. *C. Simulium
noelleri* Friederichs (Simuliidae). Male, general view. Length 4 mm. (*A. B.* after Goetghebuer; *C.* after Rubstov, 1956).

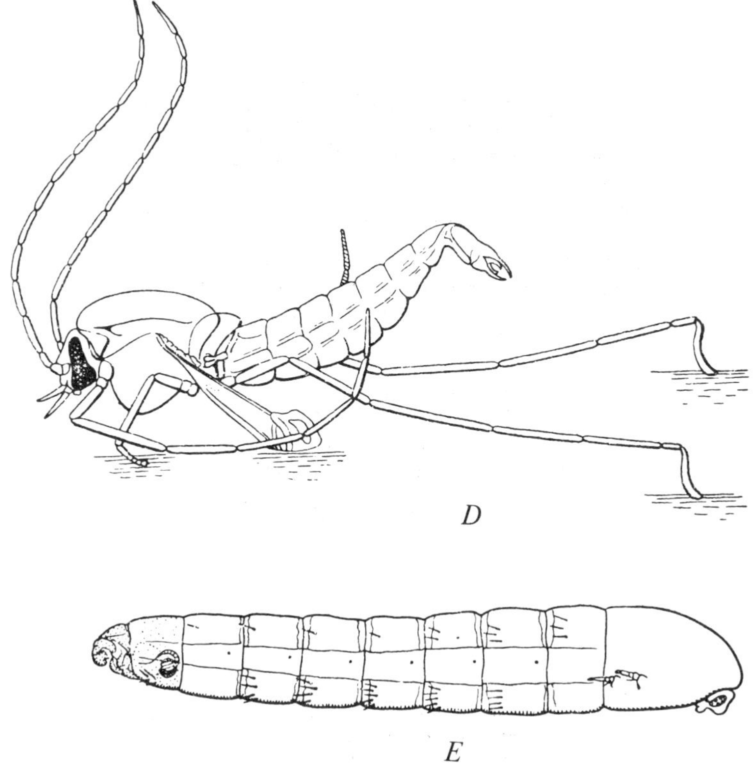

Fig. 14. (cont.) *D. Pontomyia natans* Edwards (Chironomidae). Male, insect on the surface of the water. *E.* The same. Female, general view from the side. (*D. E.* after Hendel, 1937.)

Another aspect of improvement of aquatic life in Chironomidae is their ability to live in stagnant pools. The development of a red respiratory pigment (haemoglobin) in the blood of the larvae, almost the only example in the whole Class Insecta, is undoubtedly a direct response to respiration under such unfavorable conditions.

Up to now the exact food material of the larvae has remained obscure, and it tells little to say in general terms that they are 'mud feeders', or 'predatory'. There seems to be little doubt that they have evolved an ability to utilize at least some components of mud, which is present in abundance, and this must be the most important step forward in their evolution, since it guarantees them plenty of food. Unfortunately this assumption still remains unsupported by precise facts.

Finally it is also very important to bear in mind the biocenotic changes which took place while the midges were evolving and becoming established. I have in mind the interrelation of these insects with their enemies, the many species of freshwater fish for which their larvae serve as a main item of food. It would not

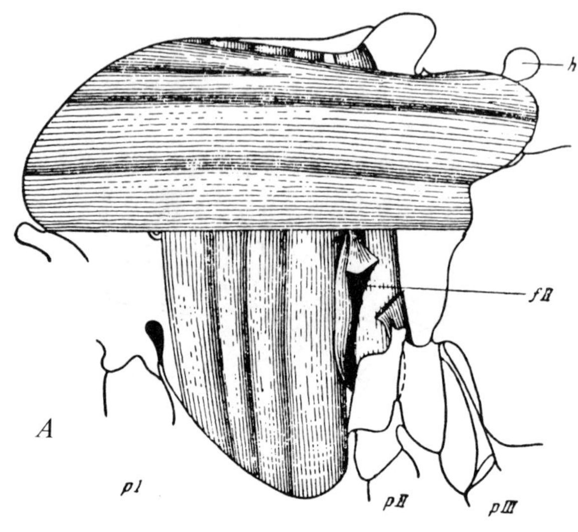

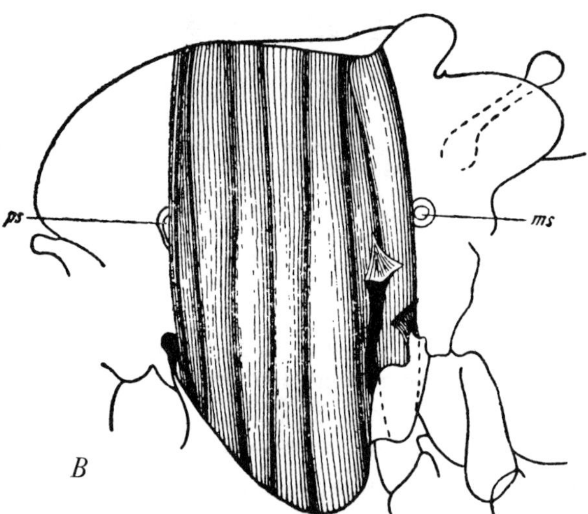

Fig. 15. *Tendipes plumosus* (L.) (Chironomidae). Male, muscular equipment of the right half of the meso-thorax. Length 3 mm. *A*. Longitudinal. *B*. Dorsoventral. (Original, diagrammatic.) (for abbreviations, see fig. 13. p. 42).

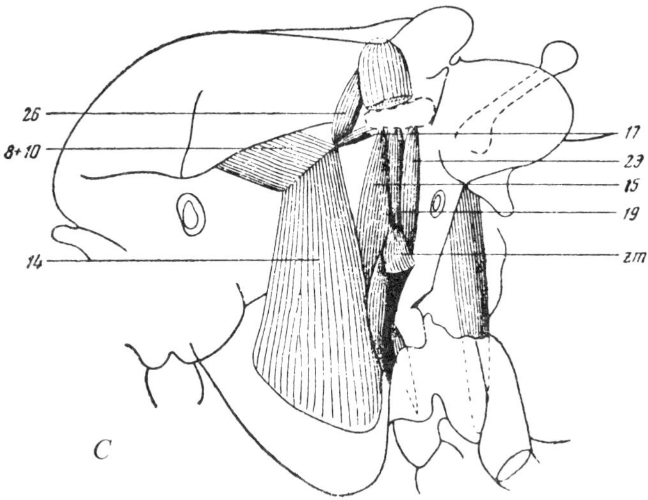

Fig. 15.(cont.) *Tendipes plumosus* (L.) *C.* Pleural muscles. (Original, diagrammatic.) (for abbreviations, see fig. 13. p. 42).

be an exaggeration to say that without the larvae of midges large populations of carp, eels and many other fish simply would not be able to exist. Such interdependence is, of course, an important part of the evolution of both groups.

The second largest family of Chironomidea, Ceratopogonidae (Heleidae) is still very incompletely known, though it is certainly a very interesting one (fig. 14*A*. 14*B*). An especially interesting feature is that one group of ceratopogonids, the subfamily Forcipomyiinae, have colonized terrestrial habitats, thus sharply isolating themselves not only from the other Chironomidea, but from the majority of Tipulomorpha. In this they approach another superfamily, Psychodidea, but it is necessary to point out that the terrestrial habits of Forcipomyiinae are evidently secondary.

Ceratopogonidae are sharply distinguished from Chironomidae in having well developed mouthparts in the adult, showing that feeding is important at this stage to Ceratopogonidae, in contrast to Chironomidae. Predation by adult Ceratopogonidae seems to be a primitive condition, and they seem to have started by feeding on insects and gone on to suck blood (Beklemischev, 1951). The importance of adult feeding is reflected in their whole organization — strong, prehensile legs, sturdy, costalized wings and integration of the whole body. Thus the Ceratopogonidae are shown to have followed quite a different evolutionary path from the Chironomidae, but, as already stated, we know so little about them as a group that it is useless to speculate about details of their evolutionary history.

It can be stated that the peculiar subfamily Leptoconopinae (fig. 14*B*) is a

definite derivative of one of the groups of Ceratopogonidae, arising apparently
following upon the evolution of extra-intestinal digestion in the larva, which has
lost the head capsule to become an acephalous larva of the muscoid type. Other
developments were ultra-costalized lifting wings and the habit of bloodsucking
from vertebrates.

The family Simuliidae is third in order of size, and is completely differently
characterized from any other Chironomidea. The larvae are strongly rheophilic,
and live a stationary life in a strong current of water. They are microphagous,
collecting minute food particles from the water by means of fan like appendages
of the head. The adults show great integration of the body, suck the blood of
vertebrates, and have developed broad, fan like wings of the type called 'strepsi-
dipterous' (Rohdendorf, 1949, p. 151).

All these characteristics reflect the influences that have acted upon the evolution of the Simuliidae. The original Chironomidea which were able to colonize the
rapids of a swift stream did so by making use of tubes, by attachment to the substrate and with the evolution of microphagia. Although this guaranteed for the
larva a habitat in the freer, less heavily populated zones of the stream, under more
sheltered conditions, it was also a habitat with insufficient food material, and with
other problems. Inadequacy of feeding of the larva called for intensification of
feeding by the adults, and so led on to improved mobility, integration of the
body, wider wings, development of the sense organs of the head, prehensile,
powerful legs, and finally bloodsucking from vertebrates. Such an evolutionary
history shows the importance of correlated changes in different parts of the body.
We must assume that this correlated evolution took place comparatively slowly,
and that Simuliidae are a relatively old group (Rubtsov, 1937, 1956).

Superfamily Dixidea (fig. 16)

There is only one surviving family, Dixidae, with about 100 species distributed
among six genera and subgenera, and clearly a rather specialized relict family. The
oldest members of this superfamily were found in the mid-Jurassic of Karatau
(family Dixamimidae, Rohdendorf, 1951; see below, p. 232).

This group of Tipulomorpha is undoubtedly closest to the ancestral Culicidea.
Until recently many authors even considered the superfamily Dixidea as no more
than a subfamily of the Culicidae, on the strength of larval similarity, disregarding
adult differences (Martini, 1929-30; Monchadski, 1936; Shtakelberg, 1937). A
similar disparity between larval and adult classification and organization is often
found among Diptera, especially among the older infraorders, and indicates that
specialization and adaptive change took place independently and at different rates
in larva and adult.

In both Dixidea and Culicidea the larvae became completely adapted to living
and feeding in the surface film of a pool, with microphagia, but in Dixidea these
steps occurred earlier and more completely than in the first Culicidea.
Consequently Dixidea progressed to adult aphagia, with a reduction in the length
of adult life (desimaginization, Rohdendorf, 1961). As a result the Dixidea led on
to the present relict family, exhibiting on the one hand some primitive characteristics (lack of scales, broad wing, simple antennae, short genital appendages) com-

bined with some specializations associated with the improvement of larval feeding, notably reduction of adult mouthparts and aphagia.

If the evolution of the Dixidea is compared with that of the Culicidea, it will be noted that the latter have gone a step further; they have evolved a coating of scales on wings and body, indicating some improvement of flying characteristics, and possibly some additional sensory function. It seems that the process of desimaginization has gone forward more quickly in Dixidea than in Culicidea, which, on the contrary, have concentrated on improvement of the winged insect.

Superfamily Culicidea

Third in point of size in the infraorder Tipulomorpha, Culicidea is sharply distinguished from two of the superfamilies mentioned above by being both monolithic and relatively monotonous (fig. 17). Only two families enter into its composition, Chaoboridae with only about 60 species, and Culicidae which includes all the rest. Culicidea are widely distributed all over the world, are an important component of nearly all the biocenoses of humid habitats, and consequently are of major significance for man. This is one of the most studied groups of Diptera, and especially the bloodsucking mosquitoes of the family Culicidae.

Culicidae is a well-known family, consisting of three subfamilies: Culicinae with about 1,700 species; Anophelinae with about 200, and Megarhininae with about 70. The most diversified subfamily is the Culicinae with about 30 genera, and the others are much smaller, Anophelinae with only three genera, and Megarhininae with a single genus, *Megarhinus.*

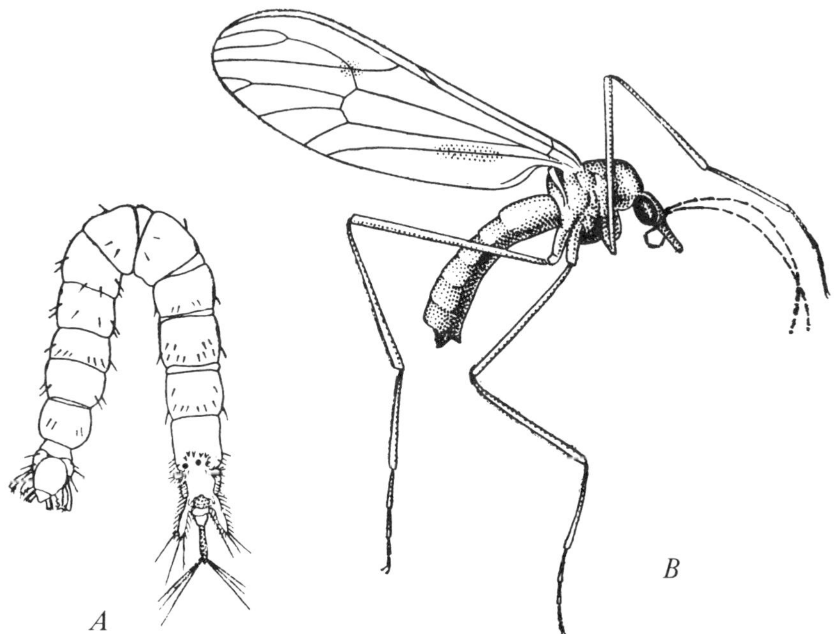

Fig. 16. Dixidae. *A. Dixa* sp. General view of a living larva. Note the characteristic curved body. *B. Dixa maculata* Meigen. Female, general view. (After Lindner, 1930.)

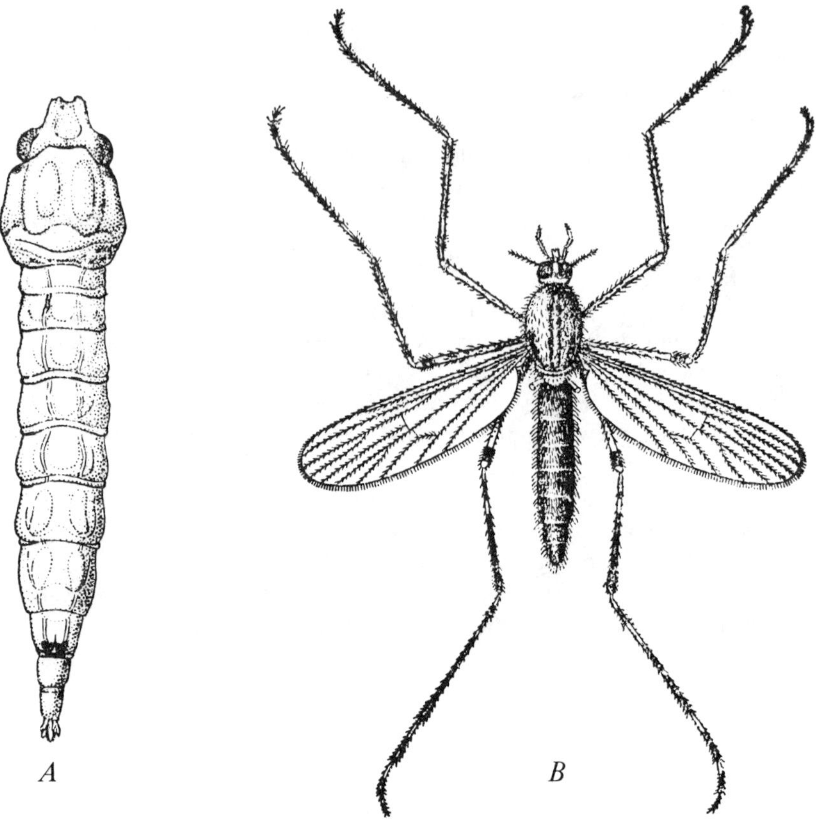

Fig. 17. Culicidea. *A. Cryophila lapponica* Edwards (Chaoboridae). General view of larva from above.
B. Mochlonyx cinctipes Coquillet (Chaoboridae). Female, general view from above. (*A.* after Monchadski,
1936; *B.* after Curran, 1934.)

The other family of Culicidea, the Chaoboridae, has been studied less thoroughly, and contains six genera of different sizes, three of which are monotypic.
Secondary groupings between genera within the family apparently exist, but are
not yet fully investigated (Martini, 1929-31).

Fossil records of Culicidea are negligible, and are limited to a few forms found
in Tertiary faunas, chiefly in Baltic amber. All the Tertiary species belong to
genera represented in the fauna of the present day. It is interesting to note that
the family Chaoboridae, poor in species at the present day, is comparatively well
represented in the fauna of the Paleocene, where six forms have been reported.
The phylogenetic relationships of the Culicidea are more or less clear: this superfamily is a derivation of unknown primitive groups of Tipulomorpha, which also
lay close to the ancestral Chironomidea and Dixidea. Culicidea approach most
closely to Dixidea, and they remain similar in their early stages, but differ sharply
in the organization of the winged adults, in particular in the peculiarly narrow,
scaly wings of Culicidea.

Examining briefly the characteristics of the Culicidea we must first note the presence of atmospheric respiration in the larvae, which are mobile, attached to the surface film, and have evolved 'microphagia'. Another characteristic is the development of bloodsucking in the adults, and an associated retention of a complete set of mouthparts, converted into a piercing proboscis. Furthermore they have narrow wings with peculiar scales and abundant pleural musculature (fig. 18). Undoubtedly the ancestral Culicidea evolved after their larvae had populated non-flowing or slowly flowing water with a quiet surface, possibly including some 'microhabitats' such as forest pools, puddles, temporary accumulations of water in the cavities of tropical plants, and so on. These waters were poor in fish, but rich in micro-organisms, and this fact shaped the evolution of the Culicidea. Improvement of larval feeding, with microphagia, coming to the surface film to breathe and generally increased activity, influenced the integration of the body, and stimulated the separation of the head and thorax of the larva in the way we see in mosquito larvae. Simultaneously the adult female became a competent bloodsucker (Monchadski, 1936, 1940; Beklemishev, 1951, 1957).

Further evolution of the Culicidea depended upon the exploitation of high-caloric animal food, and this required improvement in respiration so that the larvae could leave the surface film and feed upon planktonic organisms. For this they required to be able to breathe by gas exchange through the cuticle. This is how the Chaoboridae solved their evolutionary problem. Among the true Culicidae we see an example of this trend among the Megarhininae, which evolved predatory larvae, while the adults became phytophagous or nectar-feeding. *Mansonia* solved the problem of respiration at depth by tapping the air-sources in the stems of underwater plants. A study of the comparison of the differing solutions to this problem by different lines of Culicidae is a fascinating one, which certain Soviet investigators have attempted with some success (Monchadski, 1936; Beklemischev, 1944, 1951, 1957).

Superfamily Psychodidea

This is fourth in order of size in the infraorder Tipulomorpha. It comprises three families, of which two — Phlebotomidae, with about 290 species, and Psychodidae with about 250 — are the largest, while the Nemopalpidae, with only 11 species, are rare insects, and obviously a relict group. This superfamily is very poorly known, and information is disconnected and concerns a few groups only. The most that can be said is that so far, in spite of conspicuous differences in structure and ecology, each of the three groups mentioned above is considered to be a single family.

There are more fossils of Psychodidea than there are of Culicidea, though they are still an insignificant few. Authentic fossils of Psychodidea are all of the Tertiary, and earlier finds are very doubtful; e.g. *Mesopsychoptera dasyptera* Br. a poorly-known form from the Lias of Ust-Baleja. All the Tertiary species were found in Baltic amber, and belong to each of the three contemporary families. There is one species of Nemopalpidae, one species of Phlebotomidae, and no fewer than 30 species of Psychodidae. In addition to these fossils that can be assigned with certainty to their families there are several species that are still of

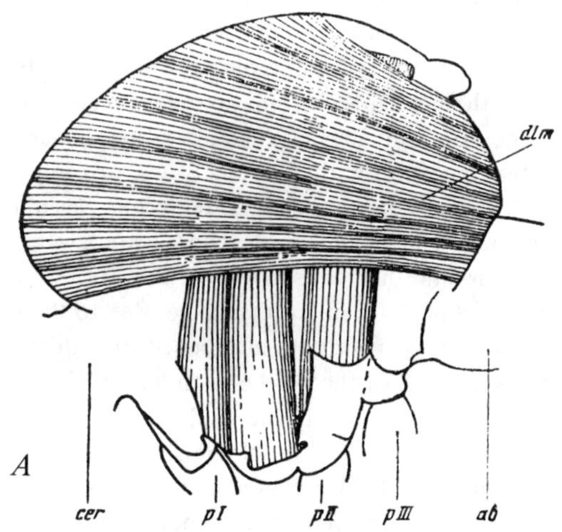

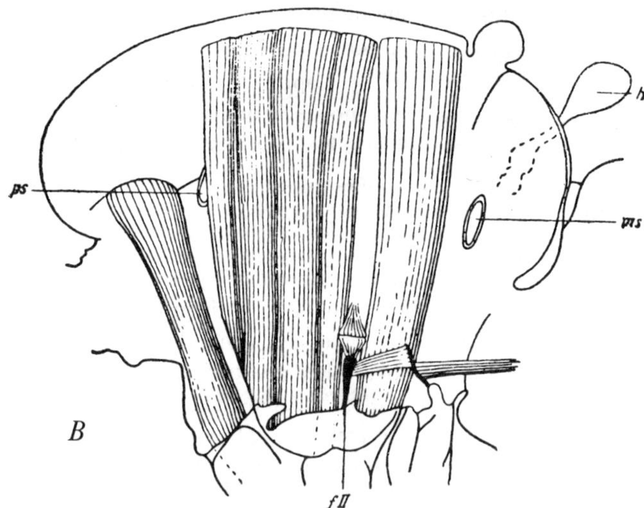

Fig. 18. Culicidae. Muscular apparatus of the right half of the thorax. *A. Aedes* sp. Longitudinal muscles of female. *B. Anopheles maculipennis* Meigen. Dorsoventral muscles of female. (Original, diagrammatic.) (for abbreviations, see fig. 13. p. 42).

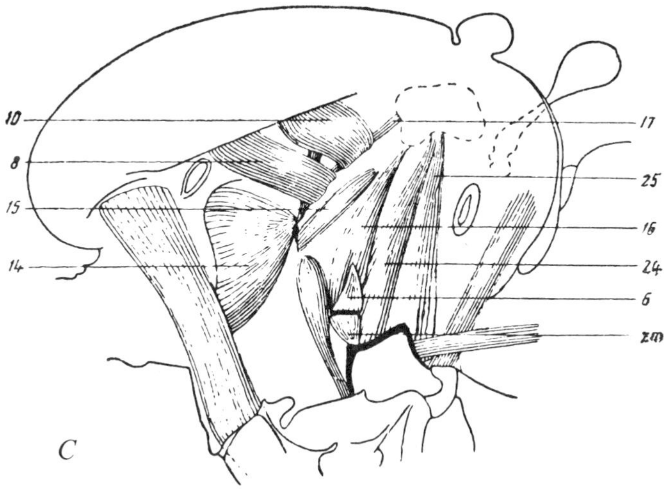

Fig. 18. (cont.) *Anopheles maculipennis* Meigen. *C.* Pleural muscles of female. (Original, diagrammatic.) (for abbreviations, see fig. 13. p. 42).

uncertain systematic position (Handlirsch, 1906-8). Of course the actual figures become out of date as time passes, but their relative numbers remain significant. For instance, species have been found belonging to the subfamilies Trichomyiinae and Sycoracinae, which are relict groups, poorly represented in the contemporary fauna. Their existence in the Upper Eocene indicates that they were formerly more abundant than they are today.

The phylogenetic relationships of the Psychodidae have not yet been studied very much. Obviously they were related to the most ancient Tipulomorpha, and perhaps even to the Dictyodipteromorpha, and had quite primitive ancestors which had not yet lost the complete mouthparts of the female adult, and had not costalized the wings.

It is interesting that these Diptera have developed a coating of macrotrichia on the body and wings, but it is quite probable that this feature is merely retained from distant ancestral forms, and is not evidence that these two groups are particularly closely related. The relationships between groups in the superfamily Psychodidea are still very little known. Although the evolution and segregation of Psychodidae and Phlebotomidae, by essentially different paths, have been more or less clarified, the third family, Nemopalpidae, is an obvious relict, and has been little studied as yet (1964).

The lack of information about the organization and history of the Psychodidea is especially annoying in view of the primitive features of the larvae which make this superfamily completely distinct from any other Tipulomorpha. An example is the occurrence in most larvae of rudimentary anterior spiracles, which points to the great age of this group: as indicated above, withdrawal into water was the

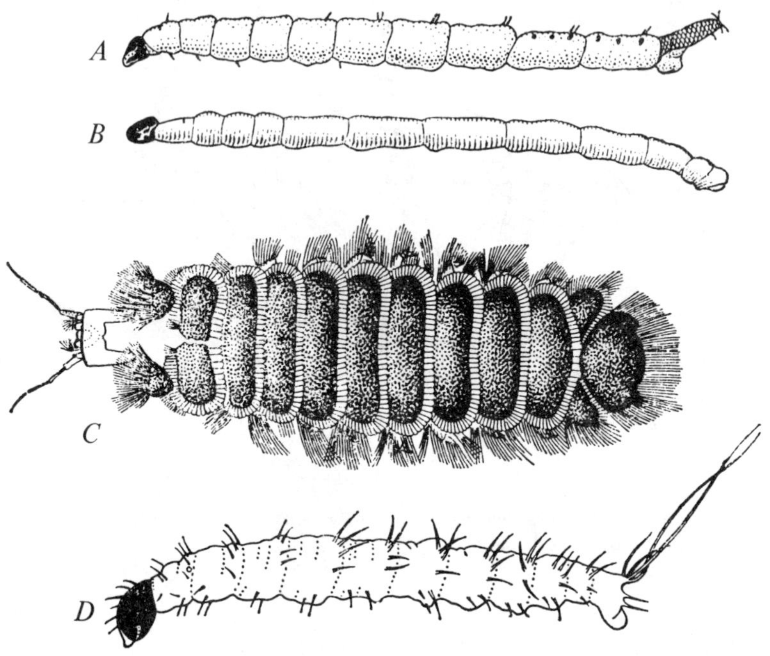

Fig. 19. Psychodidea: larvae. *A. Psychoda alternata* Say (Psychodidae). Side view. *B. Trichomyia urbica* Curtis (Psychodidae). Side view. *C. **Sycorax silacea*** Curtis (Psychodidae). Side view. *D. Phlebotomus papatasii* Scopoli (Phlebotomidae). Side view. (According to Hennig, 1948-52.)

primary solution of a problem which led to the evolution of the very first members of the Psychodidea points to their separation from the ancient Tipulomorpha at a very early stage, when they still populated only the surface layers of water, the boundaries of pools, and the damp earth of the shore.

We may now go on to describe in very general terms the organization and the problems of evolution of the Psychodidea (fig. 19, 20). The characteristic features of the moth-flies of the family Psychodidae are an aquatic larval life, either in fast-running fresh water, or in accumulations of stagnant water — but the feeding habits of the larvae remain quite unknown — together with the development in the image of running legs, of broad, hairy wings of the ancient lifting type, and of reduction of the mouthparts, which can take only liquid food. The main problem in the evolution of Psychodidea has been the deficiency of food, which they apparently solved by improving the feeding of the larval stage, which went into a moist medium, and consumed the decaying plant substances that were abundant there. The abundance of such larval food has permitted the winged phase to give up bloodsucking and change to plant food, or even to aphagia.

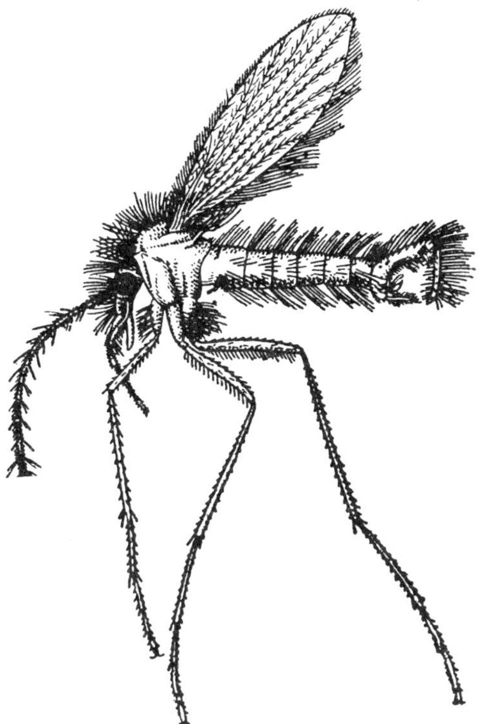

Fig. 20. *Phlebotomus papatasii* Scopoli (Phlebotomidae). Male, general view. (According to Perfilev, 1937.)

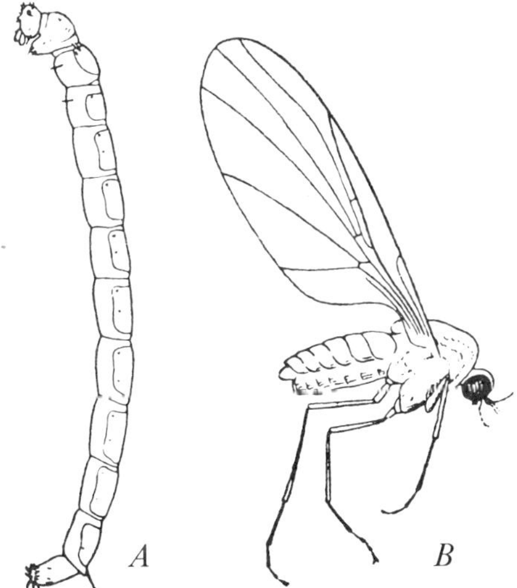

Fig. 21. Orphnephilidea. *A. Orphnephila* sp. Larva from the side. *B. Orphnephilia major* Bezzi. (According to Lindner, 1930.)

The Phlebotomidae followed a different path. They solved the problem of a damp ancestral environment and a deficiency of food by feeding on different rotting substances that were present in small quantities, e.g. bodies of insects, and perhaps by feeding on bacteria. Evidently this was not an entirely satisfactory solution, since it did not lead the adult to become either plant-feeding or non-feeding; on the contrary the adult remained predatory, and possesses piercing mouthparts. Bloodsucking on vertebrates was developed, and this, combined with an improvement in flying abilities throughout the Phlebotomidae, allowed them to become widespread. At the same time the larvae became terrestrial, almost xerophilic, and acquired the ability to live in the soil of burrows occupied by vertebrates, and to feed on a variety of decomposing animal substances (Burakova, 1931; Perfilev, 1937).

One other important feature of the evolution of Psychodidea, has been the reduction in overall size of the body. That this reduction is secondary is shown by the rich, complex veining of the wings, a reminder of the formerly large wings of their ancestors. Insects with such small bodies as those of Psychodidea and which are, furthermore, covered with plentiful microtrichiae, have no functional need for complicated veining of the wings. Psychodidae run well, and flight is of little biological importance to them; reduction in body size has been accompanied by broadening of the wings, and their partial adaptation into peculiar integumentary organs. This reduction in body size, in response to evolutionary problems, proved to be advantageous to the Psychodidea, since it allowed them to colonize 'microhabitats', and to develop 'microphagia'.

Superfamily Orphnephilidea (syn. Thaumaleidea)

This includes about 70 species of four genera of the single family Orphnephilidae and is insufficiently known to me (fig. 21). The existing literature on these peculiar Diptera illustrate very well the 'non-correspondence' of the features of the larvae and the adults which has led at various times to very different assessments of the systematic position of these flies. At times they have been considered to be a subfamily of the Chironomidae, at others as a separate family and, finally, at other times as an individual superfamily which almost shows a connection with the Brachycera (Lindner, 1930b).

Completely unknown as fossils, Orphnephilidae are distributed on almost all the continents of the earth, and are clearly mountain-dwellers. Their phylogenetic relationships can be indicated only in very general terms. They are undoubtedly a derivative of the original forms of Tipulomorpha which moved into habitats on the margins of pools, and which kept the ability to breathe atmospheric air, as shown by their amphipneustic tracheal system. Links with the Chironomidea are comparatively remote, and consist of a resemblance between the larvae of the two groups in the structure of the head, and in the pseudopods located on the anterior thorax and posterior end of the body. Almost certainly this must be assessed as convergence resulting from a peculiar method of feeding (eating plant-growth on underwater stones) and of locomotion. There are deep-seated differences between the winged phases of the two groups, which prevent them from being regarded as very closely allied. Undoubtedly the Orphnephilidea

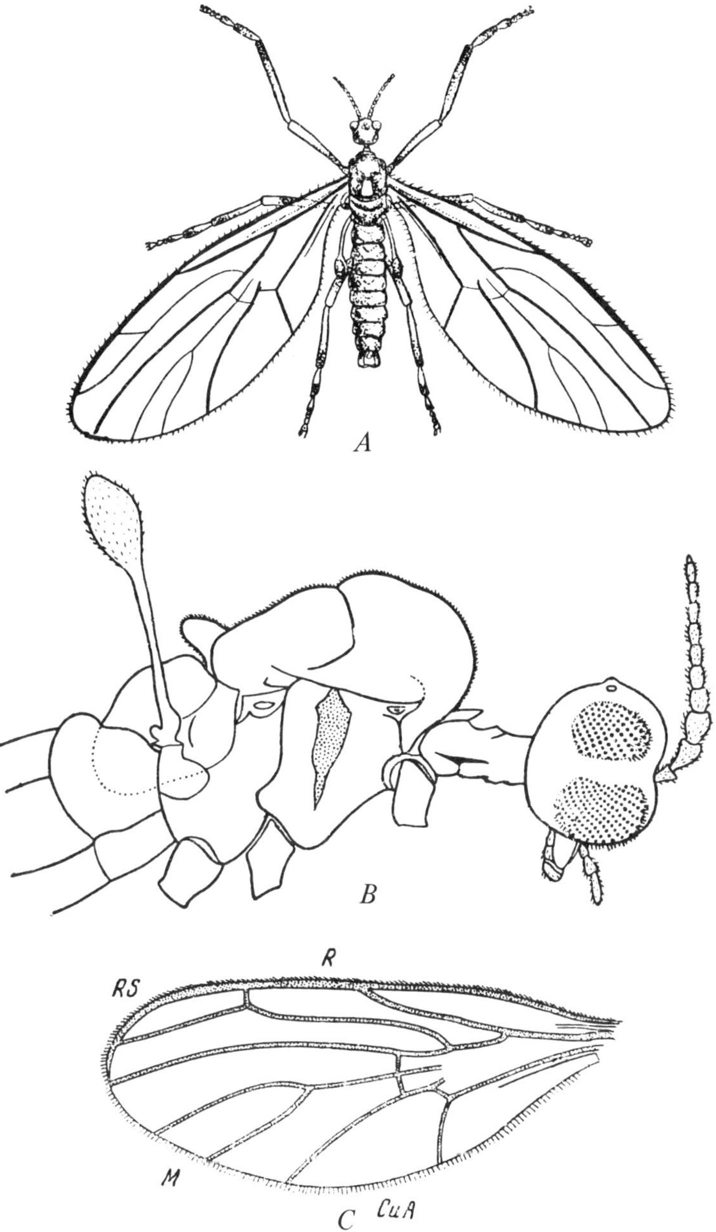

Fig. 22. *Perissomma fuscum* Colless. *A*. General view. *B*. Head, thorax and base of abdomen (legs and wings not represented). *C*. Wing. (According to Colless, 1962.)

forms an entirely separate superfamily within the infraorder Tipulomorpha, on a
level with the Chironomidea, Culicidea and others.

Superfamily Rhaetomyiidea

Until recently this superfamily was known only from the Upper Triassic of
Central Asia (see p. 161). Recently Colless described the genus *Perissomma*, with
two species from the south of Australia, and placed these in a sharply isolated
family, Perissommatidae. This family is almost certainly a relict form of the
ancient Rhaetomyiidea, a superfamily which had attained great specialization by
the Mesozoic (fig. 22) (Colless, 1962).

The Perissommatidae are characterized by the sharply isolated head of the
adult, with multisegmented, but short antennae; by the reduced proboscis; and by
the very large wings, which exceed the length of the body and have highly
reduced venation, while the halteres are extremely long and reach to the fourth
tergite of the abdomen. The well-marked transverse suture of the mesonotum, the
isolated head with long cervical section, and the branching radial sector all indicate
the relationships of this group with the infraorder Tipulomorpha; the reduction of
the anal veins, and the general specialization of the wing venation form a link with
the Rhaetomyiidae of the Mesozoic. The structure of the larvae of the Perissommatidae shows clear resemblances to larvae of a different infraorder, the
bibionomorph family Scatopsidae, but this is undoubtedly an example of
convergent resemblance, and does not indicate any real phylogenetic association.

The way of life of the larvae of these remarkable relict Australian Diptera is
also noteworthy. They dwell in the highly putrified fruiting bodies of the
mushroom *Boletus granulatus*, which are converted into a dark, semiliquid mass.
The winged adults fly in the winter. All the biological features of this insect that
have so far been described support the view that it lives in isolated refuge habitats,
a characteristic of phylogenetic relicts.

Infraorder Bibionomorpha
Hennig 1948, p. 74

Extent, evolution, systematics. — This vast complex of Diptera includes six
superfamilies: Bolitophilidea, Fungivoridea, Cecidomyiidea (syn. Itonididea),
Bibionidea, Scatopsidea and Rhyphidea (syn. Phryneidea), and no fewer than 22
families. The total number of species is about 7,400. The Diptera belonging to this
infraorder were formerly known as the Nematocera Oligoneura, but this is an
artificial taxon (see above, p. 24). The coherence of the infraorder Bibionomorpha is not open to doubt, and its phylogenetic unity is confirmed by abundant
palaeontological evidence (Rohdendorf, 1946). The bibionomorphs first appear in
the Upper Triassic in great number and variety, being represented by no fewer
than five superfamilies, two of which — Fungivoridea and Rhyphidea — have
survived until the present time. The Middle Jurassic fauna of Karatau contains
representatives of two other modern superfamilies, Bibionidea and Scatopsidea,
while some of the ancient superfamilies of the Upper Triassic have already
disappeared.

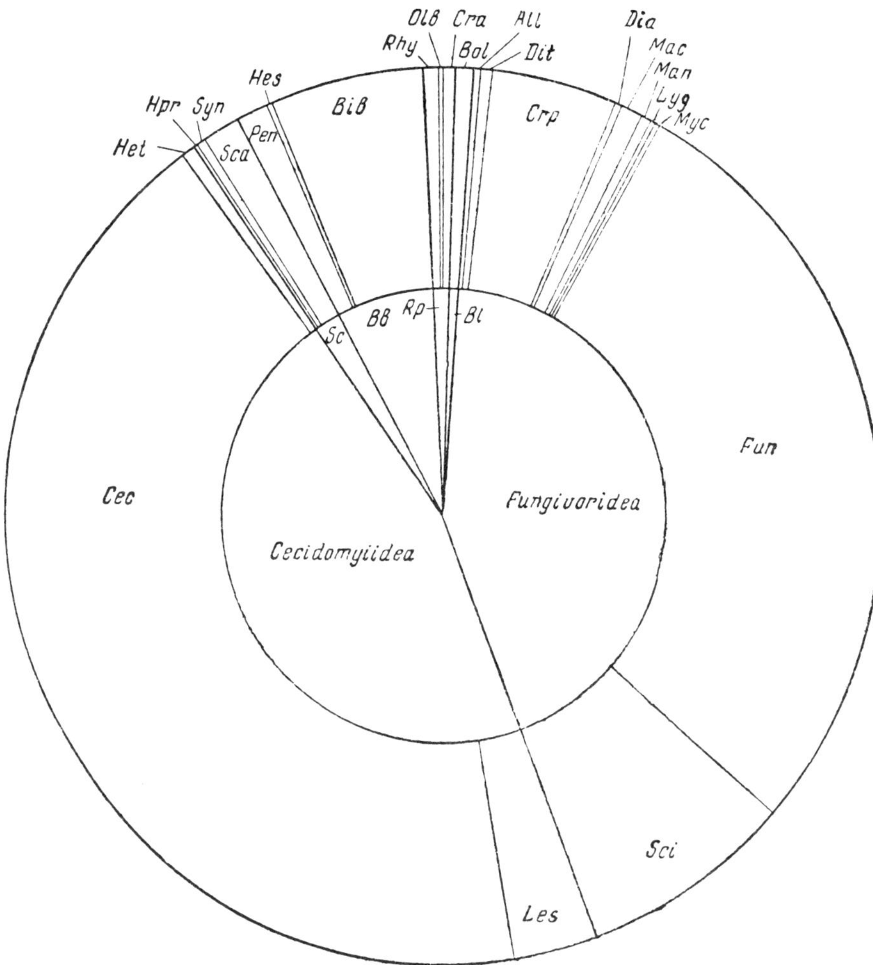

Fig. 23. Diagram of the composition of the infraorders of bibionomorphs of the present-day fauna. The super-families and families are indicated as the sectors of a circle the sizes of which indicate the number of species in the given taxon (1° of arc equivalent to 21 species). In the centre of the circle are shown the superfamilies, and on the circumference the families. (Original.) For abbreviations, see p. xiii.

The previous history of the superfamily Bolitophilidea is still obscure. These Diptera appear suddenly in Tertiary times, in the Baltic amber, although they show some similarities to more ancient forms of the Upper Triassic.

The relationships of the recent families of Bibionomorpha is satisfactorily understood (fig. 23), and the existing superfamilies are well isolated from each other. Most isolated is the relict superfamily Rhyphidea, with about 100 species of two families; less isolated are the relict superfamilies Bolitophilidea (one family with 32 species) and Scatopsidea (three families with 135 species), which are

close to some Triassic Fungivoridea. The superfamily Bibionidea is comparatively
scarce in the contemporary fauna — about 500 species in three families, two of
which, comprising about one hundred species, are clearly relics — but it is fairly
certainly a derivative of the Triassic Pleciofungivoridae, and reached its most diverse
evolution in Tertiary times.

The two superfamilies richest in species are the Cecidomyiidea (three families)
and the Fungivoridea (ten families), which together include about nine-tenths of
the present-day species of this infraorder. The two superfamilies are no doubt
related to each other, the Cecidomyiidea being descendants of one group of
Jurassic fungivoroids. The superfamily Fungivoridea shows definite connections
with the ancient Dictyodipteromorpha, as witness the peculiar upper Triassic
group the Pleciodictyidae, which is undoubtedly related to the original forms of
the Fungivoridea, and which possesses some of the structural features of the
Dictyodipteridae. The different age of these two huge superfamilies of Bibionomorpha is clearly reflected in their taxonomy. The superfamily Fungivoridea, with
3,214 species, is the smaller on the basis of number of species, and appears to be
the more ancient, including many clearly isolated families among which relict
groups are numerous: no fewer than six out of the ten families come into this
category. In contrast, the larger (3,403 species), but younger superfamily Cecidomyiidea contains only three families, among which relict groups are few.

Chief features. — The various members of this infraorder may be briefly
characterized as living in a terrestrial habitat, away from water, yet in a medium
of high humidity. The larvae live in decaying plant residues, in animal dung, or in
living plant tissues, generally in a medium saturated with water vapour. Only
rarely, as in the predatory Cecidomyiidae, do they live openly on the surface.
Very rarely larvae of Bibionomorpha live in a liquid medium, such as the juices of
rotting vegetation, or the juices present on living plants, or even in temporary
accumulations of water, but never in permanent pools. As a rule the larvae have
anterior spiracles, and may be either amphipneustic or peripneustic, but sometimes they are metapneustic, or even apneustic, as in the Ceroplatidae and some
others.

Generally the larvae have a well-defined head with biting mouthparts, but
sometimes there is external digestion, with a correlated reduction of the skeleton
of the head and mouthparts (Cecidomyiidea and some Scatopsidae). The larvae
are mainly phytophagous, but some feed on detritus.

Adult flies of this infraorder are characterized by a shortening of the body,
accompanied by an increase in size of the thorax to accommodate bigger wing-
muscles (fig. 24). Change in shape of body is reflected in changes in powers of
locomotion: the legs are only rarely elongate, and are suited to running rather
than being of the prehensile type. The wings are often reduced, with varying
degrees of costalization, which sometimes goes to extreme lengths as in the
Scatopsidae. Some species with extremely small body size have feather like wings.
The adult flies of this group do not show any specialization towards more efficient
feeding. Most of them take liquid plant food, or are aphagic; predators and blood-
suckers are unknown. Nor is there much specialization of the reproductive cycle.

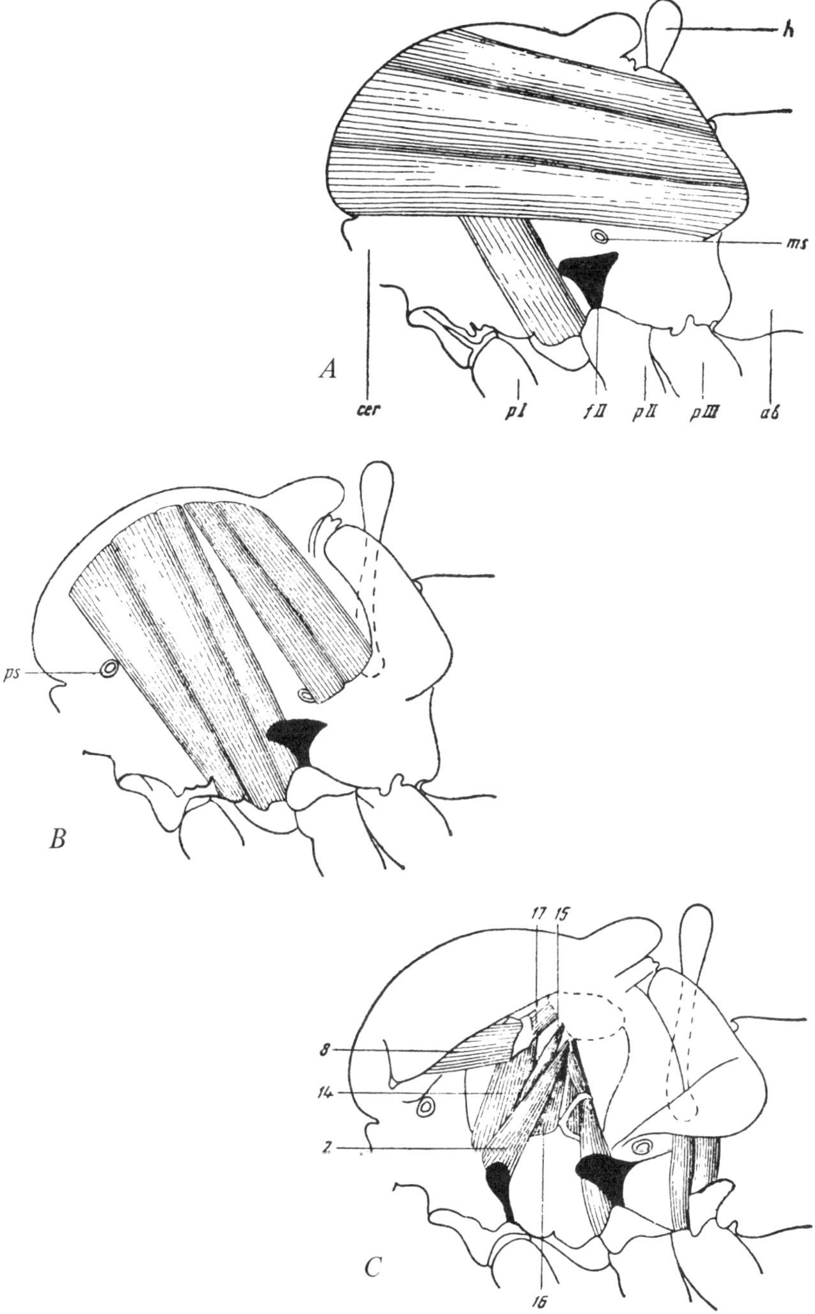

Fig. 24. *Lycoria thomae* (L.) (Sciaridae). Female, muscles of the right half of the thoracic section. Length 1.68 mm. *A.* Longitudinal dorsals. *B.* Dorsoventrals. *C.* Pleural muscles. (Original, diagrammatic.) Abbreviations as in fig. 13, (p. 42), plus *2*, and *16* which are corresponding pleural muscles.

The eggs are minute, and are laid in a suitable food medium; only rarely are the eggs concealed (Bibionidae) or laid somewhere else altogether (Cecidomyiidae).

Factors influencing the evolution of the group. — It is difficult to review the infraorder Bibionomorpha because it has had a long history, and has reached a high level of abundance of variety, but we can try to assess what its main characteristics are, and how these point to the factors which led to the isolation of this group of flies. Some of the main features are as follows: larvae living in localized, terrestrial habitats, independent of pools or other standing water, and incompletely nourished on detritus or plant tissue, so that the adult winged phase is obliged to provide supplementary nutrition, and must therefore be mobile. This insufficiency of larval food was one of the oldest problems to arise in the evolution of Bibionomorpha, and the various groups of the infraorder solved it in different ways. Mobility of the adults was achieved by the development of strongly costalized wings, with great lifting power, guaranteeing a swift take-off, while other forms developed powerful legs and became very mobile among the vegetation. Locomotion has been the most important feature in the evolution of this group, and improvements to the nervous system, feeding, and reproduction have played only a secondary role, except in a few superfamilies.

Phylogeny of the chief superfamilies of the infraorder Bibionomorpha

The different groups of these Diptera had very dissimilar histories, and to understand these it is necessary to examine the individual superfamilies in more detail.

Superfamily Bolitophilidea

The earliest records are from the Baltic amber of the Upper Eocene, and today there is only one family, Bolitophilidae, with two genera, *Bolitophila* and *Messala*, and only 32 species. They are obviously related to the large superfamily Fungivoridea, but have diverged considerably. The characteristics of the superfamily have been insufficiently studied, but they include poorly developed wings, with only slight costalization and incomplete basalar (see Rohdendorf, 1946, p. 24,) (fig. 31), legs of the thin type, and long antennae. The larvae lack the prostheca, or bristly upper surface of the mandible that is present in Fungivoridea (fig. 25), and their feeding habits and way of life are little known.

These Diptera are apparently direct descendants of the earliest forms of Fungivoridea and Bibionidea, but the costalization and improved basalar of the wing, with increasing wing-beat, had not progressed very far, and lifting power was only slightly increased. The legs lengthened and became thinner and more prehensile, while the muscles of the coxae declined. The larvae retained well-developed antennae, probably an indication that they remained mobile and polyphagous.

Bolitophilidea apparently solved the basic problem of shortage of larval food quite soon, and so they did not go on to develop adult mobility as far as it was developed in some other groups; but there is no direct evidence of this, and it remains only an assumption.

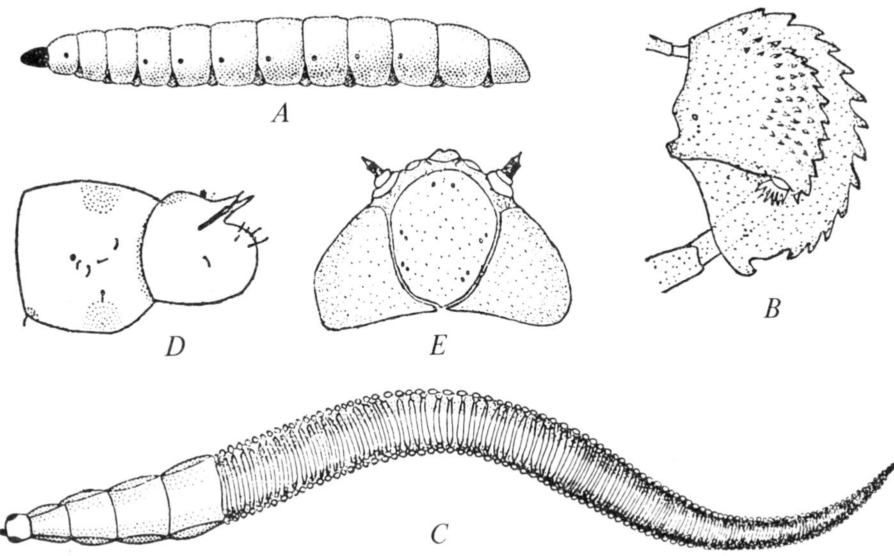

Fig. 25. Fungivoridea and Bolitophilidea. Structure of larvae. *A. Brachypeza radiata* Jenkinson (Fungivoridae).
General view from the side. *B.* The same, mandible. *C. Macrocera anglica* Edwards (Macroceridae). General view
from above. *D. Ditomyia fasciata* Meigen (Ditomyiidae). Tip of body in lateral view. *E. Bolitophila pseudohybrida*
(Bolitophilidae). Head from above. (All from Hennig, 1948-52, enlarged.)

Superfamily Fungivoridea

This is one of the most ancient groups, known from Triassic times, and widely
distributed today, with 10 families, and more than 3,200 species. A general
characteristic of this family is an active winged phase, with wide peculiar wings,
made more efficient by improvements in venation and in the basalar region.
Larvae have a powerful head and characteristically serrate mandibles (fig. 25),
but with short antennae, and feed on plant tissues, either living or decaying. These
features suggest that the chief evolutionary problem to be overcome was once
again insufficiency of larval food, leading to perfecting of the winged phase. This
last step took place in habitats among accumulations of plant residues as well as
living plants, where such nutritive substances as fungi and rotting plant material
were unevenly distributed, and subject to variation at different times of the year.
The adult fly thus needed increased mobility; indeed the determining tendency
among Fungivoridea as a whole has been the improvement of locomotion, and
larval feeding has been improved by increase in the quantity of nutritive mass
consumed rather than in more efficient digestion of what was consumed. The ten
families of Fungivoridea which survive to the present day express this tendency in
different ways (Rohdendorf, 1946).

The biggest family is Fungivoridae, which is known from the Paleogene, and
presumably came into existence not later than the Jurassic/Cretaceous, and which
is characterized by its close association with the higher fungi, in the fruiting

bodies of which the larvae live. The limited distribution of nutritive substrates in time and space; climatic features — particularly the winter season in the temperate zones which are the main habitat of the family; the character of this habitat, especially the abundance of plant residues in the upper layers of soil in forests — all these favoured the development of types which can run quickly and dig in actively, and of long-lived winged adults.

These developments, in their turn, brought new problems. The problem of overwintering actually slowed down the improvement of flying ability, as well as requiring the larva to accumulate greater reserves of fat, and so they found richer sources of food. This was accomplished by the larva concentrating its feeding upon the sporogenous tissue of the fruiting body of the fungus, which has the highest calorific value. We can observe this process in many of the youngest groups of the family.

Another group of terrestrial midges, Sciaridae (Lycoriidae), the second most numerous family at the present day, is characterized by different features. They evolved the ability to feed upon rotting leaves lying on the surface of the ground, and thus became of prime importance in soil-formation. Their reliance upon fungus-feeding is less direct than that of Fungivoridae, and is only incidental to their feeding upon the general mass of decomposing material, together with its saprophytic flora. Flight was improved by moderate costalization of the wings, but the legs were only slightly strengthened, and the body did not acquire a stream-lined form. Apparently feeding by the adult fly was of decreasing importance, and the life of the adult was shortened, with partial aphagia. In these respects Sciaridae can be regarded as a progressive family.

The other families of Fungivoridea are very much smaller, and most of them are relicts. We still know too little about their characteristics to say what problems have determined the course of their evolution with any degree of precision. In earlier work (Rohdendorf, 1946) I have noted some peculiar traits of the many relict families of this group, and I can now add new data on the evolution of the Macroceridae and Ditomyiidae which confirm my impression that these relict families have unusual histories. The structure of the larva of *Macrocera* (fig. 25*B*) is so peculiar that it must obviously be related to some very peculiar way of life, and the same may be said of some of the Ditomyiidae, which combine an unusual larva (fig. 25*C*) with a winged phase that shows quite ancient features. No useful purpose is served by trying to follow the evolution of these groups in more detail until we know more about them.

Superfamily Cecidomyiidea (Itonididea) (fig. 26)

This superfamily contains over 3,500 species, in three families, about half of the present-day content of the infraorder. It is undoubtedly a derivative of some Mesozoic representative of the Fungivoridea (Pleciomimidae). The oldest known fossil members go back only to the Baltic amber of the Upper Eocene, and belong to present-day genera. The characteristics of the Cecidomyiidea have been com-paratively well studied, and I have discussed them in an earlier paper (Rohdendorf, 1946, pp. 92-94). The particular problems in the evolution of this group was how to feed upon plant tissues, where nutritive material was abundant, but difficult for a small larva to assimilate by means of chewing mouthparts. The solution was extra-intestinal digestion, with reduced mouthparts, and more enzymes in the saliva. This allowed the larva to become even smaller, and the winged adult to develop aphagia. Thus the determining features of cecidomyiid evolution were improvement of larval feeding, and development of sense organs in the adult. The further evolution of this group of bibionomorphs is complex. There were changes in the organization of the larvae, some of which, for example, came to live freely on the surface of plants as predators of aphids, and some have even developed the capacity for parthenogenetic reproduction (paedogenesis). This evolution is reflected in the division of the superfamily into three quite sharply different families.

The Lestremiidae, comparatively poor in species, are characterized by larvae which develop in rotting wood, by rich venation of the relatively big wings, by the long main segments of the antennae; all features related to the search for more nutritive vegetable substances in the soil, or to dispersal of the adults. The structures of the legs and wings only needed to be preserved, but the organization of the larvae for wood-feeding was a new development.

The principal family, the one richest in species, is Cecidomyiidae (Itonididae), and is characterized by larvae which live in various tissues of living plants, by wings that have a reduced venation, and are often shortened and feather like, by complex protuberances of the antennal segments, and by weak legs, with shortened femora. These structures are associated with feeding upon specialized parts of particular species of plant, which have to be searched for by the winged adult.

Finally, the last family Heteropezidae is poor in species, and is characterized by the evolution of the remarkable phenomenon of paedogenesis (reproduction by larvae, without an adult stage), by reduction of the wing veins, and feather like appearance (Rohdendorf, 1949), and by the reduction of the number of tarsal segments. All these features signify favourable conditions for activity of the larvae, and a deficiency of food for the winged adults.

Comparison of the phylogeny of the three families of gall midges shows that all of them show improvement in larval feeding, as well as in the sense organs of the winged insect. Lestremiidae are the least specialized, and show little change in these features. Cecidomyiidae are considerably more specialized, and show improvement in larval feeding, together with the development of feather-like wings, and particularly in the evolution of the complex sense organs of the antennae. Finally Heteropezidae are the most specialized of the three, in which extreme specialization of larval habit (paedogenesis) is accompanied by reduction of wings, legs, and antennal sense organs in the adult fly.

In conclusion some comment is needed on the reduction in absolute size of the body that has taken place during the evolution of the gall midges. Diminution in size in itself is not necessarily a sign of evolutionary decline, as is often assumed. Consideration of the way in which the habits of the Cecidomyiidea have changed during their evolution suggests that reduction in overall size is clearly an advantage to a group which seeks out and utilizes a form of plant food that exists in very small, scattered locations.

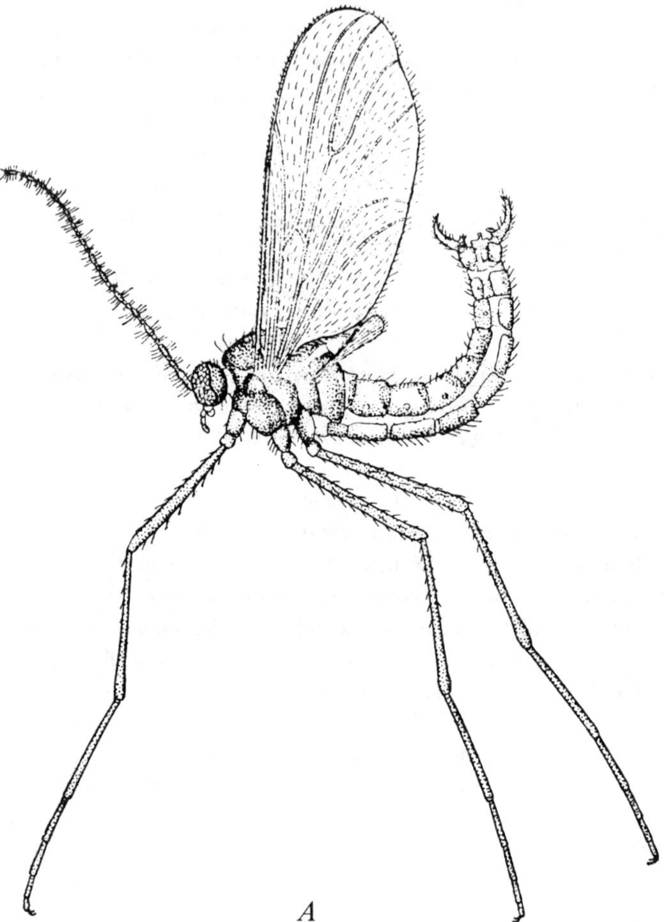

Fig. 26. Cecidomyiidea. *A. Catocha* sp. (Lestremiidae). Male, general view. (According to Curran, 1934, greatly enlarged.)

Superfamily Bibionidea

These first appeared in the Middle Jurassic, and are therefore slightly more recent than Fungivoridea. The present Bibionidea comprise three families, with almost 500 species, more than three-quarters of which belong to the family Bibionidae (about 390 species); 100 species belong to Penthetriidae, while the remaining family, Hesperinidae, with only two species, is clearly a relict group (Rohdendorf, 1946; Hardy, 1960).

This group is characterized by a sharp sexual dimorphism, with the eyes and antennae of the males being more highly specialized than those of the females. They are flies of comparatively large size, with large, costalized wings (but slow flight!), and strong, prehensile legs. The direction of evolution is known only for

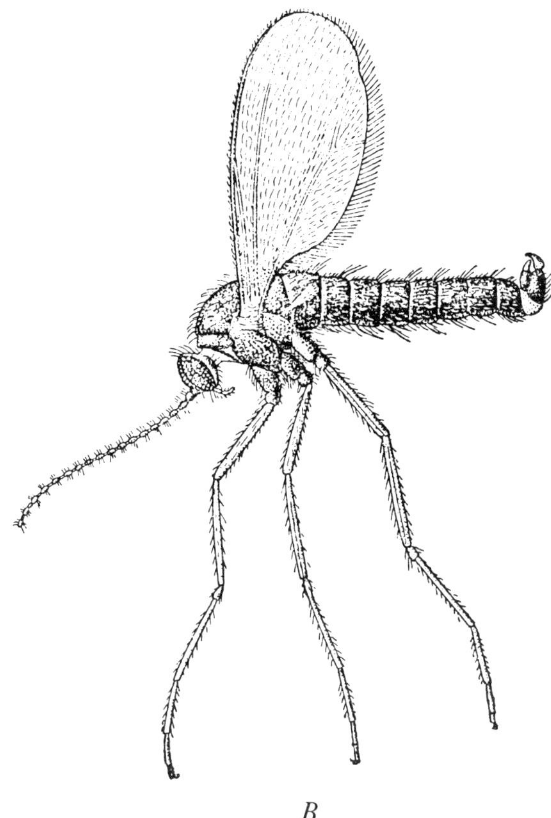

Fig. 26. (cont.) *Mayetiola destructor* Say (Cecidomyiidae). *B.* Male, general view. (Accórding to Curran, 1934, greatly enlarged.)

the two bigger families, and does not include anything very new, except for armature of the body-surface of the larvae, most particularly in Penthetriidae (fig. 27). So far the feeding habits and metabolism of these larvae remain unknown. The only information about reproduction in Bibionidea concerns the laying of eggs in the soil, during which the armed anterior legs have a part to play. Adult bibionids feed on flowers, though adult Penthetriidae apparently take no nourishment. Our total knowledge about the biology of Bibionidea is thus inadequate to speculate about the factors that may have directed their evolution, but it seems that they first diverged from the ancient Fungivoridea (presumably in Triassic times) by the development of prehensile legs and moderate costalization of the wings, accompanied by an increase in overall size of the body. It is not clear why they became bigger, but it may be linked with the exploitation of a source of food that was abundant and readily available. The evolution of Bibionidea led in Jurassic times to Eopleciidae, Mesopleciidae and Paraxymyiidae, and later on to the Cenozoic Hesperinidae and Penthetriidae, of which only the last remain today.

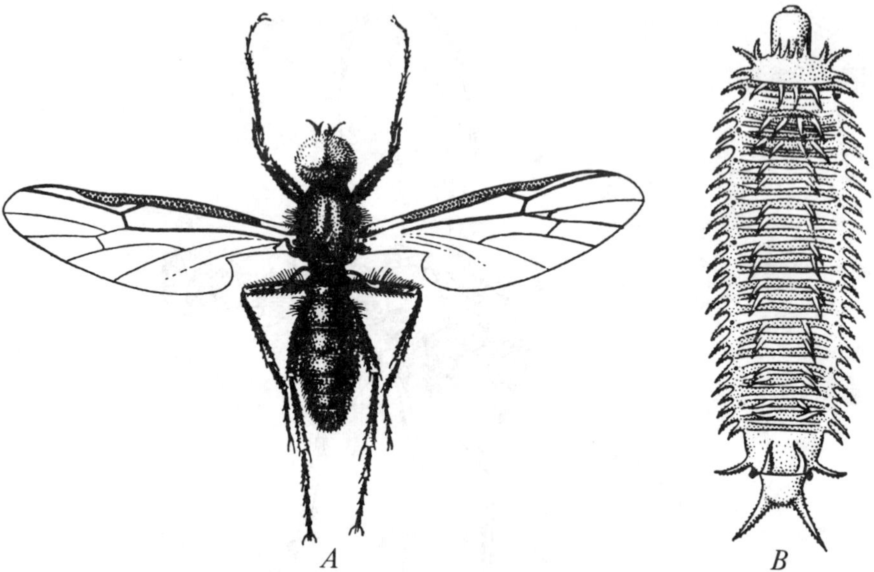

Fig. 27. Bibionidea. *A. Bibio marci* L. (Bibionidae). Male, general view. Length of body 10 mm. *B. Penthetria holosericea* Meigen (Penthetriidae). General view of larva from above. Length 12 mm. (*A.* according to Lindner, 1928; *B.* according to Hennig, 1948-52.)

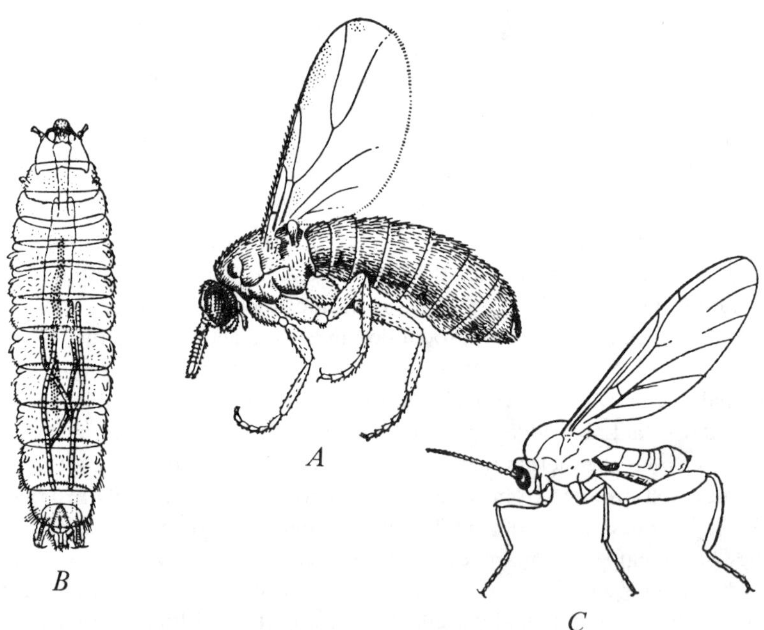

Fig. 28. Scatopsidea. *A. Scatopse fuscipes* Meigen (Scatopsidae). Female, general view. *B.* The same, larva from above. *C. Cathyoscelis antennata* Edwards (Hyperoscelididae). Male, general view. (*A. B.* after Burkova, 1931; *C.* after Edwards, 1922, enlarged.)

Very soon after the separation of the original stem of the superfamily there developed the sharply costalized, still very small Protobibionidae, which may have given rise to the Bibionidae of the Cenozoic; on the other hand they may be merely a parallel analogous group which arose considerably earlier than the true Bibionidae. We do not know what factors caused the evolution of the characteristic bibionoid features of armed, strongly prehensile legs, with highly developed appendages of the pretarsus, the sharp sexual dimorphism of the structure of head and the colour of the body, the costalized, high-lift wings, and finally the 'simple' larvae.

Superfamily Scatopsidea

This is of the same age as the Bibionidea, its most ancient fossil members, (family Protoscatopsidae) being found together with Bibionidea in the Middle Jurassic of Karatau. Scatopsidea are not common in the present day fauna, and comprise only three families. The family Scatopsidae contains about 130 species, but the other two are clearly relict groups, Hyperoscelididae with 10 species, and Synneurontidae with only one.

Characteristic features of the Scatopsidae (fig. 28) are improvement of the sense organs of the head (with enlargement of the compound eyes, and reduction of the antennae to fewer, stout segments), shortening of the body until it is small, or even minute, highly costalized wings, and strong, prehensile legs. The predatory larvae live on any kind of moist terrestrial substrate. In the imago the thoracic section is large, with greatly enlarged terga, and the sternite of the prothorax is not bridged to the pleura. The chief differences between the three families lie in the details of metamorphosis of the larva, with the most primitive condition in Scatopsidae, where metapneustic first stage larvae transform into amphipneustic second stage larvae, and these into peripneustic third and fourth stages. The larvae possess chewing mouthparts, and the posterior spiracles lie at the end of processes (fig. 28*A*). The pupa bears anterior thoracic spiracles (Burakova, 1931; Hennig, 1948-52).

The following points are noteworthy in the evolution of the Scatopsidea. The larvae lived in and on damp, terrestrial substrates of a temporary nature, any kind of decaying matter, excrements of vertebrates, or plant residues. They sometimes preyed on other minute insects (for example the larvae of other Diptera, see Burakova, 1931). These larval habits arose very early, and led to the development in the adults of a high-lift type of wing, as well as to the improvement of the sense organs for seeking out new larval habitats. The body of the adult did not become bigger, and there was no improvement in the feeding of the winged adult, which remained phytophagous. The ancestors of the Scatopsidae have not yet been found, but they almost certainly belonged to the ancient fungivorid stock.

It is easier to explain the later history of the Scatopsidea, with their fission into two families, Scatopsidae and Hyperoscelididae. These two families arrived at different solutions for the basic problem of providing an abundance of food and shelter for the larvae. The ancestral Scatopsidae solved this problem by improving their powers of flight and the acuity of their sense organs. The adults became smaller and more mobile, better equipped to find and colonize the restricted

habitats in which their larvae could live. The larvae, on the other hand, were little
modified, and developed only the one special feature of not discarding the last
larval skin, which became a primitive puparium. In contrast, Hyperoscelididae
evolved along quite different lines; larval feeding was improved by the development of extra-intestinal digestion, but the flying apparatus was not improved, the
size of the body did not decrease, and there was no formation of a puparium
(Rohdendorf, 1951, 1958-59).

Superfamily Rhyphidea (fig. 29)

This is one of the most ancient groups of Diptera, being represented even in the
oldest Upper Triassic fauna. At present the Rhyphidea consist of three monotypic
families, which have clear relict characteristics, and which include about a hundred
species — Rhyphidae (66 species), Olbiogastridae (31 species), and Cramptonomyiidae (1 species). These Diptera are of great theoretical interest since they are the
nearest forms to the original ancestral group of the infraorder Asilimorpha and its
derivatives, and hence of many younger groups of Diptera, notably Myiomorpha
and some of the parasitic families.

The chief characteristics of the Rhyphidea are numerous 'primitive' features in
adult structure and in the mouthparts of the larvae. The head of the adult has long
antennae, four-segmented palpi, eyes of moderate size, and a proboscis capable of
sucking the juices of plants. The convex thorax bears broad, slightly elongate wings,
and long legs of the thin type suitable for active running; the prothoracic sclerite is
almost completely isolated from the pleural sclerites. The wing venation is very
complete, almost without any displacements or reductions of any individual veins,
with strong basalar sclerite, and a strong phragma; a free complex shaft of the
cubital and anal veins, reduced second anal, with microtrichiae, and lastly, a well-
developed vannal lobe to the wing (Rohdendorf, 1946, p. 29), (fig. 39). The
abdomen is cylindrical, and consists of eight free segments. Larvae with sclerotized head with a complete set of chewing mouthparts (including prostheca);
amphipneustic, elongate, with body-segments subdivided, anterior part of each
splitting off. Larvae live in various damp media, and ar polyphagous. The pupae
are elongate, with short anterior thoracic spiracles. The known species of the
genus *Sylvicola* (=*Rhyphus*, *Phryne*) have a very rapid larval development.

The differences between the surviving families of Rhyphidea are not yet well
understood, and we do not know enough to draw conclusions about phylogeny
within the group. Some light is thrown upon the chief evolutionary problems of
the superfamily by considering the main characteristics of the principal family,
Rhyphidae; these are, on the one hand, a rapidly developing and polyphagous
larva, and on the other hand, an adult with large wings with widely isolated basalar
sclerite, long, thin running legs, and mouthparts suitable for feeding on the juices
of plants. These characteristics explain why the larvae live in widespread, damp
substrates, rich in decomposing plant substances, but of a widely variable nature,
such as the shorelines of flowing pools, the soil of temporary pools in tropical
forest, and so on. The polyphagous habits of the short-lived larva called for an
adult of only limited mobility in damp, shady habitats, and favoured the retention

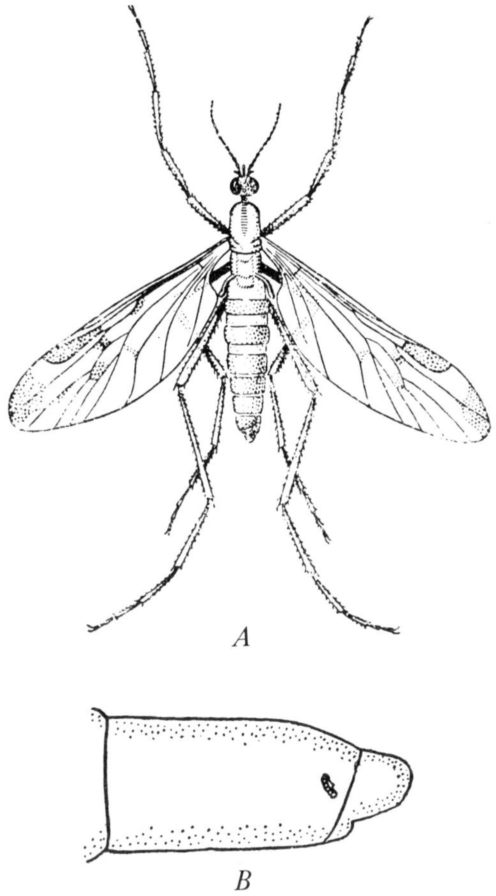

Fig. 29. Rhyphidea. *A. Rhyphus fenestralis* Scopoli (Rhyphidae). Female, general view. *B. Olbiogaster africana* Edwards (Olbiogastridae). Larva, posterior end of body from the side. (*A.* after Lindner, 1930; *B.* after Hennig, 1948-52.)

of a low-speed, high-lift type of wing, with no pressure towards costalization, while the legs retained considerable running powers. This suggests that Rhyphidae probably evolved under conditions of high temperature and humidity with an absence of direct sunlight, such as are found in tropical forests.

It is noteworthy that the Rhyphidea have evolved comparatively few 'new' features, while a number of organs have changed little or not at all, a picture that indicates a 'primitive relict' group (Rohdendorf, 1946, 1959). This is part of a wider problem of relict forms, and we still do not know precisely what factors govern this type of evolution.

Infraorder Asilomorpha
Rohdendorf, 1961b, p. 156

Extent, evolution, systematics. — This infraorder contains 23 families, grouped into five superfamilies: Tabanidea, Stratiomyiidea, Asilidea, Bombyliidea, and Empididea. The total number of species is very large, about 22,500. These families comprise all those which formerly were included in the suborder 'Brachycera Orthorrhapha' of earlier systematic schemes, with the sole exception of the peculiar family Lonchopteridae. The removal of this sharply isolated family leaves the rest as a real, single, systematic group.

No representatives of the Asilomorpha have been found in the early Upper Triassic fauna of central Asia, and the first species come to us only from the Liassic deposits of Germany. This most ancient of asilomorphs belonged to the peculiar family Protobrachycerontidae of the superfamily Stratiomyiidea, and there are more varied asilomorphs in the Middle Jurassic fauna of Karatau. Among these are Stratiomyiidea of three extinct families: Archisargidae, Palaeostratio-myiidae, and Protocyrtidae, as well as seven representatives of the Tabanidea (four genera of the family Rhagionidae and two genera of the extinct families Eostratio-myiidae and Rhagionempididae, with one genus of Asilidea from the extinct family Protomphralidae). Diptera collected from deposits of the Upper Jurassic (Malm) of western Europe include one representative of the Tabanidea (one genus of the extinct family Protohirmoneuridae), and one species that cannot be allocated to any known systematic position. In total the Jurassic fauna of Asilomorpha comprises no fewer than 12 species of seven families and three super-families, (Handlirsch, 1906-08; Hennig, 1954).

The Tertiary fauna of Asilomorpha shows great similarity to that of the present day, with representatives of all the superfamilies now known, and of the vast majority of the families. Families specially abundant in the Tertiary were Rhagionidae, Nemestrinidae, Stratiomyiidae, Asilidae, Empididae and Dolichopo-didae. On the other hand Apioceridae, Scenopinidae, and Hilarimorphidae have not yet been discovered as fossils.

The ancestral group of this infraorder is still uncertain, but apparently it was related to the early Rhyphidea, family Protorhyphidae. The Asilomorpha are therefore descendants of the first Triassic bibionomorphs. The closest to the original forms are species of the superfamilies Tabanidea and Stratiomyiidea, in particular of the families Coenomyiidae, Tabanidae, and Rachiceridae. All other Asilomorpha are either direct or indirect derivatives of the Tabanidae. Such are Asilidea, which presumably arose from forms related to the Rhagionempididae. Bombyliidea and Empididea developed at the base of the ancestral forms of the family Therevidae (Asilidea). The separation of the first Bombyliidea was more ancient, but the Empididea split off later; these two are the most recently evolved superfamilies of Asilomorpha, and besides them the ancestors of the peculiar Phoromorpha also split off from ancient Asilidea.

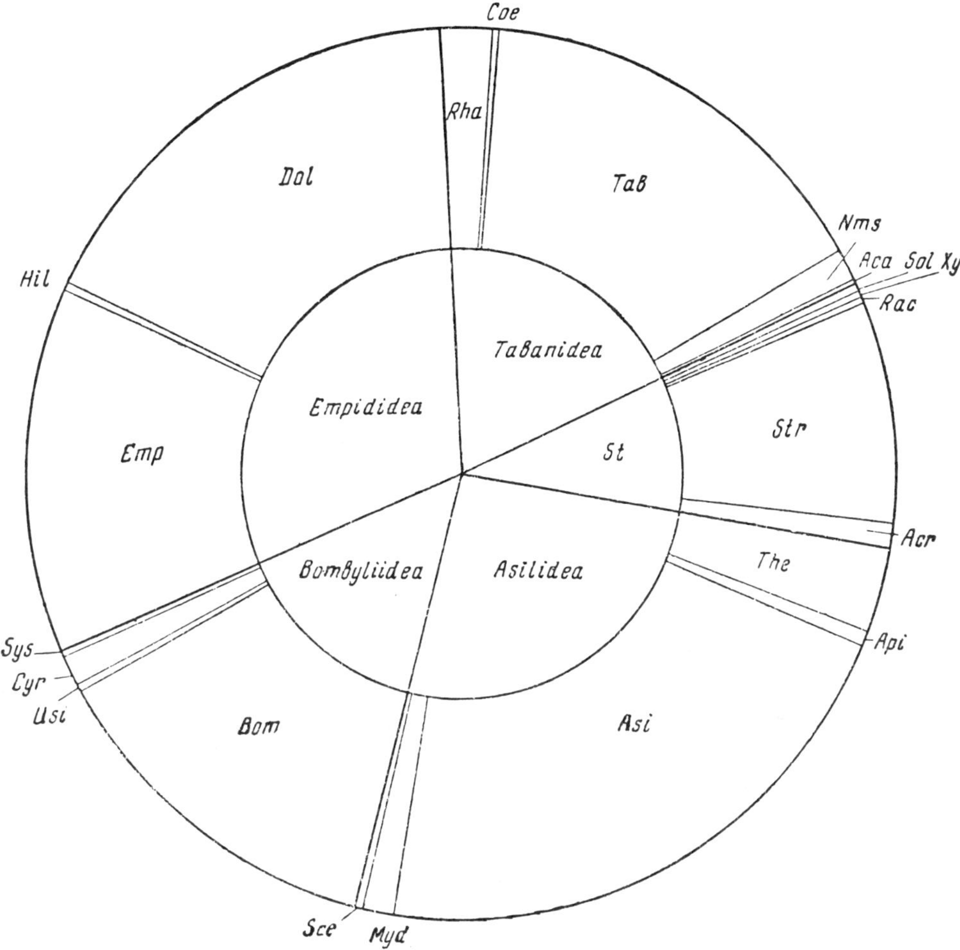

Fig. 30. Diagram of the composition of the infraorder of asilomorphs of the contemporary fauna. The superfamilies and families are indicated in the form of sectors of a circle, the sizes of which correspond to the number of species of the given taxon (1° of arc is equivalent to 63 species). In the centre of the circle are shown the superfamilies, and on the circumference the families. (Original.) For abbreviations, see p. xiii.

The systematics of present-day representatives of the series are comparatively well understood (fig. 30). The main superfamily, Tabanidea, which contains about 4,300 species, is divided into five families of extremely unequal extent. The largest family of horseflies (Tabanidae), to which belong five-sixths of all the species of Tabanidea, is subdivided into 11 subfamilies, many of which are relict groups; there are more than 40 genera, some of which include several hundred species (for example, *Tabanus*). The distribution of the family is very wide, embracing all the continents and it is represented most diversely in the tropical zone (Kröber, 1925; Curran, 1934; Olsufev, 1937). The remaining families of the

Tabanidea have a more or less obviously relict character. The most diverse, Rhagionidae, has more than 400 species distributed among four subfamilies of which some (for instance the Vermileoninae, known from Jurassic time) are ancient relict groups. There are about 15 genera distributed universally. The next family of Tabanidea according to size, the Nemestrinidae, which includes about 250 species, is divided into two related subfamilies. There are about 20 genera. The two final families, Acanthomeridae and Coenomyiidae, are the most obviously relict groups, and include only about 40 species: the acanthomerids, which have about 30 species of three genera, and the coenomyiids, which have about 10 species belonging to no less than five genera, many of which are monotypic (Lindner, 1924-25; Sack, 1933; Hendel, 1937).

The second superfamily, also an ancient one, Stratiomyiidea, includes the smallest number of species, approximately 2,230; there are five families, most of which are poor in species. Only the main family, Stratiomyiidae, distributed universally, contains more than 1,800 species, and these are distributed among 11 subfamilies and more than a hundred genera. The second family according to size, the Acroceridae, includes only somewhat more than 280 species, distributed among approximately 30 genera and three subfamilies, of which one is clearly relict (Philopotinae). The remaining families of the Stratiomyiidea are in large part ancient, relict groups. The Solvidae contains about 70 species of several genera; the Xylophagidae with 20 species of three genera and the Rachiceridae with 15 species of a single genus are residual groups which contain uncommon species restricted to definite, very local regions (Lindner, 1936-38).

The third superfamily, Asilidea, apparently younger than the first two, contains about 6,000 species distributed among five families of which only one, the Asilidae, is distributed universally, and this embraces more than 4,800 species. This family is subdivided into four very clearly separated subfamilies to which more than 200 genera belong. The remaining families of the Asilidea are considerably less diversified. Only one of them includes several hundred species, namely the Therevidae, which is evidently homogenous and not divided into secondary groupings but this needs further study; there are about 20 genera. The next family on the basis of size, the Mydaidae, contains upwards of 250 species of 25 genera distributed in tropical and desert regions; systematics of the family have been comparatively little studied and it has not yet been divided into subfamilies. The last two families, Scenopinidae with 85 and Apioceridae with 60 species, are clearly relicts. The scenopinids contain about 10 genera and are distributed very locally; the apiocerids are rare tropical Diptera distributed partly in subtropical and desert zones and include about five quite isolated genera in two clear subfamilies (Engel, 1930-38; Kröber, 1924-26; Sack, 1934; Hendel, 1937).

The fourth superfamily, Bombyliidea, is a derivative of unknown, presumably Lower Cretaceous Diptera, close to the original forms of the Asilidea. It contains over 3,300 species spread among four families, of which the Bombyliidae, distributed universally, embraces the vast majority of species (about 2,950) and is divided into a series of subfamilies and tribes (Engel, 1932-1937). The family Cyrtosiidae, also widespread, contains about 250 species and is divided into four subfamilies. The families Usiidae and Systropodidae are quite poor in species, each including approximately 60 species of few genera; the usiids are distributed chiefly in arid subtropical regions and the systropodids in humid tropical areas in the Oriental and Neotropical regions. It should be noted that until recently the group Bombyliidea was considered to be only one family of the Asilidea and its system remained little studied. The scheme of division of the bombyliids into four families, as I have proposed, is purely preliminary and must be reasoned out in detail by a special investigation.

The fifth and last superfamily, Empididea, is apparently the youngest of all the groups of asilomorphs and contains upwards of 6,700 species in three families of very unequal size. The largest and at the same time the youngest family, Dolichopodidae, which includes more than 3,700 species, is distributed among no less than 11 subfamilies also of very unequal size. Dolichopodids are found all over the world though some subfamilies show a local distribution; such, for example, are the Plagioneurinae and Stolidosomatinae which are discovered only in the Neotropical region, and the Sciapodinae, found chiefly in tropical regions. The second family of Empididea is also a very big one, the Empididae, which includes about 3,000 species in eight subfamilies of which the largest are the Empidinae (with upwards of 1,600 species), the Corynetinae (with over 700) and the Noezinae (with about 300). These are distributed universally. A few subfamilies are poor in species and have a local distribution. Such, for example, are the Ceratomerinae which contains about 10 species in the Neotropical region. The last family of Empididea is clearly a relict group, poor in species. There is, for instance, Hilarimorphidae, which includes only five species of the single genus *Hilarimorpha*, the systematic relationship of which was established only recently.

Chief features. — The infraorder of asilomorphs as a whole is characterized by a sharp integration of the body, which leads in the first place to enlargement and shortening of the head and increase of the thoracic section. These morphological processes are an expression of the improvement of the central nervous system and the muscular apparatus of the thorax (fig. 31). Associated with this is the complication of the function of the sense organs (especially the antennae and eyes), more complete (accelerated) nervous regulation, and increased locomotion of the insect, especially in flight. This improvement of the winged form is inseparably connected with changes in the organization of the larvae, which were primarily predators and which developed the capacity for extraintestinal digestion. This development required complication of the enzymes of the saliva and changes of the head and mouth parts for the sucking in of liquids. Another important trait of the original asilomorphs is the medium or large size of the body, which apparently was a very ancient acquisition in their historical development: a decrease in size happened repeatedly in different groups of the infraorder but is proved to have been only a secondary process. The larvae of representatives of the series are found in various ecological conditions, chiefly in terrestrial media, — soil, but also in rotting or living plant tissues — and in aquatic media, remaining as air-breathing insects. For the whole infraorder, therefore, a eurybiont nature is characteristic.

The principal structural features of asilomorphs also influenced the

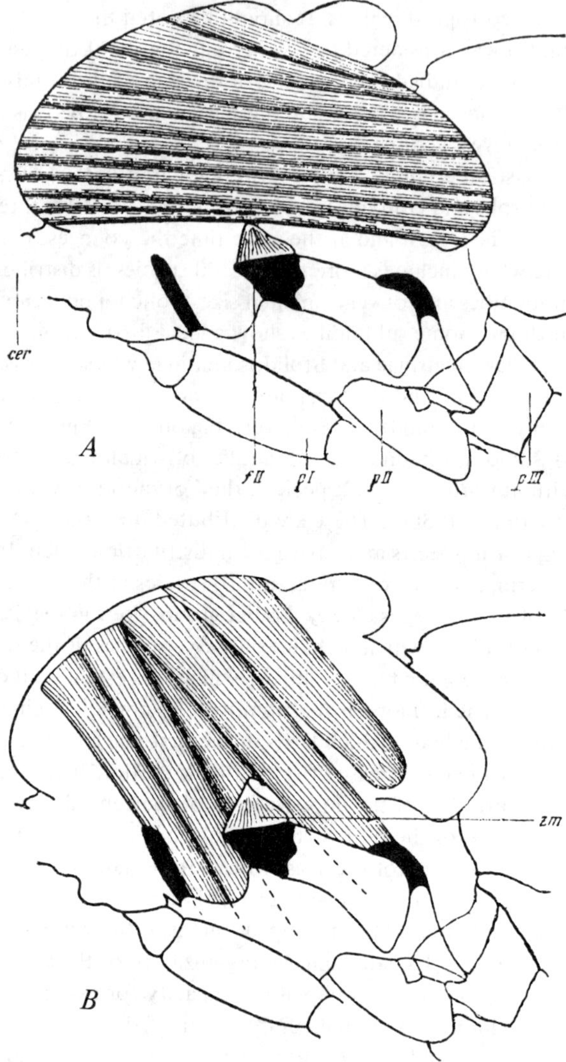

Fig. 31. *Rhagio scolopaceus* (L.) (Rhagionidae). Female, muscular apparatus of right half of thoracic section. Length 3.5 mm. *A.* Longitudinal dorsals. *B.* Dorsoventrals. (Original.) (for abbreviations, see fig. 13, p. 42), plus *7, 20, 24* – corresponding pleural muscles.

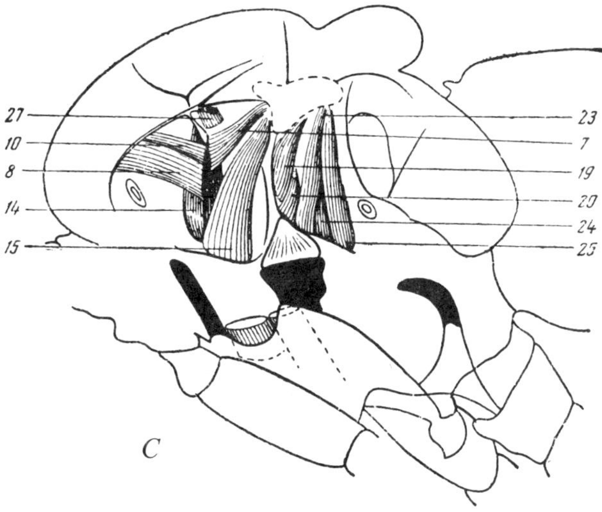

Fig. 31. (cont.) *Rhagio scolopaceus* (L.) *C.* Pleural muscles. (Original.) (for abbreviations, see fig. 13, p. 42).

development of many other separate traits which are characteristic of these Diptera. Thus, it is characteristic for the larvae to have a strong development of the internal skeleton of the head capsule (vertical plate – internal protuberance of the facial shield, protuberances of the posterior part of the head – outgrowths of the posterior rim of the epicranial plate and finally of the tentorium) which permit the development of a powerful muscular apparatus to move the anterior end of the body and the mouth parts. The paired mandibular sclerites are not capable of the usual gnawing function of the jaws of an insect but move parallel to one another and carry out the grasping, tearing, puncturing or rasping of nutritive material. The respiratory system of the larvae is usually amphi- or metapneustic. Most characteristic of asilomorphs is the structure of the winged phase, which possesses comparatively short wings, not exceeding the length of the body, and which are provided with a rich venation covering almost the whole wing. The phenomenon of costalization occurs in only a few groups and always bears a secondary character (fig. 35*D*). The legs as a rule are prehensile and belong to the running type only in a few of the most ancient groups. Especially noteworthy is the structure of the head of these Diptera, on which very large eyes often occupy a great part of the anterior surface. The antennae nearly always show far-reaching processes of reduction; sometimes, but rarely, they are elongate as, for example, in some Stratiomyiidea (Rachiceridae, Solvidae) with many segments, and Asilidea (Mydaidae, part of the Asilidae) with elongated three-segmented antennae. The mouth parts are of very diverse structure: from the complex piercing probosces of horseflies, which have almost all the structures

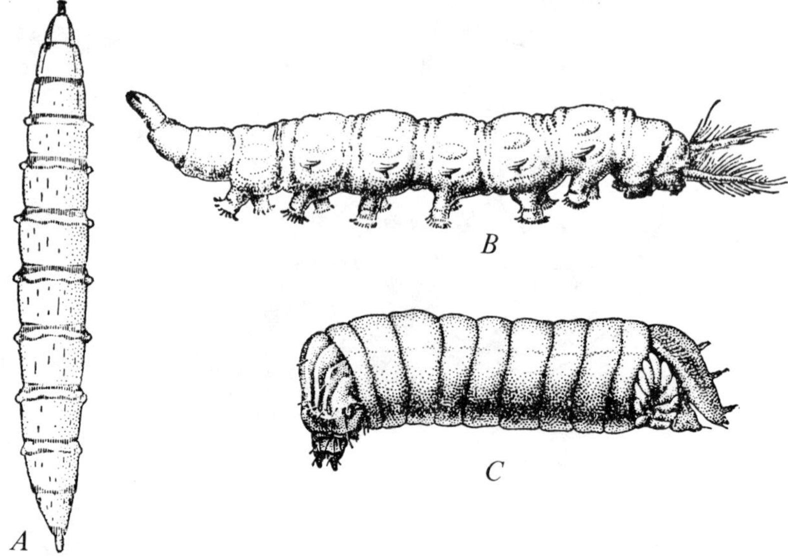

Fig. 32. Tabanidea; Larvae. *A. Tabanus autumnalis* L. (Tabanidae). View from above. *B. Atherix ibes* Fabricius (Rhagionidae). View from the side. *C. Pantophthalmus tabaninus* Thunberg (Acanthomeridae). View from the side. (*A.* according to Olsufev, 1937; *B.* according to Lindner, 1925; *C.* according to Hendel, 1937.)

characteristic of chewing insects, to the extreme reduction of the mouth parts of some aphagic species (e.g. Mydaidae) and some aberrant Tabanidae.

Conflicts and determining tendencies in historical development. — We have no paleontological data to show the derivation of the first asilomorphs from ancestral forms related to the ancient Rhyphidea, and we can only deduce the course of this evolution by examining the differences between these families and studying the direction of the changes that have taken place. It is obvious that these first asilomorphs were insects increasing in size, as witness the preservation of abundant "primitive" venation of the lamina of the wing; these veins are necessary to guarantee stability in large wings. Feeding of the original forms consisted of the consumption of highly caloric substances presumably of animal proteins; these Diptera carried on a predatory way of life, catching other insects and sucking them out with their piercing probosces, as witness the presence among asilomorphs of blood suckers with original mouth parts (horseflies and some rhagionids) and the predatory way of life of the larvae of many forms. As a result of these processes in the history of the order Diptera there arose the first asilomorphs, which are characterized by an integrated body with a powerful flying apparatus and by larvae in which the processes of acephalization combined with an improved method of feeding are indicated.

Features of the phylogeny of the separate superfamilies of the Asilomorpha. — As mentioned above this infraorder of Diptera is very diversified and already at

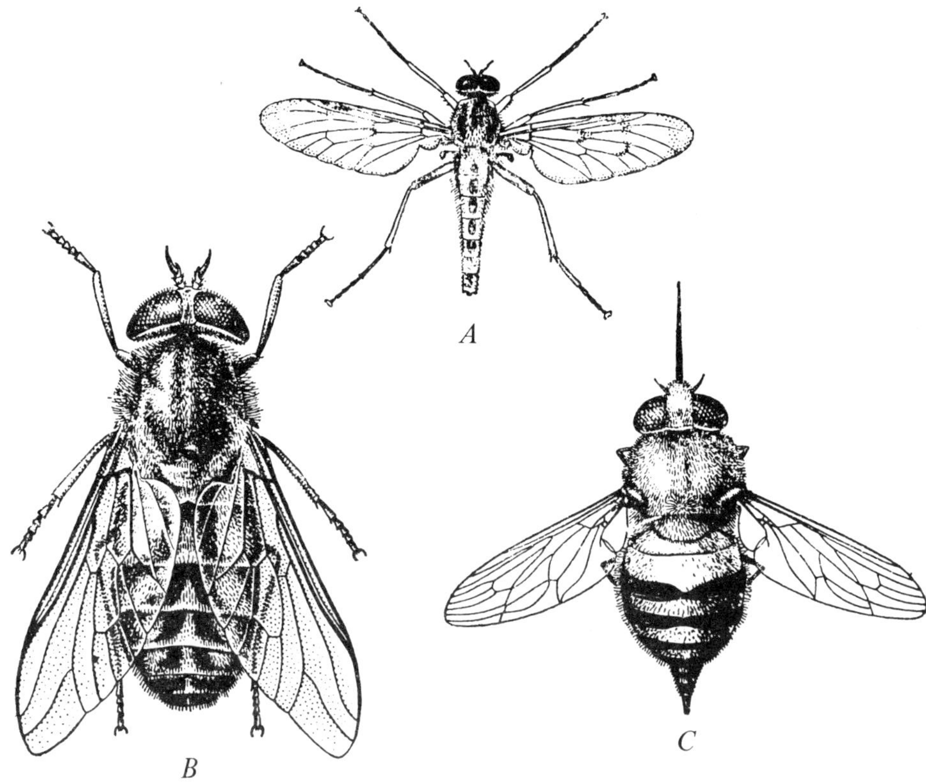

Fig. 33. Tabanidea. *A. Rhagio scolopaceus* (L.) (Rhagionidae). Male, general view. Length of body 12 mm.
B. Tabanus fulvicornis Meigen (Tabanidae). Female, general view. Length of body 15 mm. *C. Rhyncho-
cephalus fasciatus* Olivier (Nemestrinidae). Female, general view. Length of body 14 mm. (*A.* according
to Hendel, 1937; *B.* according to Olsufev, 1937; *C.* according to Sack, 1933.)

the beginning of the Cenozoic era consisted of several individual superfamilies
which included quite differently organized insects. Consequently a consideration
of the features and character of the conflicts in the phylogeny of the infraorder of
asilomorphs as a whole is necessarily very brief and general. Only an examination
of some superfamilies allows us to throw more light upon the history of these
Diptera.

Superfamily Tabanidea (figs. 32 and 33).

One of the most ancient of the infraorder, it is characterized by the absence of
costalization of the veining of the wings, by the large size of body, by a slight
development of the anterior thoracic section, and by predation or blood-sucking.
This last feature, as also the structure of the predatory larvae, may be established
only on the basis of an examination of the present-day representatives of the
superfamily, although they were probably present in the first Jurassic forms.
Nearly all these traits of the Tabanidea indicate this group to be the most
primitive in the whole infraorder. Most families of the Tabanidea are remnants,

relicts of earlier, richly developed groups and only the Tabanidae and, in
lesser degree, the Rhagionidae and Nemestrinidae show a comparatively
abundant development in the contemporary fauna. The reasons for the
preservation of these families of Tabanidea up to the present-day epoch are
apparent: the development of blood-sucking on mammals, strong flight and large
body size, perfect vision, high fertility, and a habitat for the predatory larvae in
pools (Olsufev, 1936; Rohdendorf, 1951). The progressiveness of these processes
is evident: the new determining features turned out to be those of feeding,
development of the nervous system (sense organs) and powers of flight, and the
colonization of sheltered habitats by the larvae. Another family of the Tabanidea
– the Rhagionidae (figs. 32*B*. 33*A*), still insufficiently investigated, also
evidently may be outlined as having preserved ancient traits (wings, legs, form of
body) together with improvement of the nervous system (integrated antennae and
large eyes) and presumably the development of active predation at all stages.

Third according to abundance of species is the family Nemestrinidae (fig. 33*B*)
which arose as a result of the production of parasitic features of the larvae
together with the development of thrusting flight and an increase in the size of
body. Hence these features of development were new determining traits. The
other two families, clearly relicts, are still very little known: The Coenomyiidae,
the larvae of which are inhabitants of the soil, almost unstudied, and the huge
tropical Acanthomeridae, possessing larvae which bore in the wood of tropical
forests and which have peculiar protuberances on the posterior end of the body,
presumably having a respiratory function.

Superfamily Stratiomyiidea

This is apparently as ancient a group as the Tabanidea. The chief feature of the
superfamily is the absence (as a primary process) of increase in size of the body
which led to the process of costalization of the venation. Another important trait
of the structure in the Stratiomyiidea was the powerful strengthening of the
anterior thoracic section of the body, the pleural sclerites of which increased and
united with the sternite: as a result the anterior coxae of the winged insect
became closed stable chitinous plates in front. This feature of the skeleton of the
thorax apparently reflects an unusual arrangement of the muscular apparatus
which moves the sclerites of the pleura; a more precise determination of the role
of this skeletal feature calls for further study. A process of general integration of
the body originated during the evolution of some families of Stratiomyiidea: as a
result of this process, some Jurassic representatives of the superfamily acquired
for the first time in the history of the Diptera the characteristic shortened form of
the body of a 'fly'. This process of the integration of the body is well illustrated
by the structure of the muscular apparatus of the thoracic section (fig. 34).
Another important feature of most Stratiomyiidea is the development of plant-
eating in all stages; the improvement of the anterior section of the intestine of the
larvae which contains the peculiar and still insufficiently studied apparatus which
digests the food both mechanically and chemically, has special importance. Thus
the features which characterize the Stratiomyiidea as a whole are the
strengthening of the skeleton of the thoracic section of the winged form, the

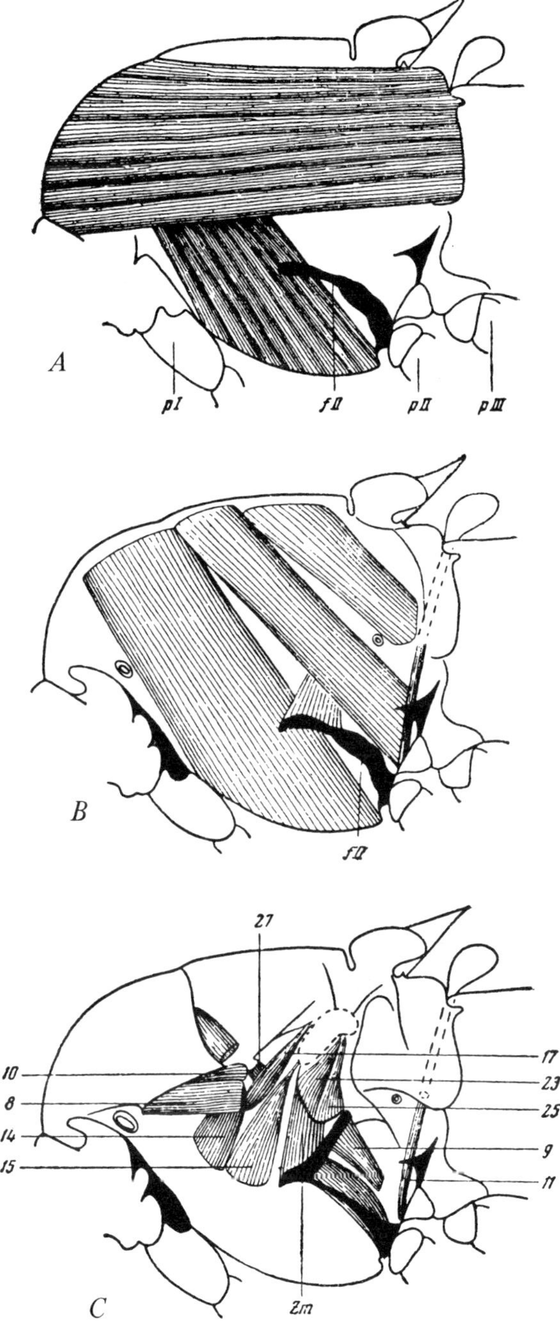

Fig. 34. *Stratiomys chamaeleon* (L.) (Stratiomyiidae). Female, muscular apparatus of the right half of the thoracic section. Length of thorax 7 mm. *A*. Longitudinals. *B*. Dorsoventrals. *C*. Pleural muscles. (Original.) Abbreviations as in fig. 13, (p. 42), plus *9* and *11* − corresponding pleural muscles.

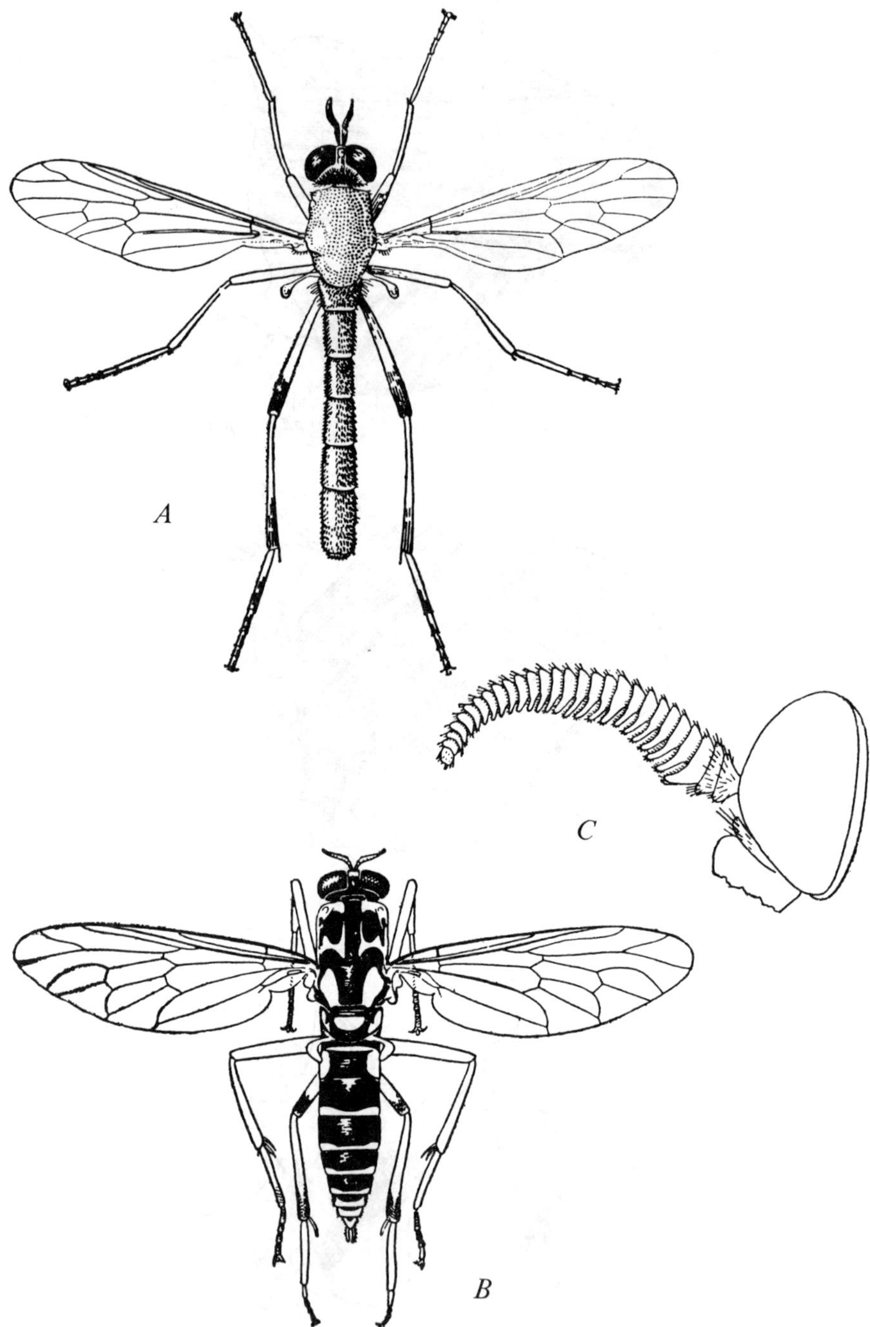

Fig. 35. Stratiomyiidea. *A. Xylophagus ater* Fabricius (Xylophagidae). Male, general view. Length of body 10 mm. *B. Solva maculata* Meigen (Solvidae). Female, general view. Length 10 mm. *C. Rachicerus* sp. (Rachiceridae). Male, head from the side. (*A.* according to Berrol, from Hendel, 1928; *B, C.* according to Lindner, 1938).

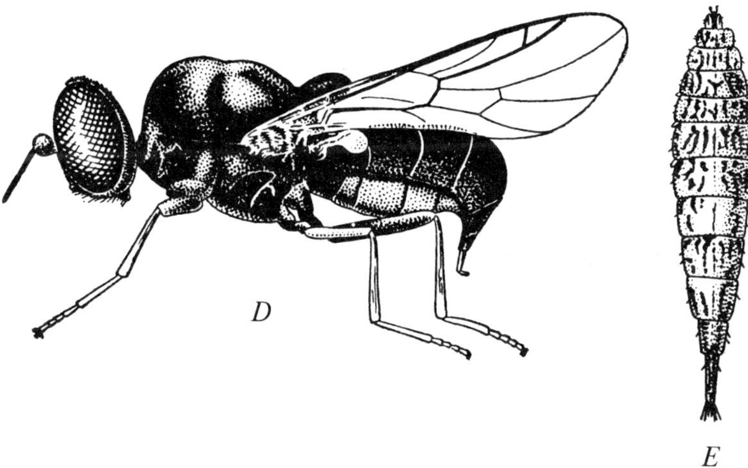

Fig. 35. (cont.) *D. Pachygaster orbitalis* Wehlberg (Stratiomyiidae). Male, general side view. Length 3.25 mm. *E. Stratiomys riparia* Meigen (Stratiomyiidae). Larva from above. (*D.* according to Lindner, 1938; *E.* according to Curran, 1934.)

absence of increase in body size and the improvement of larval feeding (Hendel, 1928; Monchadski, 1940).

Further development of the superfamily led to quite different groups which have comparatively little in common. The greatest development in the present-day fauna is attained by the family Stratiomyiidae with a large number of species of numerous subfamilies (fig. 35). This family is characterized by improvement of the larval form, which lives in an aquatic medium, and increased plant feeding; at the same time there developed a puparium which is especially important. Stratiomyiids show an integration of the body which results in the shortening of, and in fewer, apparent segments in the abdomen and in the reduction and integration of the antennae. They also show an increase of the thoracic section, strengthening of the legs and, most conspicuously, an abrupt development of costalization of the wings.

The processes of integration and costalization led to the derivation of numerous groups (subfamilies, tribes and genera) in the family but they were not the primary features which influenced the derivation of the original forms of the family. A consideration of the history of the same family is a vast undertaking beyond the frame of the present investigation. Appraising the conditions which might have led to the derivations indicated above, one can assume the colonization of warm temporary pools rich in decomposing plant substances by larvae of the ancestors of these Diptera; the temperature, temporary character of the pool, the abundance of food and, finally, the necessity for strong flight to seek this and, even more, to seek shelter from overheating — all influenced the first small body size in these Diptera.

Three relict families of the Stratiomyiidea — the solvids (fig. 35B), rachicerids (fig. 35*C*) and xylophagids (fig. 35*A*) — are poor in species and are inadequately known, yet each of them is distinguished by peculiar features which throw light on the problems that were solved during the course of their evolution. Solvidae are characterized by the development of unusual larvae, which live in the rotting fibrous inner bark of woody trunks, together with the development of relatively slight integration of the body, presumably imaginal aphagia or nectarophagia and the absence of costalization of the wings. The improvement of larval feeding is accompanied by a withdrawal into an aquatic medium, but in contrast to the first stratiomyiids which followed a similar evolutionary path, there is an absence of integration and costalization. The Rachiceridae are comparatively close to the solvids, being distinguished by multisegmented, peculiar antennae and apparently by predatory larvae. The Xylophagidae are more isolated and possess a less enlarged prothorax like the Tabanidea. Their peculiar predatory larvae live in the rotting wood of tree stumps.

The peculiar family Acroceridae is sharply distinguished from the remaining Stratiomyiidea in the immature stages: the larvae of these Diptera are parasites and live in the bodies or egg cocoons of spiders (Araneina). The winged form of acrocerids is characterized by a great reduction in the size of the head, by very great increase of the thorax with weak legs, and by small, often costalized wings; the flies feed, if at all, on the nectar of flowers. The finding in the Jurassic fauna of a form (*Protocyrtus jurassicus* Rohd.), related to the present-day acrocerids points to the ancient separation of these Diptera from other Stratiomyiidea. The parasitism of acrocerids on spiders isolates them sharply, yet at the same time it throws no light on the course of the evolution of this group. There are evident only the most general mechanisms: aphagia in the imago in association with the parasitism of the larvae, as well as a powerful lifting, flying apparatus in relation to contacts with spiders' webs (Sack, 1936).

The superfamily Asilidea first made its appearance in the Jurassic fauna; we have a single species of a family related to the Scenopinidae from Karatau. The Asilidea are characterized by a large body size, powerful legs, non-costalized elongate wings and by the elongated, non-integrated body (fig. 36). One of the most important features of the asilids consists of the improvement of predation both in larval and the winged forms. Apparently these features reflect the derivation of the first forms of this superfamily from ancestors related to the Tabanidea (namely the Jurassic Rhagionempididae and Rhagionidae). The history of the Asilidea is still very little known and will be revealed only after a study of Cretaceous faunas which have not yet been found.

The largest and most diverse family, the Asilidae, sometimes called robber flies, (fig. 36*B*, 36*C*) is characterized by the development of powerful, grasping prehensile legs which are a catching apparatus for the capture of prey, by large eyes, integrated antennae, complete piercing proboscis, big thoracic section with a powerful flying apparatus and by elongated, quite short, peculiar pulling-lifting wings with little costalization. The larvae live in the soil, rotting wood and even in the tissues of living plants, being in the main plant-eating, only rarely predatory, insects; predation almost certainly is a primary phenomenon. The development of

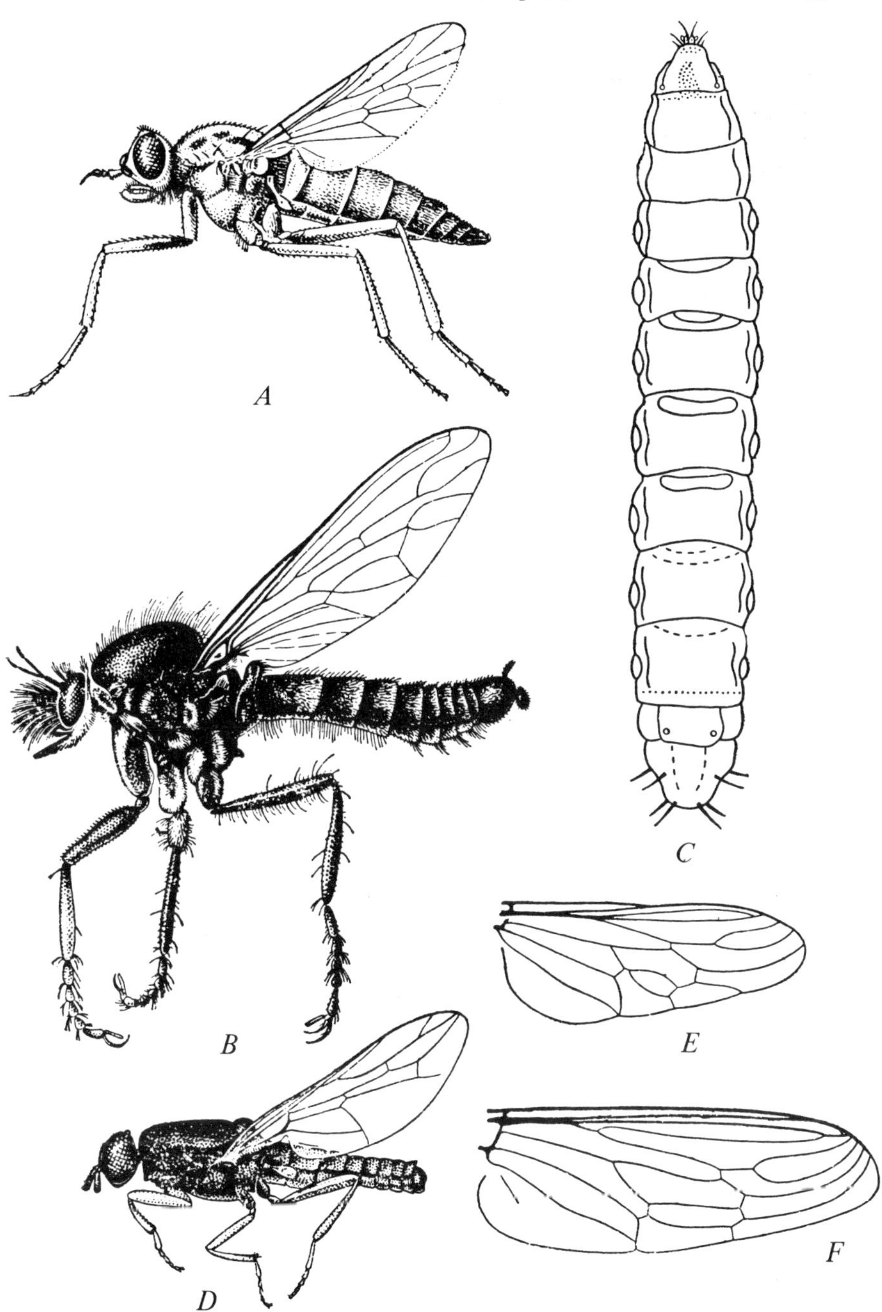

Fig. 36. Asilidea. *A. Pherocera* sp. (Therevidae). Female, general view. *B. Protophanes punctatus* Meigen (Asilidae). Male, general view. Length of body 15 mm. *C. Machimus atricapillus* Fallén (Asilidae). Larva from above. *D. Scenopinus glabrifrons* Meigen (Scenopinidae). Male, general view. Length 5 mm. *E. Apiocera* sp. (Apioceridae), Wing. *F. Rhaphiomydas* sp. (Apioceridae). Wing. (*A, E. F.* according to Curran, 1934; *B. D.* according to Lindner, 1927; *C.* according to Melin from Hennig, 1948-52.)

active imaginal predation (capture of prey on the wing), to include insects which may considerably exceed in mass and strength the predators themselves (for example, grasshoppers) together with the development of phytophagia in the larvae, throws light upon the derivation of robber flies. Undoubtedly the vastness of this whole family, which is subdivided distinctly and clearly into four subfamilies and a series of tribes, testifies to the complexity and diversity of the processes of its phylogeny, an account of which requires a special investigation and is beyond the scope of the present survey.

Next in size, but much smaller than the Asilidae, the Therevidae (fig. 36*A*), is characterized by the development of active predation in the larvae (which live in the soil and which develop a peculiar wormlike body), and by the development of powerful running legs, large antennae and a short weak proboscis on the small head in the winged form. These traits of the therevids relate to the withdrawal of the larvae into soil (from the probable moist habitat of their ancestors), and to the improvement of both their predation and the running of the winged form. The non-costalized venation of the therevids is the reason for these Diptera having been considered the original ancestral forms for all other Asilidea (Hendel, 1937). However, it is obvious that one feature of the structure of the wings cannot in itself be the basis for so definite a conclusion, which is really quite mistaken. The therevids and asilids actually are related groups, but the second are by no means descendants of the first.

The third family of the Asilidea on the basis of size is the Mydaidae, which contains large, sometime gigantic, Diptera, the bodies of which may attain 55 mm. in length. The family is characterized by a big head with elongated antennae and an enlarged third segment; little integrated, elongated body; moderately large legs of the unarmed prehensile type; thorax little enlarged; quite wide, little costalized wings, the venation of which forms a peculiar border of the membrane, free from veins, on the posterior margin; and by predatory larvae which live in rotting wood. The way of life of these insects is little known and indications of predation and the habitat of the larvae in rotting wood tells little about the phylogeny of these peculiar relict Diptera (Shtakelberg, 1948).

The tropical family Apioceridae (fig. 36*E*, 36*F*) is closely related to the therevids. It is a clearly relict group, and is even less known. The imago of the apiocerids is characterized by short antennae, moderately enlarged eyes, large thorax, quite weak legs of a reduced running type and short powerful wings with a border free of veins at the edge. The development of these Diptera is completely unknown. To pass judgement on the historical development of the apiocerids is premature (Hendel, 1937).

We still have to examine the relict family Scenopinidae (fig. 36*D*) which, out of all the contemporary Asilidea, is probably the group most closely related to the original forms of the superfamily. The scenopinids are characterized by small body size, elongated massive thorax bearing running legs, and peculiar costalized wings which are sharply distinct from those of the majority of the Asilidea. The structure of the head, which bears simplified, rather integrated antennae and very short mouth parts indicating aphagia, is curious. The larvae of scenopinids live in various restricted habitats: in mushrooms, under the bark of trees, in entanglements of the silk of caterpillars and even in human habitations; under the carpets, in mattresses or in the creases of upholstery. They are clearly predators, which attack Copeognatha, moth caterpillars and probably other small insects (Kröber, 1926). All these features allow us to assume the nature of the derivation of the relict scenopinids, which consisted of the improvement, or to be more precise, the specialization of the feeding of the larval phase, which became a predator on various small insects of sheltered, restricted habitats. This process presumably occurred in a warm medium, apparently in desert or at least in dry terrains.

The activity of man turned out to be beneficial for these Diptera, extending the habitat for their development. The chief thing in the history of scenopinids was specialization in the feeding of the larvae and the development of lifting flight.

The superfamily Bombyliidea (fig. 37) originated at the end of Jurassic times from peculiar forms of Asilidea, related to the ancestors of the therevids (and probably also to the ancestors of the Empididea). Fossils are known only from the Tertiary series and belong in the main to contemporary genera, but the chief processes of the divergence of the Bombyliidea originated during the Cretaceous period. Most characteristic of this superfamily is the development of parasitism of the larvae in very diverse insects – Hymenoptera, Orthoptera, and even Diptera. The development of parasitism in the Bombyliidea is the main governing feature which influenced the derivation of this group. All other traits of the Bombyliidea are closely linked with their parasitic way of life. These are the development of perfect pulling flight, which governs the elongation of the wings[5] ; plant feeding in the winged form and occasionally even aphagia (genera *Villoestrus*, *Oestranthrax*); and the peculiar development of an abundant covering of hairs or scales, apparently connected with the thermal insulation of the body, which has an important value as a result of these insects living in open, highly isolated habitats. With the development of powerful pulling flight there is connected a weakening of the legs, which in the majority of the Bombyliidea belong to the reduced running subtype.

Bombyliidea have not been studied sufficiently to allow discussion of the phylogeny of its individual groups in more detail (Engel, 1932-37).

The last superfamily of asilomorphs, the Empididea, appears first only in Tertiary faunas, if one does not count the unreliable indication in the Upper Jurassic fauna of Europe of the problematical form *Empidia wulpi* Weyenbergh (see p. 271). This superfamily is characterized by the small, or more rarely medium, size of body; by reduction of the head section of the larva, which is capable of being withdrawn into the thorax; by the development of quite strong costalization (undoubtedly connected with the small size); by active predation in the winged form and the larva; and by the strong development of running legs (fig. 38*A*. 38*B*). All these traits of the Empididea influenced their separation from ancestral forms (a peculiar, still unknown group of Asilidea, the original group for both the Empididea and the Bombyliidea). The larvae of the Empididea remained

5. One of the features of the venation of the Bombyliidea is the constant absence of the fourth branch of the medial vein, by which these Diptera differ from the majority of the Asilidea, among them the related therevids. This feature of their venation indicates that the original forms of the bombylioids, before they acquired pulling, elongated wings, were distinguished from their ancestors by the development of some costalization of the venation (the production of the features of lifting flight). It is possible that the ancestral forms of the Bombyliidea were at the same time the starting point for the complex of Empididea in which costalization went much further.

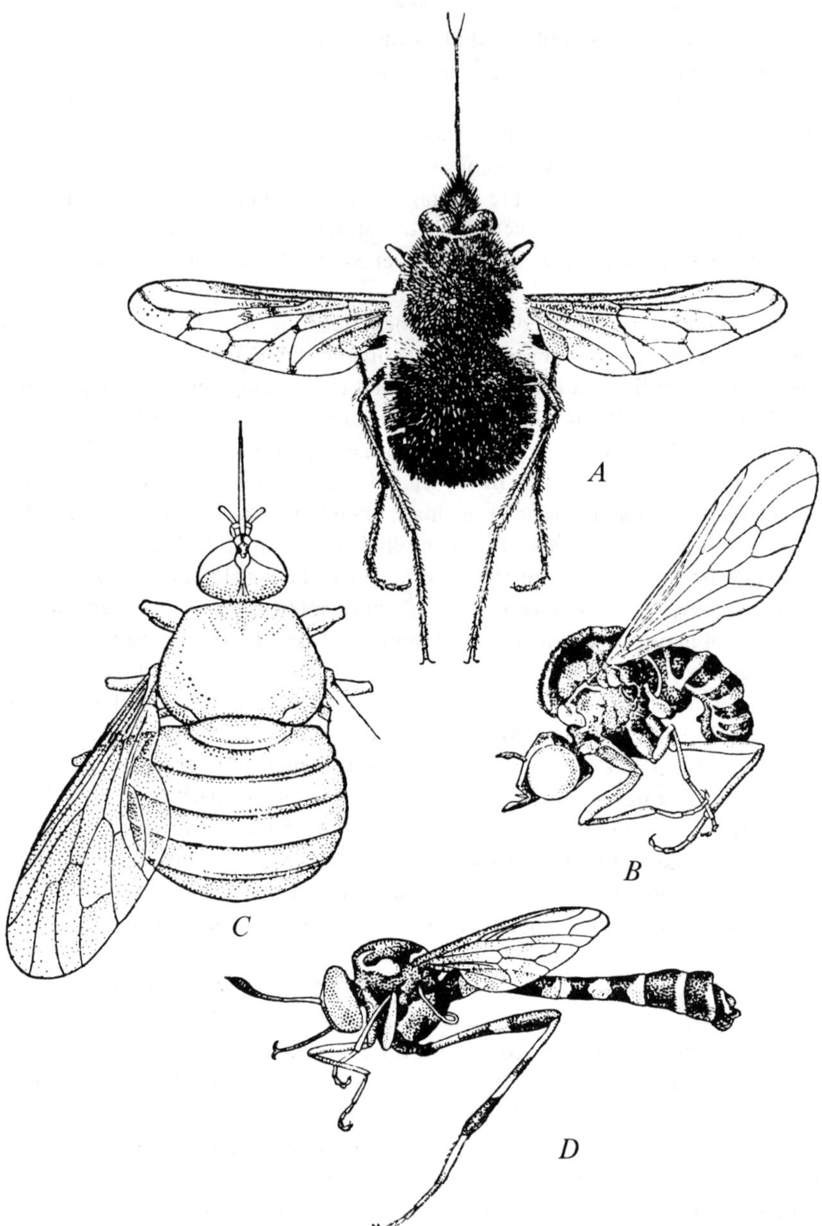

Fig. 37. Bombyliidea. *A. Bombylius discolor* Mikan (Bombyliidae). Male, general view. Length of body 13 mm. *B. Platypygus chrysanthemi* Löw (Cyrtosiidae). Female, general view. Length 4 mm. *C. Usia lata* Löw (Usiidae). Female, general view. Length 6 mm. *D. Systropus annulatus* Engel (Systropodidae). Male, general view. Length 13 mm. (*A.* according to Lindner, 1928; *B. C.* according to Engel, 1932-37.)

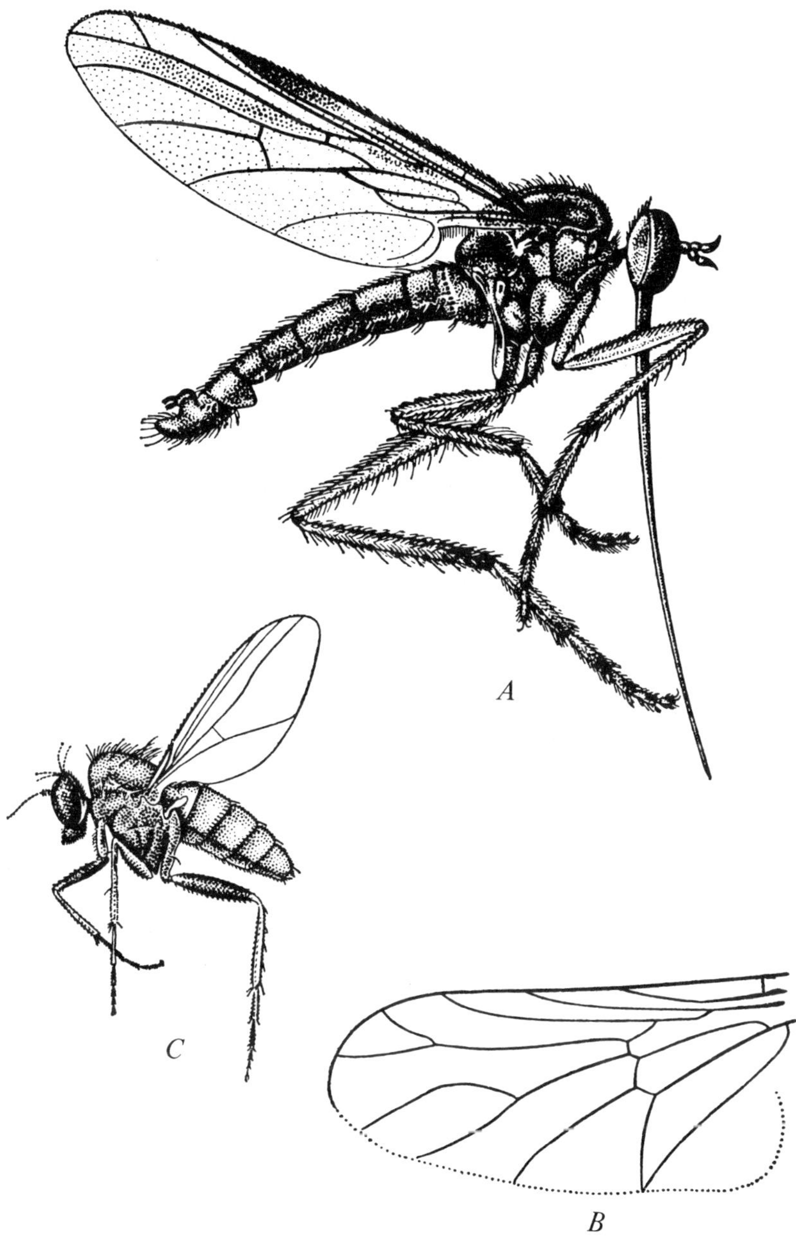

Fig. 38. Empididea. *A. Empis lindneri* Oldenberg (Empididae). Male, general view. Length of body 9 mm. *B. Hilarimorpha singularis* Schiner (Hilarimorphidae). Wing. Length of body 3 mm. *C. Chrysotus neglectus* Wiedemann (Dolichopodidae). Male, general view. Length of body 4 mm. (According to Lindner, 1924-28.)

free-living insects, did not change into parasites, and acquired new, useful features
in the structure of the head section. Very characteristic for this group in contrast
to the majority of other asilomorphs is the absence of increase in body size; there
has even been a reduction in size that in this infraorder of Diptera is a rare
phenomenon.

The family Empididae, one of the largest, is characterized by very different
forms of development of lifting flight by the development of different stages of
costalization of the wings (fig. 38*A*). An important characteristic of empidids is,
moreover, the development of peculiar powerful, prehensile legs; further, this
family is characterized by slight integration of the body: the head is quite small,
the thorax is usually little enlarged, the abdomen is elongated and frequently
cylindrical. The feeding of empidids apparently may be characterized by the
frequent development of phytophagia, in its way some sort of polyphagia:
although basically predators these Diptera also feed on nectar, often finding it on
flowers. It is interesting that a metallic color is rare in empidids; light yellowish
tones are the rule. The larvae of empidids live in various media and evidently are
predators. The governing features of empidids, we may say, were the improvement
of imaginal feeding, its polyphagous nature and the development of prehensile
legs.

The largest family of the Empididea, the Dolichopodidae, is the descendant of
still unknown forms of empidids and is characterized by well developed lifting
flight with very strongly costalized wings; by powerful, well developed running
legs, nearly devoid of prehensile characters; by clear integration of the body,
which consists of a massive thorax (including a corresponding muscular apparatus
which moves the costalized wings and big powerful legs); short head with enormous
eyes and finally by the shortened, more or less conical abdomen (fig 38*C*).
Another feature of many dolichopodids is their close association with pools, on
the shores of which, among plant materials and on the exposed sand, these insects
run in a lively manner. The larvae live in the soil, sometimes highly moistened, in
rotting wood, in discharging wood sap, more rarely in the tissues of living plants,
and most of them are predators. Governing features in the history of the
dolichopodids thus were the perfecting of the nervous system, as witness the
integration of the body; enlargement of the eyes, and development of the
antennae on the one hand, and improvement of the organs of motion, and
enlargement, strengthening of the running legs, costalization of the wings, and
increase of the thorax on the other.

The phylogeny of the empidids and dolichopodids is still very little known and
for its explanation requires a special review of the voluminous data on their way
of life which is beyond the frame of the present investigation.

The last family of empidideans, the Hilarimorphidae (fig. 38*B*), is very little
known; it is distinguished from the Empididae and Dolichopodidae by the
anterior coxae which are free in front (i.e. by the absence of fusion of the anterior
thoracic pleura and the anterior thorax); by the absence of bristles on the body
and by the intermedial transverse vein together with the presence of a long cubital
cell. These Diptera are unknown to me in nature and I have no hypothesis

concerning their phylogeny. Until recently the genus *Hilarimorpha*, as a result of similarity of habits, was considered a representative of the Tabanidea, namely of the family Rhagionidae. Its true position in the system was established by Hendel (1937).

Conclusions. — The asilomorphs came into being by a concentration of the nervous system which ensured the integration of the body of the insect (enlargement of the head by an extreme increase in the size of the eyes, and of the thorax by the powerful muscular apparatus) closely connected to and interdependent upon the improvement in feeding, chiefly by the larva. These primary processes having given a start to the first representatives of the infraorder Asilomorpha, indicate the progressiveness of this group of Diptera. Changes in the 'secondary' systems of organs (wings, legs, protective structures) in themselves did not influence the emergence of asilomorphs, that is, the separation of the original representatives of this series from the ancestral forms of bibionomorphs — some kind of Rhyphidea.

A considerably more complex problem is the appraisal of the historical development of the different stems and branches of the whole vast group of asilomorphs, represented in the present-day fauna by no less than five huge superfamilies which became distinct in the main as early as the Jurassic period. Almost all these superfamilies, after their long history, experienced complicated and diverse changes.

On the other hand an explanation of the causes of the derivation of each superfamily as a whole turns out to be a very difficult undertaking because of the ancient nature of the superfamilies which changed very much and were converted into vast, abundant groups of young forms with complicated relationships. The most ancient superfamily, the Tabanidea, which has the greatest number of relict groups living today including many forms which seem to be already disappearing, appears also to be the most 'difficult'. The chief obstacle is our ignorance not only of their paleontology but also of their biology. Especially little known are the development, ecology and anatomical structure of the rare relict forms, which always include in their organization a mixture of primary and secondary features.

The genesis of the Stratiomyiidea, a most ancient side branch of the Tabanidea, in the present-day fauna represented by a majority of relict families (five or six) may be indicated only hypothetically. A common feature of all the Stratiomyiidea is the good development of the prothorax and the enlarged empodia; the latter trait is undoubtedly primary (it is observed in ancestral groups, all Tabanidea and Bibionidea). Other features which one may consider common for the whole superfamily consist of an improvement in larval feeding, which cannot yet, however, be demonstrated for all families of the Stratiomyiidea. The degree of progressiveness of the superfamily as a whole is difficult to determine as yet.

The third superfamily of asilomorphs, the Asilidea, is also quite an ancient group. New, recently developing features which influenced the initiation and evolution of the Asilidea as a whole, undoubtedly were the improvement of feeding in the winged form, which developed the capacity for active predation by attacking the most diverse range of other insects, together with an increase in size of the insect. Changes in the organs of motion, in the first place of the legs, undoubtedly played an important role in this process. Integration of the body and simultaneously concentration of the nervous system had a very limited development. On the whole the process of formation of the Asilidea cannot be considered solely as progressive. Only in the later history of some families of the Asilidea did the more progressive processes (the perfecting of development and larval feeding in the Asilidae and Therevidae) have a place.

The fourth superfamily — the Bombyliidea — is the chief example in the Diptera of the development of parasitism on the larvae of other insects. This peculiar trend resulted in the development of complex instincts and guaranteed favorable conditions of larval life. It indicates the original progressiveness of the whole superfamily. However, in the later history of the Bombyliidae, as a result of the inevitable processes associated with parasitism and with the habitat of the larvae in particularly 'narrow' and comparatively constant conditions, several groups of these insects underwent narrow specialization approaching regression.

Finally, the last superfamily of asilomorphs, the Empididea, the youngest and most abundant in species, is distinguished sharply from the others by the improvement of the larvae (which developed a mobile, retractile cephalic end of the body indicating improved mobility and perfected digestion), by the absence of increase in body size and by an improvement in lifting flight and running in the imago. All these features point to the progressive character of the phylogeny of the Empididea. The further history of the Empididea followed various paths and led both to regressive groups (for example, the Corynetinae, Hemerodromiinae) and to highly progressive groups namely the vast stem of myiomorphs, which originated as a result of further improvement in development as well as integration of the body.

Summarizing all that has been said concerning the superfamilies of Asilomorpha one can say that determination of the degrees of progressiveness of the different superfamilies is difficult because of our lack of knowledge; the conclusions drawn can only be preliminary. The Empididea are progressive in great measure and least progressive of all, apparently, are the Tabanidea and Asilidea. An evaluation of the progressiveness of the superfamilies as a whole is hampered by the difference in their ages: less progressive but more ancient superfamilies had a long time to develop different secondary improvements and adaptations in their representatives (for example, the Tabanidae and Stratiomyiidae in the corresponding superfamilies). As a result of this the named families are richly and diversely represented in the present-day fauna, and are characterized by a different progressive but secondary and comparatively recent features in the history of these groups. These new traits often surpass in importance the original features of the superfamily by leading to great progressiveness of the given groups (for example, the perfecting of development and integration of the body in the Stratiomyiidae). A quite different situation is observed in young groups, the representatives of which are less varied and little removed from the original forms of the superfamily.

Infraorder Musidoromorpha
Rohdendorf, 1961b, p. 157.

Extent, history, system. — This infraorder contains a single peculiar family of sharp-winged little flies, the Lonchopteridae, with 24 species of one genus *Lonchoptera (Musidora)*. Until recent times the relations of these Diptera remained vague, and in different systematic schemes they were sometimes placed in the 'suborder' Orthorrhapha Brachycera and sometimes in the Cyclorrhapha: the extreme isolation from the remaining groups of the order and some purely superficial similarity to the Empididea, Syrphidea and some other Myiomorpha led to the inclusion of the genus *Musidora* in the above-named higher 'suborders' at times as a superfamily, at others only as a family. The abrupt isolation of the Lonchopteridae definitely indicates the position of this family in the system of the order as a peculiar, clearly characterized infraorder Musidoromorpha, remote from all others.

Paleontological records consist only of the remnants of some species of the genus *Musidora* in the abundant fauna of Baltic amber; Mesozoic fossils are unknown. The phylogenetic relations of the musidoromorphs are very obscure; apparently these Diptera are descendants of some kind of primary forms of myiomorphs.

The system of the whole infraorder is very simple: the little better than 20 species of the single genus *Musidora* known are related one to another and do not form any groupings of a subgeneric character.

Chief features. — These flies are quite common insects in the forest and taiga zones and are often an important component of the biocenose of humid meadows. They live in weeds of grassy vegetation and avoid exposed habitats directly illuminated and heated by the sun. The larvae dwell in the upper layers of the soil, in accumulations of decomposed plant residues such as fallen leaves, and in refuse on the shores of ponds (Czerny, 1934).

The structure of the larva is extremely peculiar (fig. 39*B*, 39*C*). The whole body is highly expanded, depressed, and made up of sclerotized segments (meso- and metathoracic and six abdominal); the head and prothorax form a soft, movable protuberance. Mandibles are lacking; the cephalic end bears large, soft angular blades which project at the sides; it has a wide, complexly constructed pharynx which contains a peculiar apparatus for processing the swallowed food. The central nervous system is united into a single mass. Spiracles are situated at the anterior and posterior ends of the body. The pupa is enclosed in a true, tough puparium produced as a sheath by the mature larva; the puparium is opened during the escape of the winged insect by a peculiar T-shaped opening, not by a round aperture.

The winged insect is not less peculiar (fig. 39*A*). The whole body is covered by numerous strong bristles, particularly large on the head and legs. The antennae are three-segmented, short; the proboscis is short and soft, the eye is of moderate size; the ptilinum is not developed. The lower margin of the head and frons above the antennae have sturdy bristles. The legs are of a running, powerful type. Especially interesting is the structure of the thorax which contains the muscular apparatus

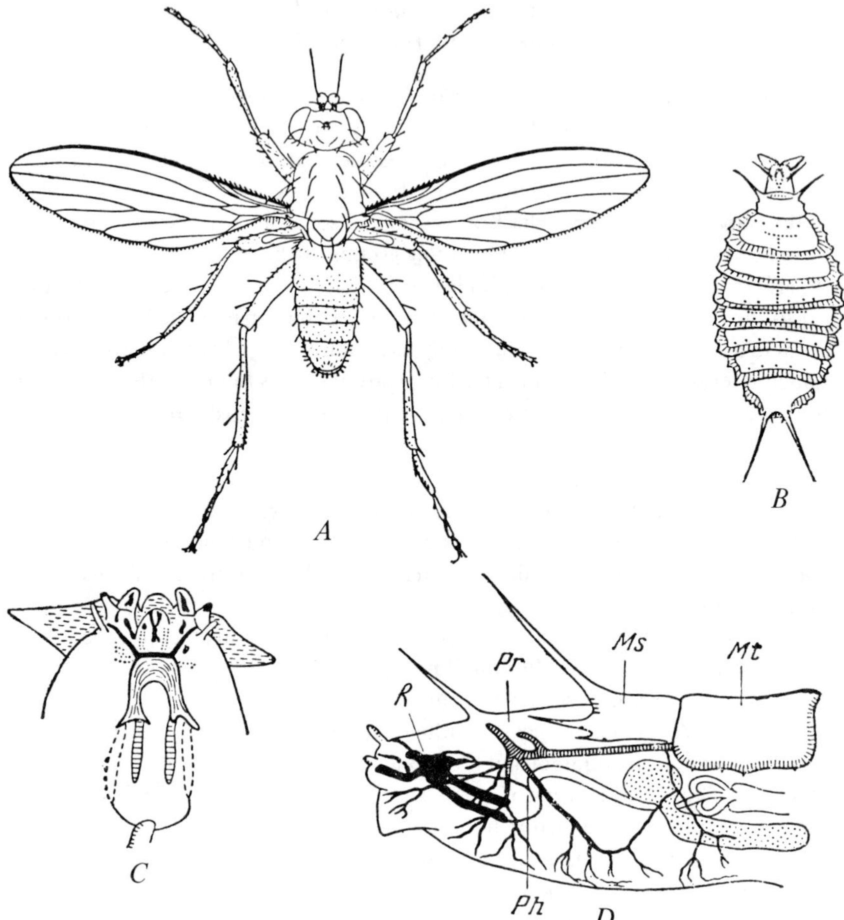

Fig. 39. Musidoromorpha. *A. Musidora lutea* (Panzer) (Lonchopteridae). Male, general view. Length of body 2.5 mm. *B.* The same, larva from above. *C.* The same, cephalic end of larva from above. *D.* The same, anterior end of body of larva from the side. (*A.* according to Verrall from Hendel, 1928; *B, C, D.* according to de Meyer from Czerny, 1934.) Abbreviations: *K* – jaw apparatus; *Ms* – middle dorsal; *Mt* – posterior dorsal; *Ph* – pharynx; *Pr* – prothorax.

which is made up of obliquely disposed dorsoventral muscles (fig. 40*B*); the back of the thorax is slightly arched. The wings are of the lifting non-costalized type (Rohdendorf, 1951, p. 101); their venation is peculiar and is characterized by a uniform distribution of veins over the whole membrane of the wing, not shifted to the anterior margin, and by the absence of transverse veins – vein *RM* lies obliquely and has the appearance of a short longitudinal vein in the base of the wing. Traces of costalization are apparent in the absence of the costal vein at the posterior edge and the reduction of the subcostal and first radial veins. The clear sexual dimorphism of the venation which consists of the fusion of the cubital and

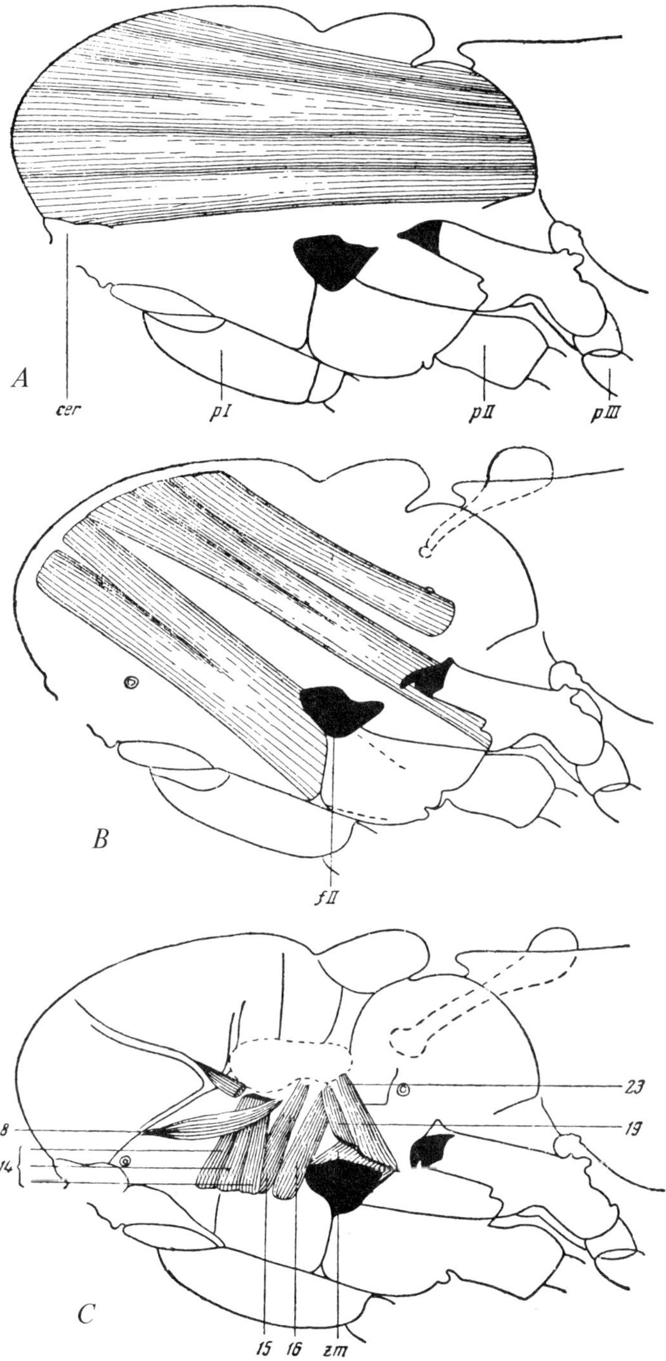

Fig. 40. *Musidora lutea* (Panzer) (Lonchopteridae). Muscular apparatus of the right thoracic section. Length of thorax 1 mm. *A.* Longitudinal. *B.* Dorsoventral. *C.* Pleural muscles. (Original.) For abbreviations see fig. 13.(p. 42).

the last medial veins in the wings of females and the free emergence of these veins in the posterior edge of the wing in males is of interest. The abdomen is moderately contracted and consists of six segments: the genital appendages of the male are peculiar and differ sharply from those of the asilomorphs and myiomorphs.

Conflicts and determining tendencies in historical development. — The larva possesses a tough sheath in the form of sclerotized tergites of the thorax and abdomen; a shortened and expanded body, together with a movable and soft anterior end (head and anterior thorax) bearing the different sensory protuberances; weak, small mouth parts; and the development of a large pharynx and bristles on the body. All this indicates the important role of an improvement of development in the formation and evolution of these Diptera. This assumption is confirmed by the development of a puparium — thickenings of the exuviae of the mature larva, which is not discarded during the formation of the pupa, but makes an ideal protective coat. The winged insect shows a peculiar improvement of the sense organs and presumably of the central nervous system, as witness the development of large, diverse bristles on the head, legs and thorax, integration of the antennae and increase of the dimensions of the head capsule as a whole. The food of the winged insect has not been investigated; judging by the structure of the mouth parts, these little flies probably are predatory on small arthropods (Collembola, mites, perhaps small Pterygota). Without knowing the original or closest forms of this group it is difficult to clear up the primary traits which determined the separation of these Diptera.

Conclusions. — The known features of the lonchopterids permit us to express an assumption concerning the progressive character of these insects. The new developing features of the lonchopterids consisted of an improvement of the feeding of the larva and of the nervous system at all stages. The positive antiquity of the whole infraorder, the poverty of species and the relict character of the group is all in conflict with the observed progressiveness of the organization of these insects. It is natural to draw the conclusion that the formation of the complete larva and the nervous system is a comparatively recent acquisition in the history of the Diptera. This explains the abundance of individuals and their broad distribution in all of the Holarctic and suggests that they are only relatively "relict" and at the present time are undergoing an evolutionary renaissance. This is in agreement with the observed ecological geographical variation of the most widely distributed species *Musidora lutea* (Panzer).

Infraorders Phoromorpha and Termitoxeniomorpha
Rohdendorf, 1961b, p. 158.

Extent, history and system. — These two infraorders of Diptera undoubtedly are phylogenetically related to one another. The chief original group, the Phoromorpha, includes three families: the main one, the Phoridae (with approximately 1,500 species), the Thaumatoxenidae (with 10 species), and the Aenigmatiidae (20 species). The second infraorder, the Termitoxeniomorpha, which is a direct but very altered descendant of the Phoridae, includes the single family Termitoxeniidae, with approximately 30 species of several genera.

There are paleontological records only for the history of the phoromorphs. The occurrence in the Jurassic fauna of Karatau of a species of the family Palaeophoridae which is apparently not an ancestor of the Phoridae, throws no light on the history of this infraorder. Besides this discovery, a series of species of true Phoridae [11 species of *Phora* (Lev *in* Handlirsch, 1906-08, p. 1026)] were discovered in the Paleogene fauna of Baltic amber. The original forms of the phoromorphs were probably related to the most ancient myiomorphs, descended from the first Platypezidea, namely the Sciadoceridae, and appeared in the Cretaceous. The Jurassic Palaeophoridae, which I put close to the Phoridae earlier, have no relationship to them as is shown by Hennig (1954). Fossil termitoxeniomorphs are unknown.

The systematics of the phoromorphs are not yet clear. All the families are closely connected with one another, forming a single phylogenetic complex; the Phoridae is the largest group and has served as the source of two other families of parasitic Diptera, the Aenigmatiidae and Thaumatoxenidae, for the most part codwellers with ants and termites. It should be noted that the systematics of Phoridae are still insufficiently known and the practice of dividing the Phoridae into two subfamilies (Phorinae and Metopininae) may turn out to be incorrect. There are more than 50 genera of Phoridae with approximately 20 belonging to the Phorinae and 30 to the Metopininae.

Chief features. – The main infraorder Phoromorpha (fig. 42) is characterized by sharp integration of the body, small dimensions, the presence of bristles on the body, very powerful running legs, rarely without bristles, big thorax with peculiar muscular apparatus (fig. 41) and highly costalized wings (often completely wingless); the head is covered by peculiar paired bristles, has very short three-segmented antennae with dorsal, rarely terminal arista, and a proboscis which is usually short and weak, rarely elongated and sturdy. The larvae are characterized by a reduced head, the presence of a frontal depression and mouth hooks directed forward capable of tearing and crushing food – different organic tissues; the secretion of the salivary glands has a complex collection of enzymes permitting the digestion of protein (and perhaps other) substances. The larva has very peculiar long and sturdy bristles by which these insects differ very sharply from all other Diptera, but which partly connect them with the Musidoromorpha. Particularly characteristic for the phoromorphs is the widespread development of winglessness in connection with the parasitic and symbiotic way of life; usually the winglessness develops only in females. The phenomenon of winglessness itself is only a part of the general regressive processes of the change in organization resulting from living in the nests of termites and ants and which are expressed also in a reduction in size of the thorax, eyes, and antennae, and in enlargement of the abdomen which loses distinct segmentation.

The other infraorder, the Termitoxeniomorpha (fig. 42*F*) is distinguished from all other Diptera by a series of peculiar features which reflect its way of life, living in the nests of termites. The head is elongated and bears a piercing proboscis,

Rohdendorf

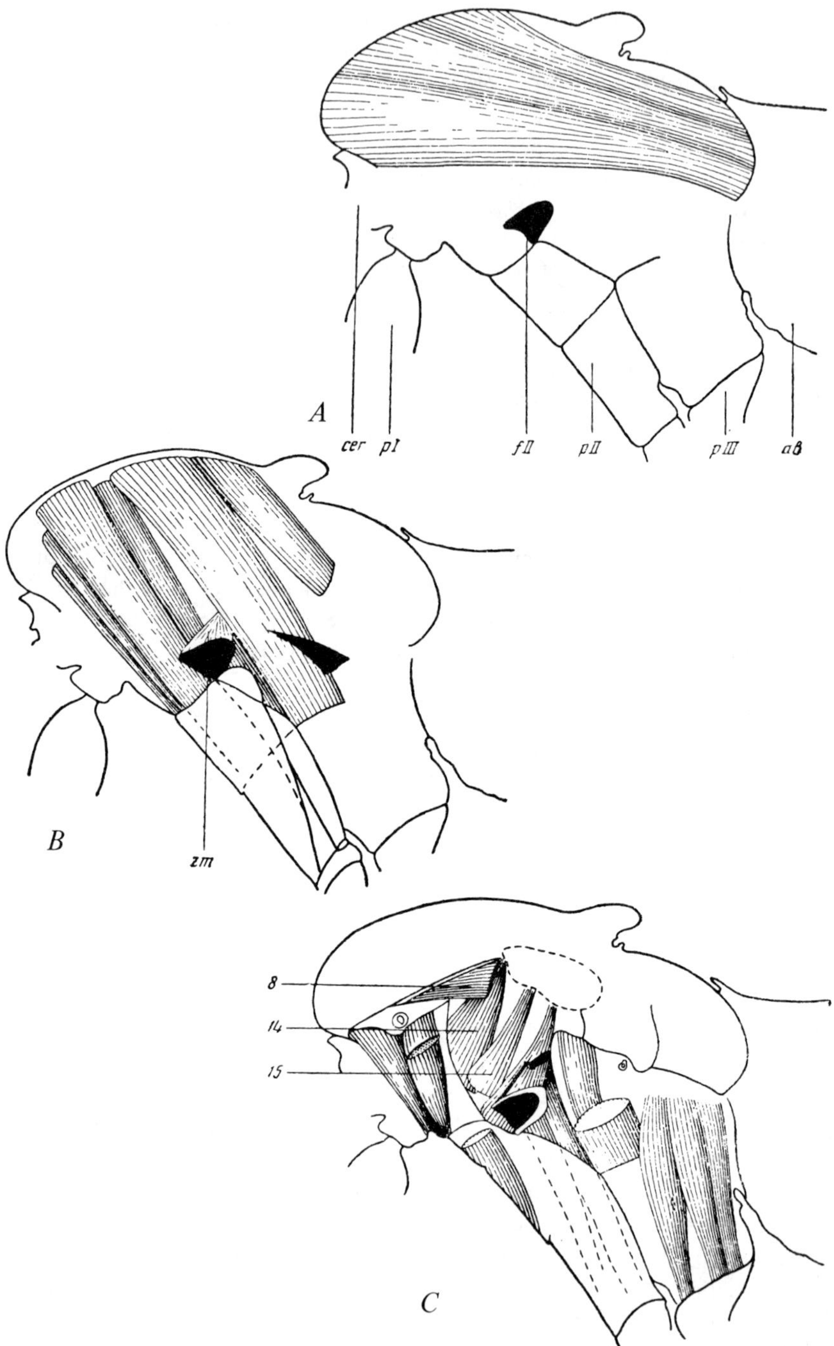

Fig. 41. *Aphiochaeta* sp. (Phoridae). Muscular apparatus of the right thorax. Length of thorax 0.9 mm. *A.* Longitudinal. *B.* Dorsal. *C.* Pleural muscles. (Original.) for abbreviations, see fig. 13, (p. 42).

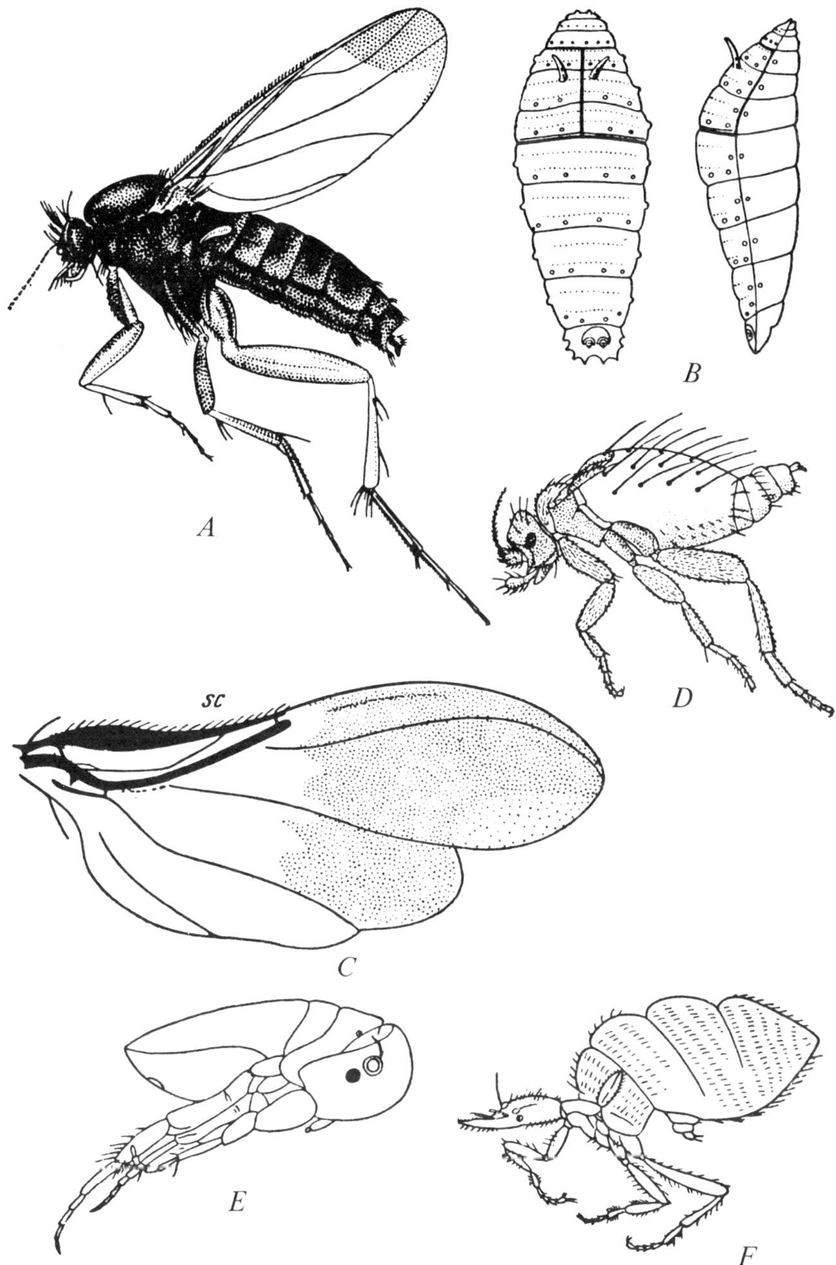

Fig. 42. Phoromorpha and Termitoxeniomorpha. *A. Phora thoracica* Meigen (Phoridae). Male, general view. *B. Megaselia rufipes* Meigen (Phoridae). Puparium from above and from the side. *C. Coniceromyia vespertilia* Borgmeier (Phoridae). Wing. *D. Ecitomyia* sp. (Phoridae). General view. *E. Thaumatoxena wasmanni* Breddin (Thaumatoxenidae). General view. *F. Termitoxenia heimi* Wasmann (Termitoxeniidae). General view (*A, B.* according to Lindner, 1925-28; *C, D.* according to Curran, 1934; *E, F.* according to Hendel, 1936.)

short antennae, and simple eyes and compound eyes which are reduced in
dimensions; the thorax is very small, equal in size to the head, and bears
prehensile legs and insignificant rudiments of wings; the abdomen is membranous,
with isolated tergites and sternites which increase in size during postembryonic
growth. Reproduction and development are very peculiar; apparently all the
representatives of this group are hermaphroditic or parthenogenetic. Males are
unknown up to now; and a free-living larval stage is completely absent. The
females either deposit very large eggs from which emerge not larvae but an
imaginal form, which is always small (the Indian genus *Termitoxenia*), or else they
bring forth adult insects, the development of which has taken place in the body of
the mother (African genus *Termitomyia*).

Conflicts and Determining Tendencies in Phylogeny. — In the Phoromorpha the
development of extra-intestinal digestion, the strong cuticle of the larva, a
puparium, powerful running legs, with bristles on the body and legs, extreme
costalization or loss of wings, small size, a connection with the ground, rotting
substances and nests of social insects — all this suggests the population by larvae
of aggregations of animal remnants, presumably corpses of small vertebrates or
large insects. A not less important, interdependent process was the development
of the highly active imago which acquired very powerful running legs (suggesting
the legs of a muscular subtype of mycetophilid) and a strong lifting flying
apparatus. These features of the winged form developed during a reduction of size
of the insect, that turned out to be one of the important conditions of the
derivation of super costalized wings and resulted from the habitat of the larvae in
a high caloric medium, which existed for a very short time (small, drying corpses).
A consideration of the later history of the phoromorphs is beyond the scope of
this investigation; one can only note that they appear to be derivatives of some
kind of primary myiomorphs, presumably related to the ancestors of the
platypezids, which also appear in the contemporary fauna in restricted habitats.

The features of the peculiar Termitoxeniomorpha point to a quite different
phylogeny, namely to the production of the pupiparous state, i.e. the
development of the larvae in the body of the mother and parthenogenesis on the
one hand, and to the formation of peculiar piercing mouth parts of the winged
insect which dwells in the nests of termites on the other.

Conclusions. — Decisions concerning the specializations of the phoromorphs
and termitoxeniomorphs can only be preliminary. The nature of new developing
features of the phoromorphs and improved locomotion, together with integration
of the body (and a concentration of the nervous system) indicate the primarily
progressive path of this group of Diptera. However, the further phylogenesis of
phoromorphs led them to diverse forms of parasitism and in connection with this
to the regressive development of many groups of families. Within the infraorder
Phoromorpha we observe examples of definitely progressive groups (for example,
the subfamily Phorinae), and some less progressive forms (for instance, the
Aenigmatiidae and Thaumatoxenidae). Appraising the features of the
Termitoxeniomorpha one must conclude that they are regressive; new developing
traits were the reduction of many sense organs and the wings, together with a

narrowing of habitat into the nests of termites. The improvement of reproduction which appeared, probably secondarily, was the sole progressive feature which guaranteed the existence of the insects, albeit as a relict group which has few forms.

Infraorder Myiomorpha
Rohdendorf, 1961b, p. 156

Extent, history, system. — This very large infraorder includes not less than 19 superfamilies: Platypezidea, Syrphidea, Conopidea, Somatiidea, Trypetidea, Psilidea, Heleomyzidea, Sapromyzidea, Borboridea, Drosophilidea, Chloropidea, Gastrophilidea, Anthomyiidea, Muscidea, Sarcophagidea, Oestridea, Tachinidea, Glossinidea and Hippoboscidea, which are made up of 55 families, including over 32,500 species (fig. 43). The components of this infraorder of Diptera entered into Brauer's well-known suborder Cyclorrhapha together with some other parasitic forms, which have been separated by me into special infraorders: Phoromorpha, Termitoxeniomorpha, Streblomorpha, Nycteribiomorpha, Braulomorpha. Thus, I treat the infraorder Myiomorpha as a single realistic whole.

Paleontological records are confined to Tertiary fossils and have been little studied. The most ancient fossil myiomorphs have been found in Paleogene Baltic amber and represent many present-day superfamilies: we have as yet no knowledge of the original, most ancient forms of this largest infraorder of Diptera in the contemporary fauna.

On the whole this infraorder shows a relationship with the Asilomorpha, being descended from an ancestor close to an unknown group of the superfamily Empididea. The nearest to the original forms of the myiomorphs are species of the superfamily Platypezidea. Undoubtedly during the second half of the Cretaceous ancient forms of this superfamily were ancestral to the Syrphidea and Phoromorpha on the one hand and, on the other, the general complex of remaining superfamilies which, very quickly presumably, divided into two large groups of superfamilies. These were the so-called acalyptrates (Acalyptrata or Holometopa or Haplostomata) and the calyptrates (Calyptrata or Schizometopa or Thecostomata). From the original stem of this vast complex the branch Gastrophilidea had already separated out as the Cretaceous passed into the Cenozoic; it is possible that the most peculiar Nycteribiomorpha and Braulomorpha also separated off from this same stem. Presumably still in Cretaceous times another branch of the calyptrates split off to give the peculiar parasitic Hippoboscidea. There remain the parasitic Streblomorpha, very little-known Diptera, presumably formed in the nature of derivatives of the original ancient myiomorphs which were not yet divided into basic superfamilies; this does not exclude the possibility that the streblomorphs will turn out to be descendants of some kind of asilimorphs not possessing connections with the myiomorphs. The origin of the infraorder Musidoromorpha is still not clear: it is possible that these peculiar Diptera are derivatives of still more ancient forms of the original myiomorphs just separated from the empidideans.

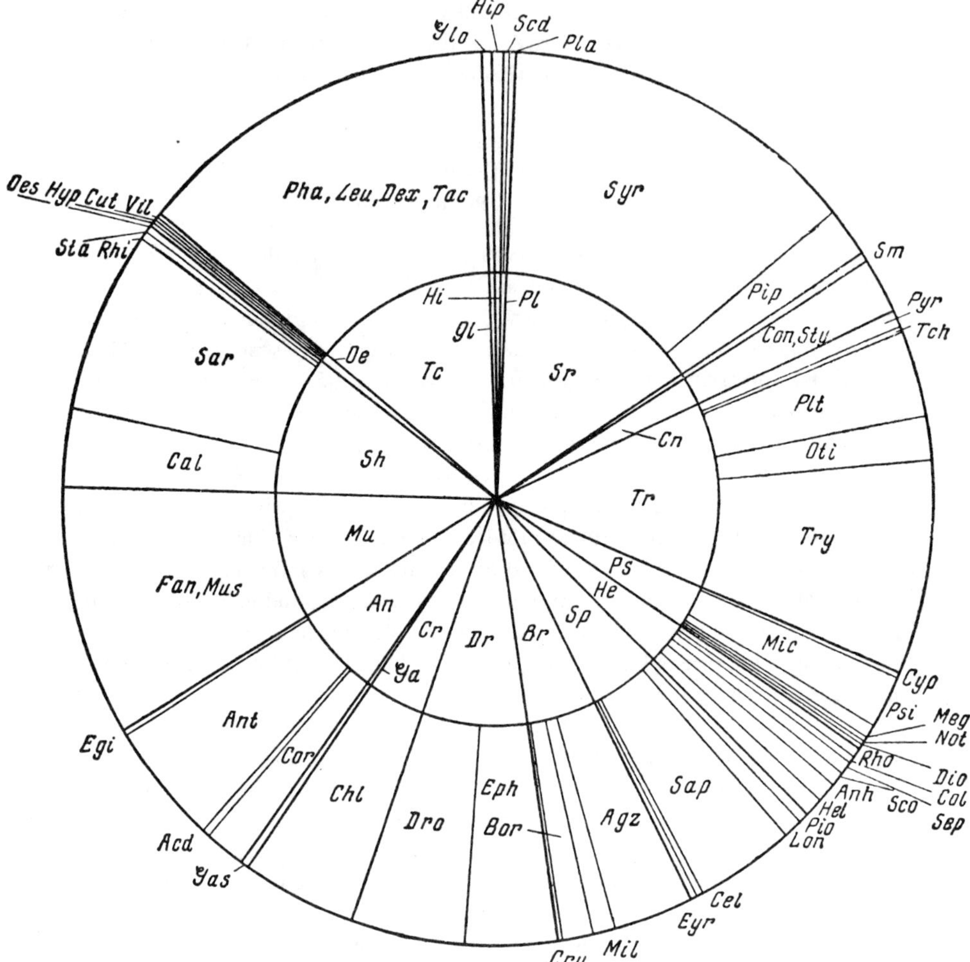

Fig. 43. Diagram of the composition of the infraorder of myiomorphs of the contemporary fauna. Superfamilies are indicated as the sectors of a circle, the sizes of which correspond to the number of species of the given taxon (1° of arc equivalent to 90 species). In the centre of the circle are shown the superfamilies, and on the circumference the families. (Original.) For abbreviations, see p. xiii.

The general systematics of the present-day Myiomorpha are known quite well although there is no general agreement in regard to the appraisal of the ranks of families and subfamilies. Until very recent times this series of Diptera was divided into a number of large categories, the rank of which was quite vague. The series as a whole (i.e. the former 'suborder' Cyclorrhapha) is divided into two groupings — the non-suture-bearing Aschiza (i.e. without a ptilinum), and the suture-bearing Schizophora. To the first grouping belonged representatives of the superfamilies Syrphidea and Platypezidea and from outside the myiomorphs the infraorders

Phoromorpha and Musidoromorpha. The second grouping, the suture-bearing, cyclorrhaphous Diptera in its turn was subdivided into acalyptrates, ones that have a second antennal segment without a split (Holometopa, Haplostomata or Acalyptrata), and the calyptrates, or those that have a split second antennal segment (Schizometopa, Thecostomata or Calyptrata of the aperture bearing Cyclorrhapha). The ones having a second antennal segment without a split include representatives of the superfamilies Conopidea, Gastrophilidea, Trypetidea, Psilidea, Heleomyzidea, Sapromyzidea, Borboridea, Drosophilidea and Chloropidea; the ones with split second antennal segment include the Anthomyiidea, Muscidea, Sarcophagidea, Oestridea and Tachinidea. All these mentioned categories only partially reflect the real interrelations of the different groups of myiomorphs, and in fact only illustrate the comparative anatomical features of the structure of the antenna. Thus, for example, the group Aschiza includes the most heterogeneous families of Diptera while the Schizometopa and Holometopa comprise a whole series of phylogenetically related superfamilies.

Another feature of the systematic arrangement of the myiomorphs is the vagueness of family groupings of some superfamilies of the infraorder. Until recently the whole complex of calyptrates and even more especially the acalyptrates were subdivided into a great number of groupings which nearly always, for no very good reason, were given the rank of family; such families, for example, in the systematic scheme of Hennig (1958) numbered not less than 60. The majority of these groupings are genuine taxa within the systematic grouping of Diptera; all that is wrong is their treatment as families when this is not justified by the scale of difference between them. A final clarification of the acalyptrates cannot yet be made; it is possible to indicate only a preliminary scheme in the present investigation, which includes eight superfamilies and approximately 30 families. This problem is made easier by the quite high level of the systematic study of this group of Diptera. To assemble all this group into one superfamily 'Conopidea' as proposed by me earlier (Rohdendorf, 1961), is incorrect; I became convinced of this after the investigations of many authors and particularly that of Hennig (1958).

Still more complex is the clarification of the real systematics of the calyptrates or 'muscoids', which have been studied less than the acalyptrates. The principal difficulty however only partly arises from the lack of sufficient study of the calyptrates which are chiefly widespread in the tropics. This group of myiomorphs is chiefly distinguished by its youthfulness and the direction in which its evolution has progressed. The calyptrates consist of an abundance of forms, the main differences between which are features that are hard to study: for example, reproduction and development. These features are reflected in the organization of the winged form only in details of structure of the sense organs (eyes and antennae, bristles of the body and legs), of the organs of movement (legs and wings) and of the genital appendages (copulating apparatus and ovipositors). As a result of such features our knowledge of the real system of the calyptrates is very incomplete; it is enough to say that on the one hand the concepts of Girschner (1896) assign all muscoids to only two families (Anthomyiidae and Tachinidae),

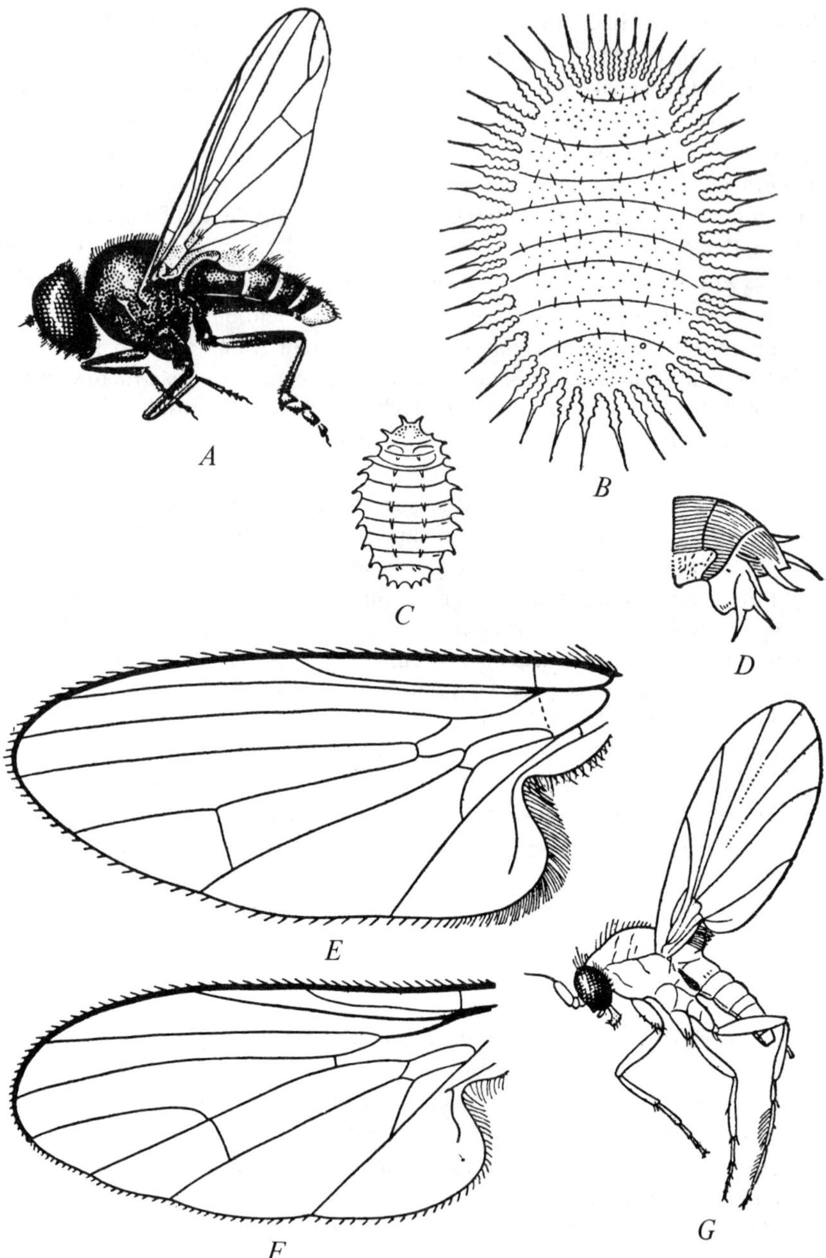

Fig. 44. Platypezidea. *A. Clythia fasciata* Meigen (Platypezidae). Male, general view. Length of body 4 mm. *B. Calomyia amoena* Meigen (Platypezidae). Larva from above. *C. Clythia dorsalis* Meigen (Platypezidae). Larva from above. *D.* The same, anterior end of body of larva from the side. *E. Agathomyia falleni* Zetterstedt (Platypezidae). Wing. *F. Platypezina connexa* Bohemann (Platypezidae). Wing. *G. Sciadocera rufomaculata* White (Sciadoceridae). General view. (*A.* according to Lindner, 1925-28; *B.* according to de Meijere from Czerny, 1934; *C. D.* according to Bergenstamm from Czerny, 1934; *E. F.* according to Czerny, 1934; *G.* according to Tillyard, 1926.)

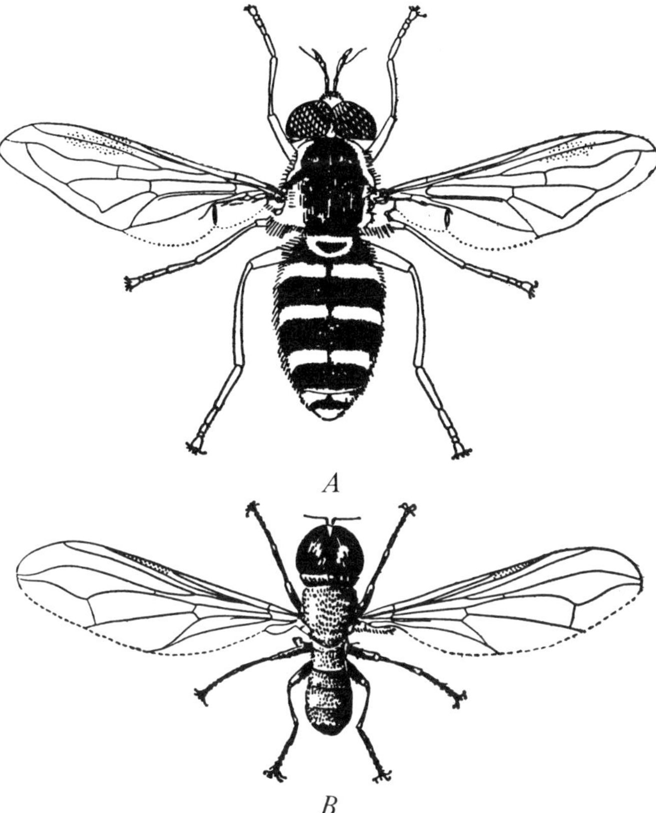

Fig. 45. Syrphidea. *A. Chrysotoxum festivum* (L.) (Syrphidae). General view. Length of body 13 mm. *B. Dorylas campestris* Latreille (Pipunculidae). General view. Length of body 4 mm. (According to Verrall from Hendel, 1928.)

and on the other there are schemes which divide the superfamily into five (Hendel, 1937; Curran, 1934 and others) eight, ten and more families. Sometimes authors deliberately did not attempt to arrange family categories in all this complex of Diptera and recorded only the presence of groups of genera; such is the scheme of Brauer and Bergenstamm (1889-94) which divided the calyptrates into 51 sections principally on the basis of features of the structure of the head. Consequently it is necessary to warn readers that the system of myiomorphs proposed by me is still far from accurate and continues to require substantial correction.

Superfamily Platypezidea

This, the most ancient superfamily, contains two families: the Platypezidae with 140 species which are distributed among approximately 10 genera, prevalent universally; and the monotypic family Sciadoceridae which bears a relict character, as to a certain extent does the superfamily as a whole (fig. 44).

Superfamily Syrphidea

This, the second superfamily, is rich in species and is divided into two sharply different families of very unequal size. The specialized, peculiar, universally distributed family Pipunculidae includes more than 500 species which belong to four genera, of which the vast genus *Dorylas* embraces an overwhelming majority of the species in the family. The remaining three genera, although widespread, yet show distinct characteristics of relict groups. The second family of syrphoids, the bulb flies, Syrphidae, contains, according to recent data, about 4,400 species. This family as a whole has been studied quite well; the family is divided into nine subfamilies of very unequal size. Together with the large groups Syrphinae and Eristalinae, which contain the vast majority of the species of the whole family, are subfamilies poor in species which bear more or less relict characters; such are the Microdontinae and especially the Cerioidinae. The total number of described genera of bulb flies exceeds one hundred; however the generic system of the family has been studied little and most genera which have been established by investigators were based often on various separate morphological features, and in reality the number of genera of Syrphidae probably is considerably less (fig. 45).

Superfamily Conopidea

The next superfamily of myiomorphs contains upwards of 660 species which are distributed between two families of very unequal size: the Stylogastridae with 30 species (fig. 46*A*, 46*B*) and the Conopidae with 630 species. This superfamily is very important for an understanding of the paths of phylogenesis inasmuch as it shows many ancient features of structure. Until recently all representatives of the superfamily Conopidea were considered as belonging to one family; a more attentive examination indicates that the peculiar genus *Stylogaster* Mcq. differs sharply from other conopids and forms a well isolated family.[6] The present Conopidae includes more than 630 species distributed among three subfamilies (Zimina, 1960).

Superfamily Somatiidea

Special interest is offered by the superfamily Somatiidea which includes a single family, the Somatiidae, with one monotypical Neotropical genus *Somatia* Schiner (fig. 46*E* to *I*). This dipterous form was for a long time considered a representative of the family Psilidae although it is sharply distinguished from the latter by many features. First and foremost one must note the large sizes of the cells M and CuP. The first is noticeably bigger than half the length of the discoidal cell (M_{1+2}). This trait sharply distinguishes the Somatiidae not only from the Psilidae but also from the majority of other superfamilies of myiomorphs. It connects with representatives of this infraorder which possess the most archaic venation, namely the Syrphidea. A further characteristic of the Somatiidea is the slight development of bristles on the head and finally the structure of the tergites of the abdomen which do not extend to the ventral side. These Diptera remained little studied until recently. Hennig in his summary of the systematics of Diptera-Schizophora (1958) does note the isolated position of the genus *Somatia* but does not draw the proper conclusion although the structure of many organs of this fly are very remarkable.

Superfamily Trypetidea

This next superfamily is large and includes approximately 4,200 species which

---

6. The Stylogastridae fam. n. differs from the Conopidae by the very short cell Cu; by the presence of bisegmented coxites (surstyli) of the male genitalia; by straight, often very long ovipositor; by the face, deprived of grooves; and the presence of bristles on the frons and ends of the tarsus. The one genus is found in the Neotropical region.

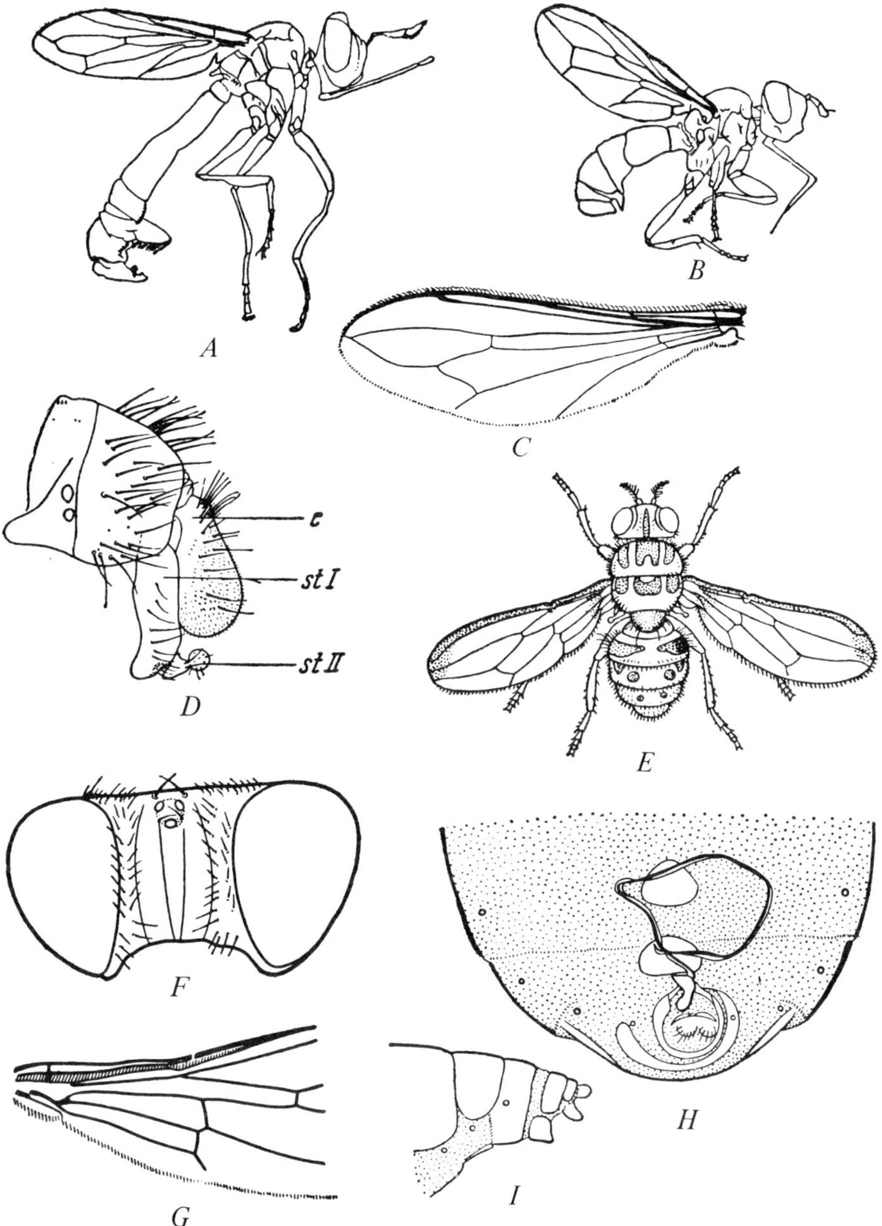

Fig. 46. Conopidea and Somatiidea. *A. Physocephala chrysorrhoea* Meigen (Conopidae). Female, general view. Length of body 10 mm. *B. Occemyia atra* Fabricius (Conopidae). Female, general view. Length of body 5.5 mm. *C. Stylogaster hirtinervis* Lopes and Monteiro (Stylogastridae). Male, wing. Length 6 mm. *D. Stylogaster breviventris* Aldrich (Stylogastridae). Male, end of abdomen. Abbreviations: *c* – cerci; *st I, st II* – bisegmented coxites (surstyli). *E. Somatia xanthomelas* Schiner (Somatiidae). General view. *F.* The same, head from above. *G.* The same, chief half of wing. *H.* The same, end of abdomen of male from below. *I.* The same, end of abdomen of female from the side. (*A. B.* according to Seguy, 1928; *C. D.* according to Lopes and Monteiro, 1959; *E.* according to Curran, 1934; *F. G. H. I.* according to Hennig, 1958.)

are distributed among five families – the Trypetidae (about 2,500 species), Platystomatidae (1,000 species), Otitidae (530 species), Pyrgotidae (210 species), and Tachiniscidae (5 species).

The largest family, the Trypetidae (syn. Trypaneidae), is clearly subdivided into three subfamilies: the widespread Trypetinae is divided into a series of tribes and subtribes, predominantly the paleotropic Dacinae and Phytalmiinae. There are more than 200 genera.

The second family according to size, the Platystomatidae, is abundant, mainly in the tropics, and is divided into two subfamilies – the widespread Platystomatinae (approximately 800 species) and the purely Neotropical Richardiinae (more than 150 species). The generic composition is large but not yet accurately established.

The rather diverse family Otitidae (syn. Ortalidae) is divided into four subfamilies of unequal size – the largest are the Otitinae and Ulidiinae, the smaller and locally widespread Euxestinae, and the chiefly Neotropical Pterocallinae. The generic composition of this family also is insufficiently investigated.

The family Pyrgotidae is almost entirely tropical, has only recently been studied and includes, in spite of the comparatively small number of species, not less than 25 genera.

Finally, the last peculiar family of quite large and bristly flies, the Tachiniscidae, contains not less than four (almost entirely) monotypical genera distributed in South America and Africa.

Superfamily Psilidea

The next superfamily (937 species) is made up of six families of very different sizes, and is of mainly tropical distribution. The largest family, the Micropezidae (syn. Tylidae), contains upwards of 600 species and is divided into four rather isolated subfamilies. The main one is the Micropezinae with 530 species and the other three are the Neriinae, Trepidariinae and Taeniapterinae. The following two families, Psilidae (156 species) and Diopseidae (about 140 species) are of very different systematic composition. The Psilidae includes three widely distributed and sharply isolated subfamilies: the Psilinae with 115 species, the Tanypezinae with 20 species and the Strongylophthalmyiinae with 21 species, while the tropical Diopseidae, widespread almost entirely in southern Asia and in Africa, is a single group not divided into subfamilies.

The last three families are poor in species and are undoubtedly relict groups. These are the tropical Megamerinidae (16 species), Nothybidae (seven species) and finally the Cypselosomatidae with five species of two clearly isolated subfamilies (Cypselosomatinae and Formicosepsidinae).

Superfamily Heleomyzidea

This has approximately the same number of species (999) as the Psilidea and is also subdivided into six families. The largest are the four families Anthomyzidae (306 species), Heleomyzidae (240 species), Sciomyzidae (220 species) and Sepsididae (160 species). The first two families are very diverse and are subdivided into a large number of subfamilies. There are seven subfamilies of the

Anthomyzidae: Clusiinae with 145 species, Asteiinae with 55 species, Opomyzinae with 50 species, Anthomyzinae with 45 species; and the Chyromyiinae, Acartophthalminae, Aulacigastrinae which are all poor in species. There are six subfamilies of the Heleomyzidae: Teleomyzinae with 205 species; and the Trixoscelidinae, Pseudopomyzinae, Choropteromyzinae, Rhinotorinae and Fergusonininae all of which are poor in species. It is necessary to note that these two big families in the main are distributed outside the tropical zone; only isolated relict groups are found in South America (Rhinotorinae) and Australia (Fergusonininae).

The second two large families are much more entire groupings: the Sepsididae which are divided into only two clear subfamilies (Sepsidinae and Orygmatinae) and the Sciomyzidae which are generally not subdivided. These families are not very diverse in the tropics either.

The two final families of the Heleomyzidae, which are clearly relict, are each divided into two clearly isolated subfamilies although they contain a comparatively small number of species. Such are the northern Coelopidae, connected partly with sea coasts. There are 48 species of the subfamilies Coelopinae, and Dryomyzinae. Of the tropical Rhopalomeridae, Rhopalomerinae has 10 species, and Heleomyzinae 15.

Superfamily Sapromyzidea (syn. Lauxanioidea)

More than 1,700 species are included in this superfamily. Of the four families, one sharply predominates; this is the tropical family Sapromyzidae (syn. Lauxaniidae) which includes approximately 1,400 species of four subfamilies: Sapromyzinae (1,260 species) Chamaemyiinae (90 species) Periscelidinae (10 species) frequently kept separate as an independent family; and the Eurychoromyiinae (one species). There are nearly a hundred genera. Second according to size is the widespread family Lonchaeidae which contains about 270 species distributed between two sharply isolated subfamilies: Lonchaeinae (220 species) and the Pallopterinae (50 species). The family Piophilidae, which includes over 80 species, is subdivided into three individual subfamilies, for the most part widespread in the tropics: these are Piophilinae with 75 species, Thyreophorinae with five species and the Neottiophilinae. Finally the tropical family Celyphidae contains upwards of 30 species of seven genera and is distributed in southern Asia and in Africa.

Superfamily Borboridea

The superfamily Borboridea is large and includes about 1,600 species in four families — Agromyzidae, Milichiidae, Borboridae and Cryptochaetidae. The first family is widely distributed and predominates markedly. It includes approximately 1,000 species distributed between two subfamilies which are sharply isolated and of unequal size: the Agromyzinae having about 930 species and the Odiniinae about 20. Second according to size is the family Borboridae which is also widely distributed and comprises over 330 species. The third family, the Milichiidae, includes approximately 260 species in four subfamilies which are very unequal in size: Milichiinae (180 species) Carninae (15 species) Tethininae (50 species) and the Conaceinae (15 species). There is finally the

Cryptochaetidae, a very isolated specialized family containing five species of two genera.

Superfamily Drosophilidea

This very large, widely distributed superfamily is made up of approximately 2,400 species in two big families: the Drosophilidae (1,340 species) and the Ephydridae (1,050 species). The first family is divided into three subfamilies and the chief, the Drosophilinae with 1,240 species, considerably exceeds in number of species the other two which are relict groups — the Diastatinae (30 species) and the Curtonotinae (70 species). The latter is most abundant in the tropics. The characteristic family Ephydridae is divided into two subfamilies. The chief subfamily, Ephydrinae, is made up of many tribes and a great number of genera, and the second, Camillinae, is a relict group and contains in all about 20 species.

Superfamily Chloropidea

This superfamily has a single family, the Chloropidae, which is divided into two subfamilies: the Chloropinae, which includes approximately 60 genera, and the considerably more diverse Oscinellinae with about 130 genera. This family is widely distributed and contains numerous tropical genera.

Superfamily Gastrophilidea

A peculiar superfamily of myiomorphs, this includes the intestinal gadflies which belong to a single, at present widely distributed family, the Gastrophilidae with about 30 species, for the most part of the genus *Gastrophilus*: the other two genera with a few species occur in Africa.

The remaining superfamilies of myiomorphs are united under the name Schizometopa, or Calyptrata, which includes an enormous number of species.

It is possible only to distinguish the most general features of the systematics of these myiomorphs which have been made clear as a result of the many works of Mesnil (1934, 1944-50), Roback (1951), Rubtsov (1951) and Hennig (1958). Obviously the original systematic scheme of Girschner (1896) although basically accurate in that it correctly indicated the real relation of the different groups of calyptrates, possessed an important deficiency — the levels of the taxa indicated in it were not compared with those of related myiomorphs, first of all with the vast complex of acalyptrates. During attempts at a comparison of these two big groups a dilemma at once arose: even though we accept the real scheme of classification of the acalyptrates proposed by Hendel (i.e. about 50 families) then Girschner's system of the clyptrates (two families) clearly included in itself taxa of another level, although in both cases they were designated 'family'. As I have already indicated the extent of investigation of the systematics of acalyptrates is incomparably greater than the investigation of the calyptrates; therefore there is no reason to doubt the reality of the established groupings of the acalyptrates, while there is every reason to affirm that Girschner's families of calyptrates are in reality at least taxa one or two steps higher, namely superfamilies or even groups of the latter. Taking into consideration the results indicated above of investigations in recent years and the rather changing systematic schemes put forward by authors, I accept for the whole complex of calyptrates the presence of five superfamilies (see above, p. 103). The first superfamily, Anthomyiidea, with over 2,000 species, includes four families: the Anthomyiidae, Cordyluridae, Eginiidae and Acridomyiidae. The first is a widely distributed family to which belong approximately 1,500 species. It contains two subfamilies, Anthomyiinae and Fucelliinae, of which the former is large, consisting of approximately 1,350 species of 40 genera, and the latter smaller with the species of ten genera. The family Cordyluridae (about 550 species) includes approximately 45 genera and several close subfamilies. The last two families are clearly relict groups which contain very few species (Rukavishnikov, 1930; Zimin, 1951).

Superfamily Muscidea

The superfamily Muscidea is made up of two families — the Fanniidae and Muscidae; the latter includes the majority of the species and is divided into a series of subfamilies. The total number of species in the superfamily reaches 3,000; the extent of the families and subfamilies is not known to me (Zimin, 1948, 1951; Hennig, 1955-62).

Superfamily Sarcophagidea

A large superfamily, Sarcophagidea has about 3,300 species and is divided into four isolated families. The biggest, a widely distributed family, the Sarcophagidae with about 2,300 species, is divided into three subfamilies: Sarcophaginae (about 1,600 species) Miltogrammatinae (approximately 600 species) and Macronichiinae (about 100 species) which include numerous tribes (Olsufev, 1929; Rohdendorf, 1937). The second, also a widely distributed family, the Calliphoridae, contains about 800 species of principally tropical insects which are distributed among five subfamilies of very unequal extent. The chief one, the Calliphorinae with about 400 species, is mainly tropical (of the eastern hemisphere). The Rhiniinae with about 280 species is also tropical. Then there are the probably relict groups Ormiinae (approximately 35 species) and Mesembrinellinae (about 20 species). Here it is necessary to refer to the relict group Mimodexiinae (about 15 species) which is widespread in the deserts of Asia. The third family, the little studied Rhinophoridae, is distributed chiefly in the Holarctic and contains more than 100 species. Finally there is the highly specialized family Stackelbergomyiidae, known until now only by a single desert species (Rohdendorf, 1948b).

Superfamily Oestridea

This includes four families of parasitic Diptera which are clearly phylogenetic relicts, consisting of a few species of relatively numerous isolated genera. The first, largest family, the Oestridae with less than 10 genera, are divided into two sharply isolated subfamilies: the Oestrinae (24 species of six genera) and the Cephenomyiinae (11 species, two genera). The second family Hypodermatidae (subcutaneous gadflies) is of approximately the same extent. It is subdivided into four subfamilies: the Portschinskiinae (seven species of one genus), Oestrodermatinae (one species of one genus), Oestromyiinae (five species of one genus) and Hypodermatinae (23 species of seven genera) (Breev, 1950; Grunin, 1947-62). The third family, Cuterebridae (American subcutaneous gadflies), contains around 20 species of three genera. Finally, one must refer to the peculiar desert genus *Villeneuviella* Austen (of this same superfamily) which forms the special family Villeneuviellidae (about 10 species).

Superfamily Tachinidea

This superfamily is most widely represented in the tropics and is exceptionally rich in species. It includes not less than four or five families, the interrelations of which have not yet been completely investigated. The largest family is the Tachinidae (syn. Larvaevoridae) with close to 4,000 species. The other families, Dexiidae, Phasiidae and others, are much smaller in size and have been insufficiently studied.

Superfamily Glossinidea

Formed by a single family of tsetse flies, the Glossinidae with about 30 species of one genus, this superfamily is at present distributed in tropical Africa.

Superfamily Hippoboscidea

Although it contains only one family with about 120 species, the Hippoboscidae, this superfamily is yet markedly diversified. It is possible that later on the component genera of this family (about 15) will be divided among separate subfamilies.

Chief features. — An unusually extensive eurybiont nature is characteristic for the whole infraorder of myiomorphs. These insects populate all habitats and geographic terrains where the life of insects is at all possible (Shtakelberg, 1937, 1948, 1950, 1953). They tolerate wide ranges of temperature and humidity. As well as the inhabitants of open aquatic media (some Syrphidea and Drosophilidea) there are forms living in the extreme dryness of the desert. The high polar latitudes are populated by many Anthomyiidea, Muscidea, some Heleomyzidea, Drosophilidea and a few Syrphidea and Tachinidea, but the variety of myiomorphs is largest in tropical conditions.

The features of feeding are very characteristic; together with the development of complete extra-intestinal digestion by the larvae, which possess large salivary glands producing a secretion with the most diverse collections of enzymes, feeding of the winged insects became almost entirely vegetarian. Only in some groups did predation (Cordyluridae, some Sapromyzidea and others) or blood-sucking (Muscidea) have a place. The extra-intestinal digestion of the larvae influenced a great variety of feeding. Together with the phenomena of detritophagia (apparently the original feeding of the first forms), phytophagia reached a wide development in the form of feeding on the tissues of living fungi as with Platypezidea, some Borboridea, Anthomyiidea, Drosophilidea, Heleomyzidea and even Syrphidea, and on the tissues of vascular plants (roots, stems, leaves and fruit) as with numerous acalyptrates, Trypetidea, Heleomyzidea, Borboridea, Chloropidea, Sapromyzidea some syrphideans and anthomyiideans. Animal feeding by the larvae became even more diverse. Many muscoids, some sarcophagideans, muscideans, heleomyzideans and sapromyzideans feed on microorganisms; coprophagy is the habit of some syrphideans, *Rhingia*, many muscideans and sarcophagideans, some borborideans, and trypetideans; predation at the expense of other dipterous larvae is practised by some muscideans, of aphids by many Syrphidae, and some Sapromyzidea, of caterpillars and pupae of lepidopterans, and egg cocoons of grasshoppers by some Sarcophagidae, and of egg cocoons of spiders by the genus *Trimerina* of the Ephydrinae. Furthermore animal feeding is accomplished by blood-sucking on birds and mammals by some sapromyzideans (Neottiophilinae), which are blood-suckers of young birds, and by sarcophagideans (tropical calliphorids, namely the genus *Auchmeromyia*, which attacks man). Finally there are those who have developed the most diverse forms of internal parasitism: Pipunculidae of the syrphideans, Conopidae, Pyrgotidae, some Sapromyzidae, Drosophilidae and numerous tachinideans and some anthomyiideans, of which a series of families contains forms with larvae which lead an entirely parasitic way of life in insects (Acridomyiidae, Phasiidae, Tachinidae), and mammals (gadflies — Oestridae, Hypodermatidae and Gastrophilidae and some Sarcophagidea — genus *Booponus*).

Concluding this condensed description of the features of feeding in the myiomorphs it is necessary to note on the one hand the various examples of co-dwelling and of predation of the larvae in the nests of solitary Hymenoptera (Miltogrammatinae of the Sarcophagidae) and social Hymenoptera (some Milichiidae of the borborideans; the genus *Brachycoma* of the Sarcophagidae; the genera *Microdon, Volucella* of the Syrphidae) and on the other the living of winged insects as external parasites on the body of birds (some Milichiidae, genus *Carnus*; and some bood-suckers, Hippoboscidae; and on mammals, Hippoboscidae). The extraordinary variety of nutritive features of myiomorph larvae have no analogies among other groups of insects. This variety of the nutritive relations of myiomorphs is true, in fact, only of the larval phase; the winged insects are considerably less varied and exhibit some examples of reduction of the function of feeding down to even complete reduction of the mouth parts. Such, first of all, are the gadflies of the family Oestridae, Hypodermatidae, some Sarcophagidae (for example the genus *Africasia* Rohd.), some Tachinidae (for instance the genus *Trixa* Meig.), the superfamily Gastrophilidea, some Trypetidea, for example the genus *Tauroscypson* Curr. of the Pyrgotidae, and the genus *Neoestromyia* Ôuchi of the Platystomatinae.

The structure of the larvae of myiomorphs forms a very important part of the characteristics of this series of Diptera. The chief trait is the extreme modification of the cephalic end of the body which results in a pulling in of the head segment into the anterior thorax; in the complete disappearance of free, protruding labrum and labium and of thickened sclerites of the head capsule; in the formation of a characteristic cephalopharyngeal sclerite and of a free movable oro-pharyngeal apparatus not connected with the wall of the segment. This process reflects the extreme concentration of the nervous system, the strong displacement of the anterior pharyngeal ganglion backwards and the fusion of it into one common, solid complex of the central nervous system which is made up of adjacent and fused body ganglia. The respiratory system is metapneustic in young larvae and amphipneustic in adults and the posterior spiracles are definitely predominant complex organs. The structure of the larvae of myiomorphs is very diverse but the general plan is a common one; the larvae of the superfamilies Platypezidea and

Syrphidea are especially different (Hennig, 1952).

The other most important traits of myiomorphs are features of transformation, which differ from the asilomorphs first of all by the constant development of a puparium, i.e. by the absence of one of the embryonic stages in the mature larva. As we already know, the development of a puparium had a place in the history of dipterous insects repeatedly: in the Scatopsidae and Cecidomyiidae of the bibionomorphs, in the Stratiomyiidae of the asilomorphs, and in the musidoromorphs and the phoromorphs. In all these cases the last larval skin is preserved during the transformation but this is very different in all of these groups and the production of a puparium was an improvement which arose 'independently'. Only in the phoromorphs and myiomorphs did the formation of a puparium and the processes of acephalization of the larva coincide. This led to the very great similarity in the development of these Diptera, which in fact are remote from one another.

The larvae of myiomorphs possess great mobility in the most diverse media; they are, moreover, protected by tough cuticle with very little permeability, able to withstand reduced moisture and also to free themselves of liquids and any kind of chemical substances (acid and alkaline media, products of the decomposition of organic substances and so forth) which are harmful to most living tissues.

The mature form of the myiomorphs is characterized by high integration of the body; abrupt enlargement of the head by virtue of the increase in the size of the compound eyes with decrease in the length of the antennae (always three-segmented, in the vast majority of cases with a dorsal arista); large broad thorax with prehensile legs, moderately costalized wings and powerful muscular apparatus; the prothorax develops slightly and the prothoracic coxae are large, considerably larger than those of the mesothorax; abdomen as a rule shortened, with four to six, rarely seven, and sometimes three developed segments. The structure of the head, furthermore, is characterized by the high position of the antennae, greatly removed from the anterior lower border of the head, leaving a highly isolated 'face' (morphologically corresponding to the clypeus and frons); the base of the proboscis forms a highly developed labial section (haustellum) and the remaining mouth parts are almost completely reduced, among them the mandibles and maxillae (sometimes in the syrphideans there are well preserved thin lower maxillae).

The thorax is formed of a very highly developed mesothorax whereas the sclerites of the pro- and metathorax are noticeable only on looking at the insect from the side (only in some psilideans is the prothoracic dorsum more highly developed). The wings have reduced venation and belong to the traction-lifting (syrphoid) and lifting (muscoid) types (Rohdendorf, 1951).

Conflicts and determining tendencies in phylogeny. — Under which conditions could the original forms, the ancient empidideans, produce puparia? Apparently under conditions requiring pupation in a drier medium, in an insufficiently humid atmosphere and at high temperatures. These are only abstract assumptions. The reasons which influenced the preservation and consolidation of the cuticle of the final larvae, in fact, are still very little known; presumably they are linked with the processes of total strengthening and thickening of larval cuticles (for example in the stratiomyiids, and partly in the scatopsids).

The derivation of the original myiomorph wing apparently came about in the following manner. The closest to the first forms of wings of some platypezideans (for example the genus *Clythia*) are the syrphideans and somatiideans; the first differs little from the empidoid type and is characterized by the elongated cubital cell and by the presence of the upper anterior branch of the medial vein, particularly in its expanded form. The features of the wing of such a genus allow us to assume that the myiomorphs originated during the absence of any tendency towards a reduction of body size. These insects probably grew in size during their phylogeny in sharp distinction to all the remaining empidideans, most of which diminished in size. The great importance of flight activity in the first myiomorphs, besides their increase in size, delayed the reduction of venation and permitted the preservation of the ancient features (large cubital cell, anterior branch of M) while the rest of the empidideans improved their wings under other conditions (decrease of size, often small importance of flight, powerful legs).

The described processes of improvement of development (production of a puparium) and of flight (large wings) suggest the habitat of the first myiomorphs in relatively warm regions with development of their larvae in conditions of abundant but local high caloric food suitable for use only for a limited time. Such conditions might have been found in dry subtropical terrains where the medium for larvae might have been the fruits of specific plants, (perhaps ephemeral?) corpses of animals, or the excrement of vertebrates.

It is possible to analyse the phylogeny only of large groups, not touching on the history of the separate families, owing to the vastness of the whole infraorder Myiomorpha which has over 32,000 species in more than 50 families.

The first superfamily Platypezidea (fig. 44) is closest to the original forms of myiomorphs and is a relict group, as was evident from the statistical data of its systematics (see above). Ecologically these Diptera are linked with forest terrain — the larvae live in the fruiting bodies of fungi and are characterized by an expanded and depressed body, provided with characteristic protuberances; the puparium differs little in form from the mature larva. The winged insects are dark colored, often with a contrasting pattern, with large eyes especially in males; antennae with a terminal arista; weak, soft proboscis, large, wide wings, and as a rule with enlarged posterior tarsi.

The next superfamily, the Syrphidea (fig. 45), is not clearly relict although it contains only two families. The widely distributed syrphids live in very diverse habitats and terrains, their larvae are also quite diversified in their feeding: there are known detritophages, consumers of rotting plant residues, true coprophages, predators and parasites and also inhabitants and consumers of living plant tissues. The features of larval organization are little known but they are peculiar and quite variable. A general trait of all syrphideans is the union of the posterior spiracles into one, on an unpaired protuberance which sometimes reaches large dimensions. The close union of the posterior spiracles, which is observed in the most diverse forms of syrphideans, whether they live in a liquid medium, in decomposing plant residues, freely in plants or as parasites, compels us to assume this to be an ancient feature indicating the manner of larval life of the original forms of this superfamily. A much more peculiar and more definite trait of the winged form of syrphideans is the development of large, comparatively long wings with rich venation and large body size that are interdependent phenomena (Rohdendorf, 1951, p. 41). Another trait of the syrphideans is the development of plant feeding by the winged insects which feed principally on the pollen of angiosperms (Shtakelberg, 1950). Everything points to the transition of ancestral forms of this superfamily into an aquatic, or at least a liquid medium with a strong growth of body size and increase of the speed of flight. This by no means reveals the chief causes which govern the phylogeny, which are hidden in the features of ontogenesis and probably in the little known features of larval feeding.

The acalyptrates are characterized by a further improvement of development, namely the production of a frontal bladder, the ptilinum, a peculiar adaptation for opening the puparium and for movement of the young fly in the mass of substrate. In addition to the bladder, the improvement of development in the acalyptrates depends still on the production of a characteristic oro-pharyngeal apparatus which acquires powerful mouth hooks (mandibular sclerites), an important organ for crushing (tearing, puncturing) the nutritive substrate and also as a supporting organ for the movement of the larva. Up to now we do not have knowledge of these first forms related to the platypezideans and which are distinguished from them by the presence of the frontal bladder and the development of "muscoid" mouth apparatus of the larvae. All known acalyptrates (among them both Conopidea and Somatiidea), moreover, are characterized by the well-known reduction of the wing venation, lesser concentration of the central nervous system, by the depressed structure of the proboscis of the winged insect and usually by little integration of the abdomen. This distinguishes them from the calyptrates, the superfamilies Anthomyiidea, Muscidea, Sarcophagidea, Oestridea, Tachinidea which definitely also are descendants of the first, original forms common for these two complexes of superfamilies. There is no doubt that the closest to the original forms of the acalyptrates in the contemporary fauna are species of the superfamilies Conopidea and Somatiidea (fig. 46), already highly specialized groups. This probable process of divergence of two groups of superfamilies of ptilinum-bearing myiomorphs was governed by the following causes. Some forms intensively improved and accelerated individual development by means of a decrease in body size — such were the Acalyptrata in which small size led to many characteristic features; simplification of the venation, little mechanical improvement of the head and less integration of the body. Such a path of improvement of the organization proved to be for these Diptera highly progressive. Although by itself dwarfness can be treated as regressive, this is obviously incorrect here; a reduction of size for insects is to a great extent a beneficial acquisition which opens up a wide possibility of colonization of very diverse media.

In the further history of the Acalyptrata we detect very different trends. It is curious that a reduction of size, in a series of cases were secondarily 'exchanged' and led to the appearance of the diametrically opposite trait – increasing size. Such are the different parasitic forms: Conopidae (fig. 46*A*, 46*B*), and Pyrgotidae, the larvae of which live in the body of large insects (stinging Hymenoptera and beetles) which stimulated increase of size by the abundance of caloric food. Other, comparatively uncommon large trypetideans, for instance Tachiniscidae, some heleomyzideans (Sciomyzidae, Dryomyzinae and Rhopalomeridae – Berg, 1960) are known insufficiently to allow us to indicate the causes of their gigantism. We see the clearest example of such 'abolition' of principal new determining features in the history of the intestinal gadflies, the Gastrophilidea, the different traits of organization of which indicate the secondary nature of their large size. Quite another thing took place in the history of the Calyptrata, derivatives of the ancient original forms of the Heleomyzidea, which had not yet begun to decrease intensively, to reduce the venation of the wings, or to simplify the proboscis. The original forms of the sarcophagideans arose presumably as a result of improvement of development in the direction of production of another form of monophagia, that is, as a result of a habitat of larval phases in any definitely high caloric but locally disposed nutritive media. Different parts of plants (fruit, tuber, root), excrement, corpses of animals, and finally, living animals, might be such media. The chief feature of all these substrates of larval feeding was its local and highly nutritional nature. This influenced the special importance of the winged form which was required to guarantee the finding of the given substrate. Therefore it had to have perfect long range flight, sense organs and nervous system. Consequently a reduction in size could have no place in the first calyptrates; improvement of the sense organs (first of all the eyes, then the intricate system of bristles on the body or dorsal shield); concentration of the nervous system (and directly combined with it, integration of the body); and, finally, the development of a complete flying apparatus (among them thoracic scales, a complete protective shield for the buzzing organs, of nervous regulators and stimulators of flight) – all this appeared in the phylogeny of the original Anthomyiidea and Sarcophagidea.

At present the system and phylogenesis of these Diptera have still not been finally explained. Even though the vast complex of Acalyptrata was intensively studied by Hennig (1958), the other group of myiomorphs, the Calyptrata, have been less studied even in regard to the faunistic composition and relationships of the taxa of the family groups. All this forces us to set aside a consideration of the paths of historical development and features of the governing traits in the evolution of these myiomorphs, although to a great extent they are peculiar and significant in a practical way.

It is possible now to discuss the features of only two superfamilies of myiomorphs, which are poor in species and which contain parasitic Diptera, the intestinal gadflies, the Gastrophilidea, and the blood-sucking flies, the Hippoboscidea. The gastrophilid larvae live in the tissues and intestinal tract of mammals, the winged forms are aphagic; their mouth parts are always reduced. The organization of these Diptera points to no direct phylogenetic connections with other groups of gadflies of the superfamily Oestridea. On the other hand a series of features (structure of veining, scales, head) clearly indicates a relationship of the gastrophiloids with the Acalyptrata which apparently are their ancestral group. The large size of intestinal gadflies as I indicated earlier (Rohdendorf, 1951, p. 135-136) is a recent acquisition in their history. The features of reproduction have been thoroughly revealed lately by Soviet investigators (Chereshnev, 1951, 1953) and denote connections of these parasitic insects with vegetation, with grassy covering that is quite unusual for gadflies, and this simultaneously strengthens the point of view concerning connections of these flies with acalyptrates. The derivation of the gastrophilideans is presented in the following. The first small forms were free-living on the surface of plants, or terrestrial larvae which fed primitively on protein substances such as the remnants of corpses or the discharges of vertebrates which contaminated grassy vegetation or soil; they frequently got into the mouths of ungulates with food; the absence of profound biochemical distinctions between the protein media of these conditions (mucus, blood or corpse liquids on the one hand and of excretions of the mucous membrane of the mouth of the animal on the other) permitted the development of a stable biochemical connection between the larva of Diptera and the ungulate mammal. Clearly that is why the gastrophilids established a connection with the odd-toed ungulates and not with the ruminants. Intake of food by the ruminants is characterized by the very rapid swallowing of torn off plant material without previous intensive crushing, and once the food has reached the stomach it is immediately subjected to the action of digestive juices, definitely very strange and harmful for free-living dipterous larvae or the larvae connected with protein media. Something altogether different is seen in the relationship with the perissodactyls in which the phenomenon of chewing the cud is absent; the torn off plant material is subjected to intensive grinding and remains in the mouth cavity for a longer time. The smallest larvae coming into the mouth cavity of the perissodactyl would be washed off by saliva into the folds of mucous membrane in the mouth cavity and would have more time to become familiar with the new medium, which was also lacking the destructive properties found in the stomachs of ruminants.

Summarizing the paths of historical development of the gastrophiloids it is necessary to note the evidently insignificant progressiveness of these processes, which have the character of narrow specialization. The improvement of feeding allowed the larvae to develop the capacity to live in tissues and in the mouth cavity of mammals. The further development of the gastrophilideans was caused by clearly progressive processes — the improvement of development and reproduction. But the limitations of the chief path of development, parasitism, could not be removed by further progressive acquisitions which had a place in the history of the gastrophilideans, and these Diptera turned out to be in the position of a dying-out relict group, the history of which was completely determined by the fortune of their hosts, perissodactyls, which became extinct (Rubstov, 1939, 1940).

The superfamily Hippoboscidea (louse flies) was until recently united with the streblids and nycteribiids in the artificial group Pupipara (or "Epoboscidea" of

Hendel), which was characterized by advanced viviparity. (Inadequate attention to the exotic Glossinidae did not permit us to put them into this group in spite of corresponding traits of their development.) The improvement of development which results in a reduction of the insect's life stages outside the body of the female is quite a widespread progressive process in the history of the Diptera. Such are some Myiomorpha (for example, Sarcophagidea, many Oestridea and Tachinidea, the representatives of which bring forth living larvae), the peculiar Termitoxeniomorpha, the parasitic Nycteribiomorpha and Streblomorpha. Among the myiomorphs, in representatives of the Glossinidea and Hippoboscidea, this improvement of development reached the extreme. As the ancient name of this group of insects indicates, the hippoboscideans bear a puparium, i.e. a pupa enclosed in the cuticle of the mature larva; the whole development of these Diptera takes place in the body of the female. This process allows a full independence of the larval stage from any kind of nourishing substrate. It is this dependence which limits the distribution of the vast majority of other winged insects in their active mature stage.

Larval development in the body of the female might have arisen because the laying of eggs and development of the larva in a nourishing substrate could not be achieved in the prevailing conditions. A simultaneous necessary condition must have been the provision of high caloric food for the winged form. Such a condition was created during the development of parasitism in the ancestral forms of these flies in mobile vertebrates; the caloric food, blood, which was available in abundance on the one hand, and the mobility of hosts (ungulates or birds) dwelling in dry tropical or subtropical terrains on the other. Such apparently were the conditions of derivation of glossinideans and hippoboscideans. The development of these two superfamilies apparently proceeded in two different directions. There developed the phenomena of pupiparity and blood-sucking whereby the winged insect did not obtain particularly close connections with its 'host', but remained feeding on the 'substrate' which furnishes nutriment, blood. The Glossinidea are related in behaviour to horseflies (Tabanidae) and the Stomoxydini of the Muscidae, but are distinct from these Diptera in their highly progressive method of development.

It is necessary to recall the extraordinary difficulty of the struggle against tsetse flies, namely because of the absence in this case of anything in the way of a nourishing substrate for the larvae, the elimination or detoxification of which is so much a regular method of combatting insect blood-suckers. In fact the fight against glossinids can only be waged by modifying the medium of habitat of the winged form, which is very difficult. The question concerning the extinction of Glossinidae in North America, where they lived in the Oligocene, is complicated and presumably must be examined simultaneously with the problem of the dying off in that same continent of a whole series of groups of ungulate mammals (for example, perissodactyls). The derivation and distinctions of glossinids[7] from other blood-suckers may be connected with the tropical distribution of the Glossinidea which live now chiefly in the highly insolated savannahs of Africa and which suck the blood of various ungulates (antelopes and others); the great

---

7. The establishment of connections of the glossinids with louse flies and the abrupt distinctions of them from the Stomoxydini was made with complete authenticity by L.C. Zimin (1951), to whom we are obliged for the solution to this important question.

mobility of these mammals, and the absence of long wool to inhibit the development of closer connections with them by the winged insect.

Another considerably more varied and widely prevalent group (found all over the world) of hippoboscideans, the louse flies, Hippoboscidae, differ sharply from the Glossinidea by the development of the winged insects' connections with 'the hosts', vertebrates, on the blood of which these flies feed. Among louse flies we may see the most diverse examples of the development of parasitic bonds. Together with the quite well and easily flying groups of the genera *Hippobosca, Ornithomyia*, are observed forms with shortened wings incapable of flight (genera *Stenopteryx, Crataerhina*), or even those altogether devoid of wings (ovine wool insects, genus *Melophagus*). Also the grasping capability of the legs increases, prehensile, serrated claws develop, the body is depressed. All these features of louse flies point to a different solution of the conflict which influenced the derivation of louse flies in comparison with the glossinids. The main thing was a strengthening and increase of the connections with the food animals, birds and mammals. Without special investigation it is difficult to tell which vertebrates were the first hosts of louse flies; as a general consideration it is possible that the character of the coverings of vertebrates was chiefly significant during the formation of this group of hippoboscoids. The long wool of mammals, the feathery covering of birds — here were the requirements which possibly determined the derivation of louse flies. Birds on the one hand and moderate climatic zones on the other, may be considered the most probable original conditions during the derivation of these Diptera.

Infraorders Braulomorpha, Nycteribiomorpha and Streblomorpha
Rohdendorf. 1961b, p. 158.

These three groups of the order Diptera which are remote from one another are brought together by their common parasitic way of life, which deeply and closely modifies their organization. So little is known of these scarce peculiar Diptera that we can examine all three at the same time.

Extent, history and system. — The Nycteribiomorpha contains in the present-day fauna one family, the Nycteribiidae, which is made up of approximately seven genera and about one hundred species. The Streblomorpha also includes one family, the Streblidae, which is considerably more diverse; it is subdivided (according to Hendel, 1937) into three clear subfamilies (maybe even families?), which contain not less than 18 genera and approximately 75 species. The infraorder Braulomorpha comprises only two species of the single genus *Braula*.

Paleontological records for these peculiar Diptera are lacking and any conclusions concerning their history and connections are obtained only on the basis of a study of present-day forms. The abrupt differences from all known groups of Diptera do not allow us to link them with any other families. Some features provide the possibility that these three infraorders are connected with the infraorder Myiomorpha, which I reflected upon when putting forward the phylogenetic schemes of the named infraorder and of the whole order (see p. 305, fig. 81). These features result in a series of criteria of a reducing character:

abbreviation of the antennal segments and abdominal segmentation, general reduction and integration of the body and simplification of the proboscis. Some of the parasitic Diptera, for instance, the Borboridea (Borboridae), Anthomyiidea (Mormotomyiidae with the Nycteribiomorpha) or Platypezidea (with the Streblomorpha) would seem to be connected in this way. Examining the relationships of these parasitic Diptera, however, it is necessary in the first place to note the absence of related connections in all these three infraorders among themselves or with the Hippoboscidea. All these Diptera appeared independently from different original groups; the traits of similarity in organization which served for a while as the basis for the establishment of the artificial grouping of the Pupipara, are definitely an excellent example of convergent development dependent on the derivation of very similar conditions of existence — living on the surface of the body of warm-blooded vertebrates and feeding on their blood. Attempting to explain the traits of organization of all these Diptera which reflect their parasitic way of life and in that way to clear up the connections with the original, other groups of the order, one can draw the following conclusions. The streblomorphs show a similarity on the one hand with the most ancient myiomorphs, the Platypezidea, and on the other with isolated Musidoromorpha. This is expressed by the peculiar veining of the wings, which does not show the phenomena of costalization; quite an unusual trait for the majority of minute Diptera. The nycteribiomorphs are so sharply transformed that they allow us to note the traits of similarity with other Diptera only with great difficulty; only the elongated metatarsi of all legs, and the multisegmented abdomen possibly indicate connections with some Heleomyzidea (p. 110). It is still more difficult to indicate the relative connections of the braulomorphs: the egg-laying condition, the absence of strong bristles on the body, the absence of a piercing proboscis, the unusualness of the head — all these features compel us to bring these co-dwellers of bees closer to the most diverse Diptera, with the phoromorphs or some Borboridea (Milichiidae or Borboridae).

It is necessary only to add that the Streblomorpha is the most diverse (approximately 65 species of 20 genera), while the Nycteribiomorpha is considerably less diversified (150 species of eight genera, Shtakelberg, 1928).

Chief features. — The Streblomorpha and Nycteribiomorpha are external parasites of bats (Chiroptera), being closely connected only with these sharply isolated flying mammals; it is necessary to keep in mind that the chiropterans are a direct derivative of the ancient order of insectivores and that the time of their derivation may not have been later than the middle Cretaceous (already in the paleocene deposits we know of remnants of bats of contemporary families). Both these groups moreover are characterized by pupiparity — their development takes place in the body of the female and the larva feeds from special glandular structures of the uterus. These flies bring forth mature larvae within the cuticle of which are formed pupae, i.e. already prepared puparia. However, here the similarity of streblomorphs and nycteribiomorphs ends. Examining the organization of these Diptera, one discovers deep-seated differences with ease.

The streblids usually possess wings which are often very abbreviated and sometimes completely reduced; the head is directed forward, wide and large and is provided with bristles or spines and a peculiar proboscis; the thoracic section is moderately depressed with well-developed scutellar sclerites; the abdomen has a clearly separated basal segment and is devoid of individual segmentation in its greater part; legs are of a prehensile type, lacking any sharply outstanding features; only the fifth segment of the extremities and posterior coxae are enlarged. The body is small in size (Jobling, 1939-45).

The nycteribiids are always devoid of wings, with a highly depressed thoracic section and elongated legs; the first segments of extremities (metatarsi) are long and thin; there is a particularly unusual structure, the little, depressed and movable head, in rest thrown back to the dorsal surface of the thorax and placed there in a special furrow or wrinkle; mouth parts flattened but peculiar; the top of the thorax is formed by the highly enlarged dorsal margins of the pleura; scutellar sclerites are very reduced and in fact form the groove, described above, in which is placed the head which is thrown back; the abdomen, at least in males, is well segmented. A general view of the insect suggests a spider; the long prehensile legs with serrated claws and very elongated first segments which differ little in length from the tarsi, make the appearance of these Diptera very peculiar.

The Braulomorpha differ greatly from the 'pupipara' parasites of warm-blooded mammals described above. A representative of this peculiar infraorder of Diptera, the bee louse, *Braula coeca* Nitz. lives on the body of the domestic bee and feeds on the food of the host, a drop of which the bee regurgitates during irritation of it by the *Braula*. The mouth parts are soft and incapable of pricking or breaking hard bodies in any way; the head is very broad with widely placed bi-segmented antennae with dorsal arista and a pair of very small compound eyes near the antennae. The whole body is shortened, expanded and arched; the general outline of a bee louse from above is a short oval. The thoracic section is small and considerably narrower than the wide abdomen, but the latter consists of five segments of which the first unites with the middle and posterior thoracic portions. The legs are very substantial and large; the five-segmented tarsi are shortened with expanded fifth segment, devoid of individual claws and bearing special teeth and a pair of hairy pulvilli (Hennig, 1938, 1958). The sharpest difference of the braulomorphs from the first two parasitic infraorders is in the peculiarity of its development. These Diptera deposit eggs in the hives of bees and the larvae feed on the forage of the latter, possessing the capacity to perforate the wax walls of the cells of the hive.

Conflicts and determining tendencies in historical development. — This analysis of the historical development throws no light upon the true features of each of the named infraorders; it is greatly hampered by the absence of exact data on phylogenesis, by the absence of data on the organization of the original groups and especially by the paucity of our knowledge about the life conditions of streblids and nycteribiids which, in the vast majority, are exotic insects and difficult to observe. The data which is known to me permit us to assume a dissimilar way of life for these Diptera.

The Streblidae apparently developed the ability to 'hold on' to the skin of the host strongly and for a long time for blood-sucking, i.e. it was able to bring the cephalic end of its body close to the skin of the bat — in this presumably the armament on the head assisted its form, and possibly the powerful posterior coxae guaranteed support. The Nycteribiidae on the contrary developed great mobility, the ability to move quickly over the surface of the body of the host. The act of receiving food did not involve close contact with the skin of the host and presumably could be accomplished quickly while 'on the go'; the insect tilted the head forward, and plunged the proboscis rapidly into the skin. Summarizing the traits of both groups one can affirm a great improvement in protective features and running in the Nycteribiidae and a great improvement in the means of food reception in the Streblidae; it is obvious that just such a solution of conflict had a place in these groups.

Something absolutely different took place in the history of the braulomorphs for which the chief determining tendency in historical development was the production of the ability to live on the body of bees, feeding on their regurgitated nutriment. The derivation of such a way of life is known in different Diptera; there are the various blood-sucking Heleidae which live on the body and wings of some insects (caterpillars, winged Lepidoptera, gold-eyes and dragon flies) and which suck the hemolymph of the veins of the wing (for example the genus *Pterobosca* MacFie). Cases of winged Diptera living on the bodies of other insects are observed in some peculiar acalyptrates, representatives of the families Borboridae and Milichiidae (genus *Desmometopa*), which live on the bodies of beetles. Presumably the borborideans (milichiids or borborids?) could have been the original group for the braulomorphs; the strengthening of connections with social bees, the reduction of body size, increase in the grasping power of the legs, improvement of body form and the development of the wingless condition — all this indicates the character of the history of these Diptera, the progressive nature of which was definitely not great.

PART II

DIPTERA OF THE GEOLOGICAL PAST

The present Diptera appear for the first time in the paleontological record at
the boundary of the Triassic and Jurassic in coal deposits of central Asia. As
mentioned above (p. 5), indications of the presence of Diptera in Permian
times turned out to be erroneous and were the result of an inaccurate determination of remnants of representatives of the suborder of scorpion flies, Paratrichoptera, which are related to the Diptera. There are other indications in
the literature (Grauvogel, 1947) of finding Diptera more ancient than the
Jurassic, but up to now they remain unconfirmed accounts and we are compelled to assume incorrectness in the identification and confusion between these
two related orders of mecopteroid insects. The fairness of such doubt is sup-
ported especially by an examination of the Upper Triassic fauna of central
Asia in the composition of which appear highly archaic forms of Diptera which
are basically distinct from all the representatives of the order known up to the
present (Rohdendorf, 1947b, 1961a). Younger Diptera are reported from the
Upper Liassic deposits of western Europe (Germany and England – Handlirsch,
1906-08, 1937-39; Tillyard, 1933). A considerably younger dipterous fauna
was discovered in the well-known Jurassic deposits of Karatau, its age cannot
yet be considered as finally determined and more probably corresponds to
Upper Middle Jurassic time (Martynov, 1925a, b, 1938). The Cretaceous fauna
of Diptera is still almost unknown. Besides these three comparatively vast
faunistic complexes which throw light upon the composition of the Mesozoic
fauna of Diptera, there are known separate finds of a few species of Diptera
from Jurassic deposits and the lowest part of the Cretaceous (Ping, 1928; Carpenter, 1935, 1937).

Findings of a single species of Diptera from the Upper Liassic location of
Ust-Baley on the river Angara, of some species from the Upper Jurassic of
England and Germany, of one species from the Lower Cretaceous deposits of
the Transbaikal region and Mongolia and of a few species from the Upper
Cretaceous deposits of Canada have been made (Rohdendorf, 1957).

Much more material is known for the Diptera of the Tertiary period. A
detailed enumeration of all the finds of Tertiary Diptera is hardly possible;
the locations of them are numerous but very irregularly investigated. The richest fauna and, moreover, one of the most ancient, including about one thousand
species, is known from Baltic amber, the age of which even now is questionable
(Upper Eocene – Lower Oligocene). In fact the fauna of the amber is the main
one on the basis of which we judge the character of the Diptera of the Paleogene. Other Paleogenic fauna, younger, is known from the Upper Oligocene

deposits of Germany (location Rott, Statz, 1940) from which approximately 400 species were described. Quite a large number of Diptera were described also from the Oligocene deposits of southern France (site Aix). Diptera of the Neogene are known from a whole series of locations, of which it is necessary to refer to the abundant, but still not fully studied, fauna of the Miocene of Stavropol (northern Caucasus — Rohdendorf, 1939, 1940) and a series of Miocene faunas of Europe (for instance Eningen in Germany) and North America (Florissant, Scudder, 1890; James, 1937, 1939). Pliocene Diptera are known far less, and only on the basis of solitary fragments of Diptera discovered in a series of locations. Such, for example, are the finds at the river Vaenge (tributary of the northern Dvina) of a fungus culicid *Boletina* and finds of bibionids in the Akchagyl deposits of western Turkmenia.

It is possible to obtain the most complete data on the organization of fossils by investigating Diptera in amber, which are in fact natural preparations of insects in the resin of conifers comparable to the artificial preparation of contemporary insects in Canada balsam. The study of insects in amber permits us to observe almost any detail of their external structure, including very fine bristles and color. The Diptera in amber throw light upon a forest fauna first of all of the Upper Eocene of the Baltic region and Upper Cretaceous of Canada; as I mentioned above these two faunas are far from equivalent according to volume and importance. We observe another kind of preservation in cases of the burial of entire insects in very fine and laminated deposits of slowly flowing or standing reservoirs; any kind of prolonged transportation of insects during burial is practically absent. Such Diptera are from the Middle Jurassic locations of Karatau and some Tertiary locations (for example from Stavropol in the Caucasus and Rott in Germany). The preservation of these fossils is also quite good and permits us to pass judgement on the general structure of the body of the insects although of course they are far from the faultless quality of amber 'preparations'. Finally, a preservation of fossils of another kind is observed in the vast majority of locations. It can be characterized by the finding of separate parts of the body of the insect, principally of single wings. Such location of wings may be naturally explained as the transport of remnants of insects by running water. The wings as the lightest parts are easily separated; comparatively tough parts of the body are conveyed together with various small parts of plants and are buried separately from the entire corpses of insects. It should be added that the finding of single wings also occurs both in amber and in fine shales of the Karatau type; in these cases the preservation of single wings is explained by the destruction of the insect at the moment of burying and is not a result of transport and separation by a stream of water.

Another feature of fossil Diptera is their relative infrequency and, as a rule, the absence of a series of any kind of forms. In fact the vast majority of the investigated faunas of Diptera were established by a description of single specimens. Finding remnants of many specimens of one form is an exceptional rarity. Thus, for example, with a study of the Upper Oligocene fauna of Rott, Statz described 50 new species of contemporary genera of Diptera on the basis of 51

fossils (Statz, 1940). The same had a place in my investigation of the Diptera
of the Middle Jurassic of Karatau: all 40 new species described up to now were
represented only by single fossils (Rohdendorf, 1938, 1946). For the Lower
Upper Triassic Diptera described, the number of species is also nearly equal to
the number of fossils discovered. Remnants of many specimens were found; for
instance, there are three specimens of Karatau *Dixamima villosa* Rohdendorf,
a series of remnants of pupae of Chironomidea from Karatau, and some species
of the Upper Triassic genus *Rhaetofungivora* with two to three specimens of
each. Such a variety of fossil Diptera is not exceptional and is basically typical
for all fossil insects, the finding of which is rare. Only when fossils have remained close to one place where they were hunted, for example Permian Orthoptera,
do they sometimes appear in large numbers. All this testifies to the wealth of
insect faunas of former times and the fragmentary nature of our paleoentomological records.

The Upper Triassic Diptera

Until recently the most ancient Diptera were considered to be the fossils
described by Handlirsch (1906-08, 1937-39) and Tillyard (1933) of the Lower
Jurassic (Upper Liassic) of Germany and England. Among the numerous sites
of insect remnants of Lower Jurassic age discovered in the U.S.S.R., one of
them, namely Issyk-kul in central Asia (Rohdendorf, 1947b, 1957), yielded a
great quantity of remnants of Diptera which after more detailed studies proved
to belong to a different, more ancient fauna, that of the Upper Triassic. This
Issyk-kul fauna includes various Diptera, part of which were reported by me for
the first time in 1961 (Rohdendorf, 1961a). The significance of this new, most
ancient fauna of Diptera is very great; in it were discovered certain previously
quite unknown, highly archaic forms which throw light on the vague phylogenetic relations of many younger groups. Some of these, however, are peculiar,
and belong to highly unusual extinct families of Diptera.

In the composition of the Upper Triassic Issyk-kul fauna, altogether, 53 species,
31 genera, 18 families, 13 superfamilies and four infraorders are known.

List of the Upper Triassic Diptera of Central Asia

Suborder Archidiptera
Infraorder Dictyodipteromorpha
Superfamily Dictyodipteridea
Family Dictyodipteridae Rohdendorf, 1961
Genus *Dictyodiptera* Rohdendorf, 1961
D. multinervis Rohdendorf, 1961

Genus *Paradictyodiptera* Rohdendorf, 1961
P. trianalis Rohdendorf, 1961

Genus *Dipterodictya* Rohdendorf, 1961
D. *tipuloides* Rohdendorf, 1961

Superfamily Hyperpolyneuridea
Family Hyperpolyneuridae Rohdendorf, 1961
Genus *Hyperpolyneura* Rohdendorf, 1961
H. *phryganeoides* Rohdendorf, 1961

Superfamily Dyspolyneuridea
Family Dyspolyneuridae Rohdendorf, 1961
Genus *Dyspolyneura* Rohdendorf, 1961
D. *longipennis* Rohdendorf, 1961

Infraorder Diplopolyneuromorpha
Family Diplopolyneuridae Rohdendorf, 1961
Genus *Diplopolyneura* Rohdendorf, 1961
D. *mirabilis* Rohdendorf, 1961

Suborder Eudiptera
Infraorder Tipulomorpha
Superfamily Eopolyneuridea
Family Eopolyneuridae Rohdendorf, 1962
Genus *Eopolyneura* Rohdendorf, 1962
E. *tenuinervis* Rohdendorf, 1962

Genus *Pareopolyneura* Rohdendorf, 1962
P. *costalis* Rohdendorf, 1962

Family Musidoromimidae Rohdendorf, 1962
Genus *Musidoromima* Rohdendorf, 1962
M. *crassinervis* Rohdendorf, 1962

Superfamily Tipulodictyidea
Family Tipulodictyidae Rohdendorf, 1962
Genus *Tipulodictya* Rohdendorf, 1962
T. *minima* Rohdendorf, 1962

Superfamily Tipulidea
Family Architipulidae Handlirsch, 1906
Genus *Dictyotipula* Rohdendorf, 1962
D. *densa* Rohdendorf, 1962

Genus *Diplarchitipula* Rohdendorf, 1962
D. *multimedialis* Rohdendorf, 1962
D. *destructa* sp. n.

Genus *Architipula* Handlirsch, 1906
    A. radiata Rohdendorf, 1962
    A. turanica sp. n.
    A. asiatica sp. n.

Superfamily Chironomidea
  Family Architendipedidae Handlirsch, 1906
    Genus *Architendipes* Rohdendorf, 1962
      A. tshernovskyi Rohdendorf, 1962

    Genus *Palaeotendipes* Rohdendorf, 1962
      P. alexii Rohdendorf, 1962

Superfamily Rhaetomyiidea
  Family Rhaetomyiidae Rohdendorf, 1962
    Genus *Rhaetomyia* Rohdendorf, 1962
      R. necopinata Rohdendorf, 1962

Infraorder Bibionomorpha
  Superfamily Pleciodictyidea
    Family Pleciodictyidae Rohdendorf, 1962
      Genus *Pleciodictya* Rohdendorf, 1962
        P. modesta Rohdendorf, 1962

  Superfamily Protoligoneuridea
    Family Protoligoneuridae Rohdendorf, 1962
      Genus *Protoligoneura* Rohdendorf, 1962
        P. fusicosta Rohdendorf, 1962

  Superfamily Fungivoridea
    Family Pleciofungivoridae Rohdendorf, 1946
      Genus *Rhaetofungivora* Rohdendorf, 1962
        R. reticulata Rohdendorf, 1962
        R. mediicubitalis Rohdendorf, sp. n.
        R. radiimedialis Rohdendorf, sp. n.
        R. destructimedia Rohdendorf, sp. n.
        R. perreticulata Rohdendorf, sp. n.
        R. parva Rohdendorf, sp. n.
        R. radialis Rohdendorf, sp. n.
        R. quadrimedialis Rohdendorf, sp. n.
        R. curta Rohdendorf, sp. n.
        R. magniradius Rohdendorf, sp. n.
        R. maxima Rohdendorf, sp. n.
        R. magna Rohdendorf, sp. n.
        R. subcostalis Rohdendorf, sp. n.

R. simplex Rohdendorf, sp. n.
R. major Rohdendorf, sp. n.
R. amasioides Rohdendorf, sp. n.

Genus *Rhaetofungivorella* Rohdendorf, 1962
R. subcosta Rohdendorf, 1962
R. sectoralis Rohdendorf, sp. n.
R. medialis sp. n.
R. analis Rohdendorf, 1962

Genus *Rhaetofungivorodes* Rohdendorf, 1962
R. defectivus Rohdendorf, 1962

Genus *Archipleciofungivora* Rohdendorf, 1962
A. binerva Rohdendorf, 1962

Genus *Archihesperinus* Rohdendorf, 1962
A. phryneoides Rohdendorf, 1962

Genus *Protallactoneura* Rohdendorf, 1962
P. turanica Rohdendorf, 1962

Genus *Archipleciomima* Rohdendorf, 1962
A. obtusipennis Rohdendorf, 1962

Genus *Palaeohesperinus* Rohdendorf, 1962
P. longipennis Rohdendorf, 1962
P. minor sp. n.

Family Palaeopleciidae Rohdendorf, 1962
Genus *Palaeoplecia* Rohdendorf, 1962
P. rhaetica Rohdendorf, 1962

Superfamily Phragmoligoneuridea
Family Phragmoligoneuridae Rohdendorf, 1962
Genus *Phragmoligoneura* Rohdendorf, 1962
P. incerta Rohdendorf, 1962

Superfamily Rhyphidea
Family Oligophryneidae Rohdendorf, 1962
Genus *Oligophryne* Rohdendorf, 1962
O. fungivoroides Rohdendorf, 1962

Family Protorhyphidae Handlirsch, 1906
Genus *Protorhyphus* Handlirsch, 1906

P. turanicus Rohdendorf, 1962

Family Olbiogastridae Rohdendorf, 1962
Genus *Protolbiogaster* Rohdendorf, 1962
P. rhaetica Rohdendorf, 1962

Key to the infraorders, superfamilies and families of Upper Triassic Diptera

1 (2)   Wing clearly divided into an apical portion with numerous transverse veins and a main part with sturdy double veins M, Cu and A .... Infraorder Diplopolyneuromorpha (p. 139)

2 (1)   Wing not divided into two parts: distal venation not separated sharply from the remainder, longitudinal veins of main part of wing with the exception of CuA not double; very rarely M is double, but in this case the distal veining is not isolated .... (genus *Diplarchitipula* of the Architipulidae) 3

3 (8)   Transverse veins multiple in almost all fields of the wing, usually also in the anal zone. Wings elongate with wide base, little costalized ... Infraorder Dictyodipteromorpha 4

4 (5)   Anal region, as a rule, has numerous transverse veins; weak transverse veins also numerous between the chief trunks of veins R, M and CuA, main section of R bent in a wavy way without phragma or break .... Superfamily Dictyodipteridea, Family Dictyodipteridae (p. 138)

5 (4)   Anal region without thin network of transverse veins; between the chief trunks of R, M and CuA there are only single veins; main trunk of R at the base with break or rudimentary phragma 6

6 (7)   Wing narrow and long; SC weak, smaller half of wing leading into C; RS with strong caudal branch which sends branches forward to the anterior margin; there are four sturdy transverse rm's remote from one another; MP in a condition of anterior parallel branches of CuA, connected by one strong transverse vein with MA .... Superfamily Dyspolyneuridea, Family Dyspolyneuridae (p. 139)

7 (6)   Wing wide; SC strong, parallel and leading into R at considerably more than two-thirds of the wing length; RS with three parallel branches directed to the apex; rm's numerous, weak and indistinct; MP in a condition of strong posterior branches of MA, connected with CuA by many weak transverse veins .... Superfamily Hyperpolyneuridea, Family Hyperpolyneuridae (p. 139)

8 (3)   Transverse veins if also present then not distributed in the anal region. Wing of diverse form, very often narrowed in the basal part 9

9 (20)   Vein system of radial sector usually consisting of three parallel branches directed to the apex of the wing; wings usually elongate, with more or less straight anterior border .... Infraorder Tipulomorpha 10

10 (11)   Radial sector branching out from R in the base of the wing
forming the chief trunk running parallel to R and sending off into
the middle of the wing a posterior branch which forms a long fork;
very close to and parallel with CuA is located not only CuP but also
MP in the form of a fine vein in front of CuA . . . .
. . . . . Superfamily Tipulodictyidea, Family Tipulodictyidae (p. 142)

11 (10)   RS branches off from R usually in the middle of the wing and
forms as a rule an anterior branch, which gives a long fork; posterior
branch of RS usually single; MP not connected with CuA, rarely parallel with it, generally absent 12

12 (15)   RS usually has two distinct bases beginning in the basal part
of the wing and in its middle; branches of RS connected and placed
in the middle of the wing; there are a great number of faint, sometimes thick, transverse veins between the branches of R and RS; SC
very weak; the system of veins M very complex but faint . . . .
Superfamily Eopolyneuridea 13

13 (14)   Costal-subcostal field narrow, almost parallel at the outside;
both bases of RS very sturdy; CuA in front of its end has a sturdy outgrowth forward which apparently unites with MP; transverse veins between branches of RS very fine and weak . . . .
. . . . Family Musidoromimidae (p. 148)

14 (13)   Costal-subcostal field very wide, narrowing uniformly to the
apex of the wing; distal basis of RS weak, in the form of a fine slanting
transverse vein or branch of R; CuA without outgrowth in front of the
apex; MP not parallel with CuA, sometimes quite absent; transverse
veins between R and branches of RS numerous and frequently thick
although weak . . . .
. . . . Family Eopolyneuridae (p. 144)

15 (12)   RS has, as a rule, one clear base located closer to the middle:
rarely there are rudimentary veins in the form of free branches of R
at the basis of RS but in this case transverse veins between branches
of RS are very few and the system of M consists of a few branches;
rm in number one, more rarely two to three veins; SC highly notice-
able 16

16 (17)   Anal veins almost completely reduced, anal area without veins,
wing short with sharply expressed RS, R, three-branched RS, weak M
with branching in the middle of the wing and strong CuA with a CuP;
transverse veins weak and all distributed in the distal part of the wing
. . . . . . Superfamily Rhaetomyiidea, Family Rhaetomyiidae (p. 161)

17 (16)   Anal veins always present in number from one to three; wings
nearly always elongate, M strong 18

18 (19)   Veins of irregular size, larger at the anterior margin and in the
centre of the wing: posterior edge with a few veins; veins M weak,
common trunk of M is often absent or reduced; transverse rm and mcu
approximately in the middle of the wing . . . .
. . . . Superfamily Chironomidea, Family Architendipedidae (p. 156)

19 (18)   Veins of whole blade of wing of approximately uniform size; posterior margin with numerous veins; veins of medial system sturdy, common trunk of M well developed, not reduced transverse rm, mcu and fork of RS are located usually behind the middle of the wing, closer to its apex . . . . Superfamily Tipulidea, Family Architipulidae (p. 149)

20 (9)   Vein system of RS has the appearance of branches directed to the anterior edge, wing usually moderately elongated, frequently short with arched anterior edge   . . . Infraorder Bibionomorpha 21

21 (22)   Venation of wing weak; only R and CuA and partly RS are sturdy; remaining venation very delicate; conspicuous numerous regular but weak transverse veins between M and CuA . . . . Superfamily Pleciodictyidea, Family Pleciodictyidae (p. 164)

22 (21)   Venation more uniform, R and CuA not especially sharply segregated; transverse vein between M and CuA not developed 23

23 (24)   Costal field with sharp projection; RS with complex but delicate entangled branches; there is a rudimentary phragma; also two very clear rm veins . . . . Superfamily Protoligoneuridea, Family Protoligoneuridae (p. 166)

24 (23)   Costal field without protuberance; RS and rm of variable structure 25

25 (26)   There is a sturdy and clear phragma; the chief trunks of RS and M are reduced and these veins begin in the form of free branches; instead of the anterior branches of RS there is a transverse vein between R and RS . . . . Superfamily Phragmoligoneuridea, Family Phragmoligoneuridae (p. 200)

26 (25)   A phragma is almost always absent; sometimes there is a break or a sharp bend at the base of R, rarely a thin phragma but then there are strong branches of RS. The chief trunk of RS always well marked 27

27 (32)   There is a definite discal cell and sturdy, long (one or two) branches of RS   . . . Superfamily Rhyphidea 28

28 (29)   Two long branches of RS, one of which branches out proximally, the other distal to rm; SC short reaching level with the fork of M . . . . Family Protorhyphidae (p. 203)

29 (28)   One branch of RS leaving proximally to rm: sometimes there is also one short, apical branch   30

30 (31)   Discal cell very small, length of it five times shorter than the anterior branches of M; vein M_4 leaves immediately from the general trunk of M; mcu very long and straight; total size of wing very small, about 2 mm . . . . Family Oligophryneidae (p. 202)

31 (30)   Discal cell large, half as long as the anterior branch of M; M_4 branches out from transverse mcu; larger, not less than 4 mm . . . . Family Protolbiogastridae (p. 205)

32 (27)   No clear discal cell; branch RS of diverse structure . . . .
                 Superfamily Fungivoridea 33
33 (34)   Between the ends of R and C, between the anterior branch of
                 RS and the end of R there are clear transverse veins; anterior medial
                 veins indistinct, forming irregular tangled compartments; posterior
                 medial vein straight and sturdy; mcu is absent; R straight, without a
                 bend or break; wing elongate . . . .
                 . . . . Family Palaeopleciidae (p. 198)
34 (33)   Transverse veins between R, C and RS are absent; anterior
                 medial veins are structurally distinct from posterior branches; R at the
                 base with break or bend; wing usually wide . . . .
                 . . . . Family Pleciofungivoridae (p. 167)

Suborder Archidiptera

The characters of the suborder were given above in the description of Diptera
of the contemporary fauna (p. 26). The chief features of this suborder are the
form and position of the head of the pupa, size of the mesothorax and other
traits of body structure. Not all these features can be observed in fossil representatives of the suborder, but the form of the wing (peculiarly elongated with relatively parallel anterior and posterior borders), and the presence of numerous
transverse veins in nearly all fields of the wing including the anal, are characteristic.

Composition. – There are two sharply isolated Triassic infraorders, Dictyodipteromorpha and Diplopolyneuromorpha.

Infraorder Dictyodipteromorpha

Wings with numerous transverse veins in certain places forming a peculiar reti-
cular venation. Base of wing wide, without individual basialar but sometimes with
a peculiar phragma or deflection of R. Apex of wing not isolated, wings elongate,
curvature of anterior and posterior margins approximately identical. Venation
consists of sturdy, generally few longitudinal trunks and systems of numerous
lighter transverse and longitudinal veins. The wings are small: from 2 to 3 mm
long. These most peculiar Diptera are distributed among three quite sharply differing superfamilies. The first, Dictyodipteridea, includes only one family, characterized by a comparatively insignificant number of longitudinal viens together
with a rich development of a system of transverse veins. This group is quite peculiar and differs sharply not only from the remaining representatives of the series
but also from all the known Diptera on the basis of the general appearance of
the wing, suggesting the most diverse insects up to and including the orthopter-
oids. Another superfamily, the Hyperpolyneuridea, also contains one family, no
less unusual than the first and characterized by abundant veining with a great
quantity of longitudinal viens and relatively broad wings. Finally, the last group,
the Dyspolyneuridea, perhaps more peculiar than either of the former, is sharply
distinct from their structure of the medial and cubital system of veins and the
complete aerodynamic form of the wing.

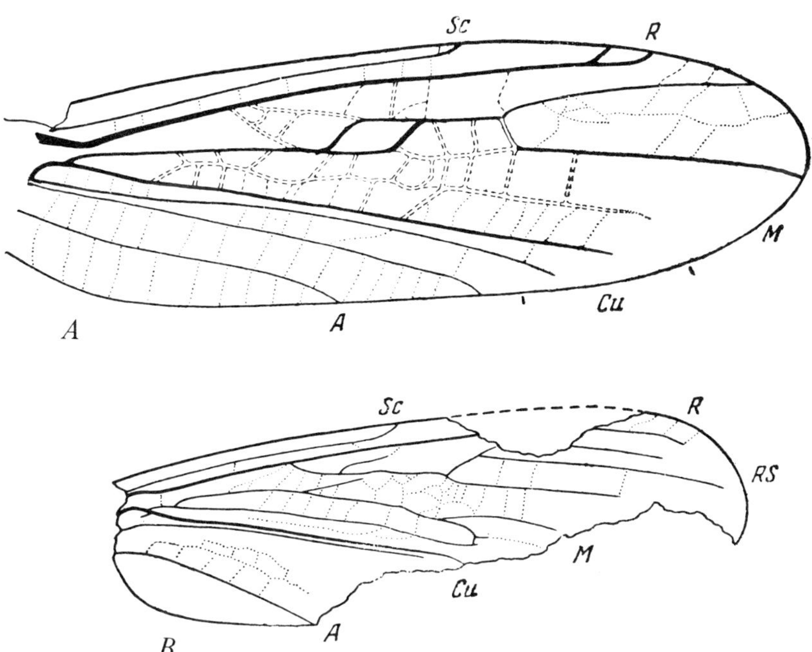

Fig. 47. Dictyodipteridea from the Triassic of central Asia. A. *Dictyodiptera multinervis* Rohd. (Dictyodipteridae). Wing. Length 2.81 mm. B. *Dipterodictya tipuloides* Rohd. (Dictyodipteridae). Wing. Length
3.44 mm. (According to Rohdendorf, 1961).

The infraorder contains so many different groups that the question arises
whether it is a valid taxon; as for any systematic category based on the structure
of one organ (in this case the wing), the possibility of revaluation is not excluded. The dictyodipteromorphs are known only on the basis of the structure of
single wings. These organs of locomotion in the order Diptera, as also in other
insects which fly well, solving different aerodynamic problems, are built very
complexly and give us the possibility of distinguishing different groups well and
of judging the character of their movement. Considering the flying qualities of
this infraorder one can sustain the conclusions that the stroke speed was slow,
the wings flexible, with an extensor mechanism only, and only partly reinforced
by the veining. In this all dictyodipteromorphs agree and are the most ancient
grouping of Diptera, serving as the origin of the formation of the majority of
other infraorders, in the first place of bibionomorphs and tipulomorphs. Furthermore, there is no doubt that many groups of this infraorder remain still unknown
to us; only the Dictyodipteridea are connected directly with the Bibionomorpha.
Two other superfamilies, apparently extinct groups, are not linked with other
younger infraorders of Diptera.

Superfamily Dictyodipteridea Rohdendorf
Superfam. Nov.

The characteristics are given above in the key (p. 133). It includes one family, the Dictyodipteridae.

Family Dictyodipteridae Rohdendorf, 1961
Rohdendorf, 1961a, p. 90

Small forms, the wings of which do not exceed 3.5 mm.

Three monotypical genera, of which *Dictyodiptera* and *Paradictyodiptera* are quite close to one another and *Dipterodictya* more isolated.

Key to the Genera

1 (2)   Between CuA and CuP in the main part of the wing delicate transverse veins are noticeable; two anal veins are almost parallel and reach to the border of the wing; between R and CuA there are only two highly noticeable longitudinal veins; the remaining veins are weak and indistinct (fig. 47*A*) . . . .
. . . . *Dictyodiptera* (p. 138)

2 (1)   No transverse veins between CuA and CuP; anal veins delicate and not parallel, connected between themselves by a network of transverse veins; between R and CuA are two larger longitudinal veins quite delicate and irregularly distributed 3

3 (4)   In the main half of the wing between CuA and M there are few transverse veins; bending of the base of R is weak; M is divided near its base into two branches between which are not less than six delicate transverse veins; RS in distal half of wing forms parallel to the anterior margin of the branch (fig. 47*B*) . . . .
. . . . *Dipterodictya* (p. 139)

4 (3)   In the main half of the wing, between CuA and M, there are numerous transverse veins; bending at the base of R very abrupt and sharp; M is divided only in the middle of the wing and does not have posterior branches running parallel to CuA. The base of M branches out from CuA noticeably proximal to the bend of R; medial vein branches out from CuA quite gradually; SC at the end with a short fork . . . .
. . . . *Paradictyodiptera* (p. 138)

Genus *Dictyodiptera* Rohdendorf, 1961
Rohdendorf, 1961a, p. 91

Type of genus: *D. multinervis* Rohdendorf, 1961. Known from a single species, the type of the genus (fig. 47*A*).

Genus *Paradictyodiptera* Rohdendorf, 1961
Rohdendorf, 1961a, p. 93

Type of genus: *P. trianalis* Rohdendorf, 1961. Known from a single species, the type of the genus.

Genus *Dipterodictya* Rohdendorf, 1961
Rohdendorf, 1961a, p. 92

Type of genus: *D. tipuloides* Rohdendorf, 1961. Known from a single species, the type of the genus (fig. 47*B*).

Superfamily Hyperpolyneuridea Rohdendorf, 1962
Rohdendorf, 1962, p. 309.

Characteristics given above. One family, the Hyperpolyneuridae.

Family Hyperpolyneuridae Rohdendorf, 1961
Rohdendorf, 1961a, p. 94

Wing not greater than 3 mm. Type of family: genus *Hyperpolyneura*. Single monotypical genus.

Genus *Hyperpolyneura* Rohdendorf, 1961

Type of genus: *H. phryganeoides* Rohdendorf, 1961. Known from a single species, the type of the genus (fig. 48*A*).

Superfamily Dyspolyneuridea Rohdendorf, 1962
Rohdendorf, 1962, p. 310.

Characteristics given in the key (see p. 133). These forms possess the most mechanically perfect wings of the whole infraorder. There is one family, the Dyspolyneuridae.

Family Dyspolyneuridae Rohdendorf, 1961
Rohdendorf, 1961a, p. 95

Sharply elongated wings which are not less than three-and-a-half times the length of the width, of small dimensions (about 3 mm). There is a real phragma at the place of a break of the base. There is a single monotypical genus.

Genus *Dyspolyneura* Rohdendorf, 1961
Rohdendorf, 1961a, p. 96

Type of genus: *D. longipennis* Rohdendorf, 1961. Known from a single species, the type of the genus (fig. 48*B*).

Infraorder Diplopolyneuromorpha

Rohdendorf, 1961a, p. 90.

Brief description. — The wing consists of a sharply isolated upper part bearing

longitudinal and transverse veins forming right-angled cells and a large main part
of the blade, strengthened by longitudinal veins, many of which form well-expressed pairs of folds. MA and MP, CuA and CuP, A_1 and A_2 are such pairs. The
wing is elongate apparently with a rather pointed apex.

Comparison. – The most peculiar fossil *Diplopolyneura mirabilis* Rohdendorf
is so sharply distinguished from all other known Diptera that it constitutes an
altogether isolated higher taxon within the limits of the order, a special infraorder.
The drawing together in pairs of longitudinal veins among related groups of
mecopteroids is unknown to me and suggests quite alien groups remote from
dipterous insects, namely mayflies (families Palinginiidae, Behningiidae, Oligo-
neuriidae). However the mechanical importance of the paired drawing together
of veins in the wings of mayflies is quite different and results in strengthening
very broad fan-shaped wings by longitudinal ribs – paired veins. In the described
Diptera, the Diplopolyneuridae, the apical part of the wing has the appearance
of a flat vibrating appendage of a blade which has been strengthened very much
by paired veins. Such a structure in part reminds one of the wings of present-
day Tipulidae, with the difference that the wings of the latter, although they
have a more or less isolated apex, do not possess such a strengthened blade bearing only the one double vein CuA+CuP usual for Diptera. The phylogenetic relations of this infraorder are more or less clear; it is a greatly changed derivative
of the ancestors of tipulomorphs, which very early obtained the high mechanical
quality of venation of the wings.

Composition. – There is a single family, the Diplopolyneuridae.

Family Diplopolyneuridae Rohdendorf, 1961
Rohdendorf, 1961a, p. 98

There is a single monotypical genus.

Genus *Diplopolyneura* Rohdendorf, 1961
Rohdendorf, 1961a, p. 98

Type of genus: *D. mirabilis* Rohdendorf, 1961. There is a single species, the type
of the genus (fig. 48*C*).

Suborder Eudiptera [8]

Infraorder Tipulomorpha

Brief description. – Wings elongate with straight anterior border and sturdy
longitudinal veins. RS forms not less than three branches, as a rule running parallel with the anterior edge to the apex of the wing. Basalar more or less isolated,
also the apex of the wing. Transverse veins scarce, but if there are many they are
almost never between the chief branches of M and CuA. Dimensions from 3.5 to
6 mm (in Upper Triassic forms).

Comparison. – A characteristic infraorder of Diptera including in the fauna

---

8. The characteristics of this suborder are given on p. 30.

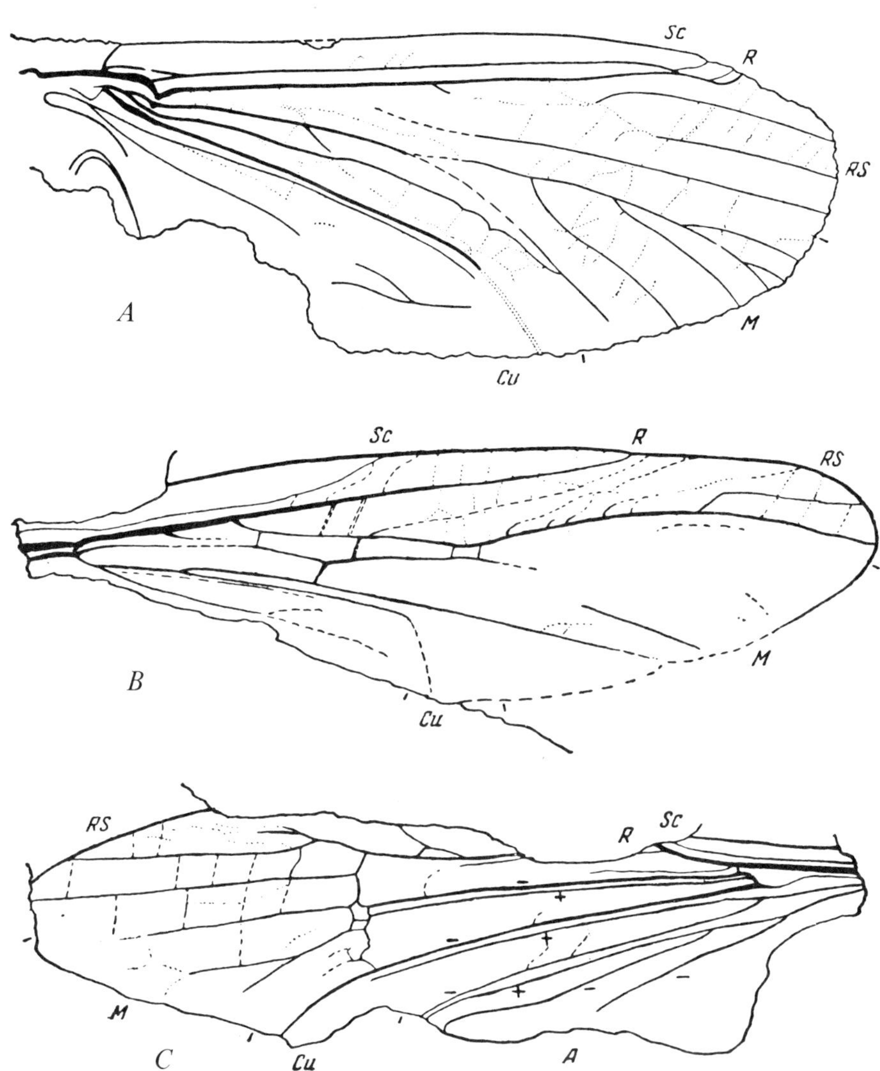

Fig. 48. Hyperpolyneuridea, Dyspolyneuridea and Diplopolyneuromorpha from the Triassic of central Asia. *A. Hyperpolyneura phryganeoides* Rohd. (Hyperpolyneuridae). Wing. Length 2.82 mm. *B. Dyspolyneura longipennis* Rohd. (Dyspolyneuridae). Wing. Length 3.06 mm. *C. Diplopolyneura mirabilis* Rohd. (Diplopolyneuridae). Wing. Length 4.38 mm. (According to Rohdendorf, 1961.)

of the Upper Triassic not less than five superfamilies, of which three are ancient, extinct groups and the other two, Tipulidea and Chironomidea, live until the present time. The most important feature of this vast and ancient infraorder of Diptera is the process of elongation of the wings which starts very early. The processes of costalization and reduction of the wings prevail little among representatives of this infraorder and are secondary, clearly developed on the basis of the already greatly lengthened 'traction' wings. Here it is necessary only to note that the extinct superfamilies, Tipulodictyidea and Eopolyneuridea, and also the Rhaetomyiidea, may be distributed between two groups. The first two superfamilies are characterized by different archaic features as for example, an abundance of transverse veins, little mechanical specialization (wide wing, little strengthening of the anterior edge, complexity of the system of M and Cu), the presence in the Eopolyneuridea of some traits of the integumentary character in the wings. On the contrary, the last superfamily, the Rhaetomyiidea (quite lately apparently discovered in the contemporary fauna of Australia — see Colless, 1962) is characterized by very high mechanical improvement, by the development of a free anal region devoid of veins, by sturdy main stems of R, CuA, and by the formation of a peculiar form of wings. On the whole this last group of Upper Triassic tipulomorphs promptly suggests different, more recent Diptera, for instance the Orphnephilidea or some Asilomorpha. On the whole *Rhaetomyia necopinata* sp. n., according to its wings, is the most perfect form among all the other Upper Triassic Diptera; the flight of this dipteran probably differs most in the rapid stroke and high maneuverability.

Among the Upper Triassic tipulomorphs there are undoubtedly forms related to the ancestors of other younger groups of the infraorder. Thus the Eopolyneuridae probably are close to the ancestors of the Ptychopteridae, Musidoromimidae to the ancestors of the Psychodidea and maybe the Dixidea and Culicidea. These relations are analyzed in more detail in the outline of the phylogenesis of Diptera (p. 291).

Superfamily Tipulodictyidea Rohdendorf, 1962
Rohdendorf, 1962, p. 310.

Description. — A peculiar isolated group which is sharply distinguished from others by the branching of RS from R in the base of the wing and by the presence of MP forming together with CuA and CuP a characteristic triple group of parallel veins. Dimensions of wing are small, about 2.5 mm.

Composition. — There is a single family, the Tipulodictyidae.

Family Tipulodictyidae Rohdendorf, 1962
Rohdendorf, 1962, p. 310

Description. — The elongate wings are small in size (approximately 2.5 mm). Phragma not developed, R is strong but uniformly curved at the base. Costal field narrow with numerous transverse veins; SC is greater than half of the wing, RS branches out from R almost from the basal bending and is made up of three well-expressed approximately parallel branches. There is a well-developed posterior branch of M which sends off from itself a small posterior branch located parallel to CuA which is sturdy and bent at the end. CuP is weak but highly noticeable and runs parallel to CuA. Veins MP, CuA and CuP form a peculiar group of parallel veins.

Composition. — The family is described on the basis of a single monotypical genus.

Genus *Tipulodictya* Rohdendorf, 1962
Rohdendorf, 1962, p. 311

Type of genus: *T. minima* Rohdendorf, 1962.

Description. — Wing is slightly narrowed toward the apex. Anterior edge to the end of SC and farther is moderately convex. C is very fine, indistinct. SC fine, almost straight, weakly and uniformly converging with C into which it goes approximately at the middle of the wing. R irregularly bent and going into C in the beginning of the apical quarter of the wing; subcostal field without transverse veins; radial field narrow with a few weak transverse veins in the apical part. RS running parallel to R and approximately at the middle of the wing there separates its posterior branch which, somewhat distal from the level of the end of SC, forms a fork of two parallel veins. All three branches of RS are directed toward the apex of the wing but the last branch of RS terminates almost strictly at the point of the apex of the wing. From the base of the last branch of RS branches out to M a sturdy straight transverse rm. The system of M is weak, the structure of it not precisely known in detail. At the border of the wing appear apparently not less than four branches of M. CuA is sturdy, especially in the basal part of the wing. CuP is thin, in its greater extent it is parallel with CuA. The anal field and its veins are unknown.

Composition. — One species which is the type of the genus.

Tipulodictya minima Rohdendorf, 1962 (fig. 49)
Rohdendorf, 1962, p. 311, fig. 974

Holotype. — Positive impression of right wing (anal region not completely preserved). Coll. PIN[*] No. 358/118, Issyk-kul, Upper Triassic (Rhaetian?) series N.

Description. — Costal field with 13 veins, apex of radial field with three (of which one connects the apex of RS_1 with C), cell RS_1 with six, cell RS_2 also with six fine straight transverse veins. In the remaining part of the wing transverse veins are inconspicuous. The length of the wing is 2.6 mm. It is described on the basis of a single positive impression of a right wing; anal region not completely preserved, medial region is not clear.

Material. — The holotype.

Superfamily Eopolyneuridea Rohdendorf, 1962
Rohdendorf, 1962, p. 311.

The characteristics are given above (p. 134). Two families, Eopolyneuridae and Musidoromimidae.

---

[*]PIN   Palaeontological Institute of the Academy of Sciences, U.S.S.R.

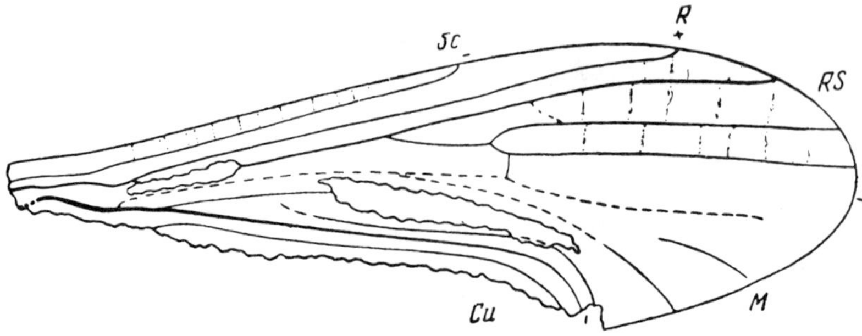

Fig. 49. Tipulodictyidea. *Tipulodictya minima* Rohd., 1962, (Tipulodictyidae). Wing. Length 2.6 mm. Upper Triassic of central Asia. (According to Rohdendorf, 1962.)

Family Eopolyneuridae Rohdendorf, 1962

Rohdendorf, 1962, p. 311

Description. – Moderately elongated wing, about 3 mm. Phragma not developed, R at base with a break, long, about two-thirds of wing. Costal field very wide, gradually tapering to the apex of the wing, without clear transverse veins. SC delicate but long, free or connected with R at the end. RS branches out from R in the basal half of the wing in the form of a sturdy and clear fork; at the middle of the wing RS splits into three branches. The medial system usually consists of finer veins than the radial; there are from four to six or more branches of M. CuA not particularly sturdy, extends beyond the middle of the wing; CuP contiguous with CuA but not parallel with it. Anal region very wide, its venation is unknown. Between branches of the radial and anterior medial veins there are straight, somewhat obliquely disposed transverse veins. The type of the family is the genus *Eopolyneura*, g.n.

Comparison and composition. – The wide costal field, the structure of the radial system of veins and the presence of numerous transverse veins distinguish this group well from all other Diptera. The family is described with two monotypical genera: *Eopolyneura* Rohdendorf and *Pareopolyneura* Rohdendorf.

Key to the genera

1 (2)   Anterior branch of RS at the end strongly connected with R and removed from the middle branch of RS; costal field very wide at the level of branching of RS, considerably wider than radial compartment; SC connected with R by transverse veins: R very thick . . . . *Pareopolyneura* (p. 147)

2 (1)   End of anterior branch of RS removed from the end of R and drawn nearer to the end of the middle branch of RS; costal field less wide, at the level of the branching of RS approximately equal to the width of the radial cell; SC at the end free, drawn together but not

connected with R by transverse veins; R is thickened only in the basal
part . . . .
. . . . *Eopolyneura* (p. 145)

Genus Eopolyneura Rohdendorf, 1962
Rohdendorf, 1962, p. 311

Type of genus: *E. tenuinervis* Rohdendorf, 1962.

Description. — Anterior edge of wing very weakly convex: apex clearly not
isolated. C is moderately reinforced. SC thin and straight, strongly drawn together with R, at the end free and not reaching the middle of the wing. R at the
base of the wing is strongly expanded and narrowed toward its break; the greater
part of it is lying in the blade of the wing, comparatively thin, nearly straight,
at the end with a small fork or short transverse vein. RS branches out from R
approximately at the boundary of the first and second quarters of the wing and
branches at the end of the second quarter sending off an anterior branch; the
posterior branch is divided into two, somewhat distal to the first fork. Two anterior branches RS and R are almost parallel; a posterior branch is parallel to the
anterior only in its basal half, in the distal it bends somewhat backwards, going
almost exactly into the apex of the wing. Transverse veins between the branches
of the radial vein thick but faint. M branches out from CuA approximately at
the level of the break of R and branches at first at the level of the middle of the
total length of RS sending off a weak posterior branch and further, at the level
of the branching of RS, at once subdivides into three branches — two anterior,
comparatively sturdy, and a weak posterior branch which unites with the main
posterior branch of M. There are not less than four radiomedial transverse veins
and several transverse veins between the two anterior branches of M. The posterior
branches of M unite with one another forming an irregular type of cell. Between
M and CuA there are two transverse veins — short between the bend of CuA and
the fork of the anterior trunk of M and long, oblique poorly expressed between
the base of the posterior branch of M and the middle of the straight portion of
CuA. CuA clearly is divided into an almost straight, greater basal part and the
smaller distal part bent backwards. Between CuA and CuP there is a weak transverse vein. The three anal veins are delicate and poorly marked.
Composition. — One species, the type of the genus.

Eopolyneura tenuinervis Rohdendorf, 1962 (fig. 50A)
Rohdendorf, 1962, p. 311, fig. 975

Holotype. — Negative impression of right wing (the extreme apex, the base and
the posterior border of the anal blade are not well preserved). Coll. PIN No. 358/
117 Issyk-kul, Upper Triassic (Rhaetian?).
Description. — Costal field with vague transverse vein in front of the fork of
R; between R and the anterior branch of RS there are six transverse veins of
which the most basal is a kind of rudimentary base of the anterior branch of RS,

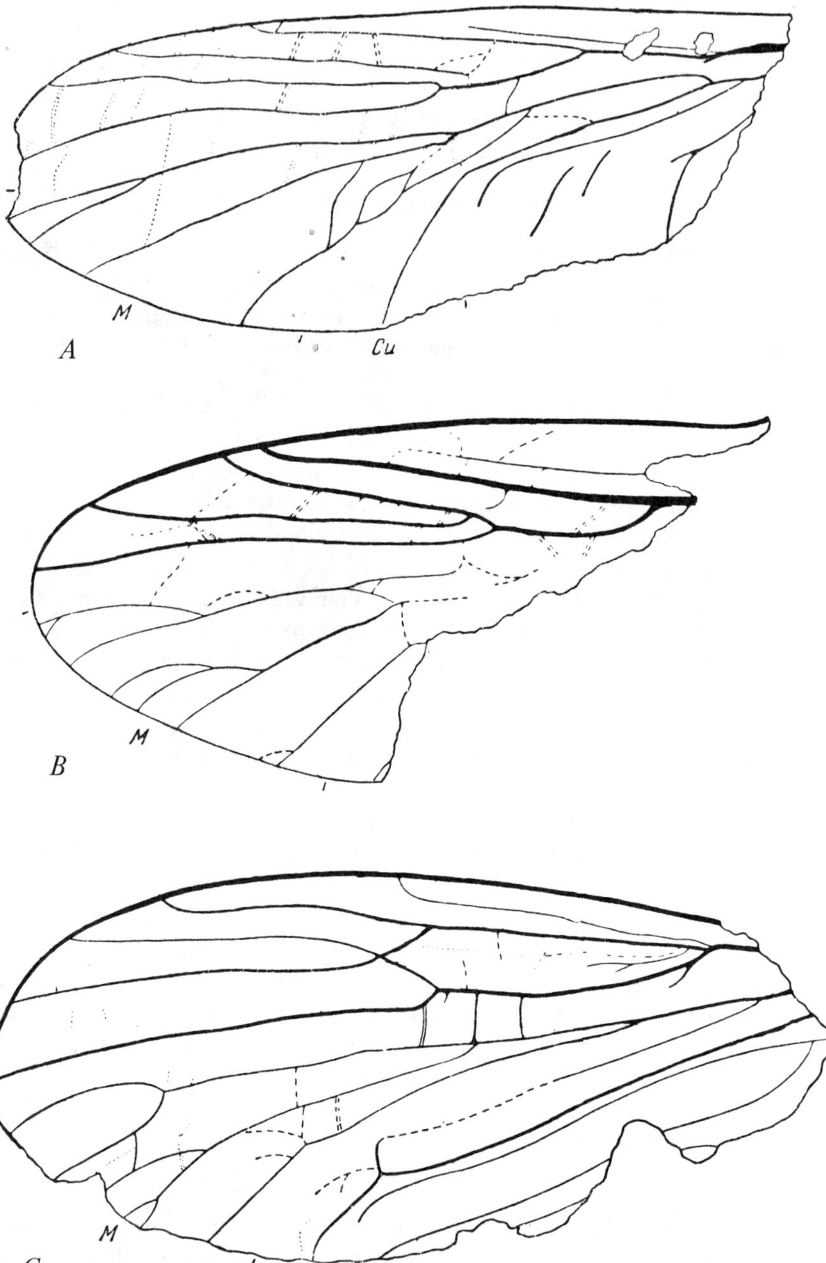

Fig. 50. Eopolyneuridae of the Upper Triassoc of central Asia. *A. Eopolyneura tenuinervis* Rohd. (Eopoly-neuridae). Wing. Length 2.94 mm. *B. Pareopolyneura costalis* Rohd. (Eopolyneuridae). Wing. Length about 4 mm. *C. Musidoromima crassinervis* Rohd. (Musidoromimidae). Wing. Length 3 mm. (According to Rohd-endorf, 1962.)

the next three are placed between R and RS and the two distal between C and a
branch of RS; between the anterior and middle branches of RS are five veins,
between the middle and posterior five to six veins. There are weakly distinguished
transverse veins between the break and the basal trunk of R and M, a strong rm
and approximately six weak transverse veins between the posterior branch of RS
and the anterior of M. The anterior branch of M at the end with a clear fork;
between the anterior and second branches of M there are not less than five weak
transverse veins. Length of wing is 2.94 mm.

Material. − The holotype.

Genus *Pareopolyneura* Rohdendorf, 1962
Rohdendorf, 1962, p. 311

Type of genus: *P. costalis* Rohdendorf, 1962.

Description. − The anterior border is moderately convex and gradually bends
to the apex, apex slightly isolated. C is strongly thickened. SC thin and strong,
moderately drawn together with R and connected with its three transverse veins
at the end apparently bending backwards and going into R; SC relatively long
and reaching the middle of the wing. R is almost straight, very sturdy and thicker
at the base, without fork at the end. RS branches off from R approximately at
that same level as also in representatives of the preceding genus; basal section of
RS curved and sturdy, and branches approximately at the middle of the wing
sending off at first a nearly straight posterior branch to the apex of the wing
and further forming a fork out of the anterior and middle branches. The anterior
branch of RS gradually draws together with R; the middle branch is parallel with
the posterior branch for most of its length and only at the end diverges from it.
Transverse branches are fewer than in *Eopolyneura*; between R and the general
section of RS there are two veins of which the distal is located at the level of the
fork of RS and has the form of a rudimentary basal section of the anterior branches
of RS. The basal division of M is unknown. There are three main distal branches
of M which form forks altogether to six or seven branches. The structure of Cu
and A are unknown.

Composition. − One species, the type of the genus.

Pareopolyneura costalis Rohdendorf, 1962 (fig. 50*B*)
Rohdendorf, 1962, p. 311, fig. 976

Holotype − Negative impression of right wing (the base, the greater part of
the medial and the whole anal region are not preserved). Coll. PIN No. 358/594,
Issyk-kul, Upper Triassic (Rhaetian?).

Description. − There are transverse veins between the anterior and middle
branches of RS to the number of six weakly-expressed branches, between the
middle and posterior branches of RS two oblique veins, between the main section
of RS and M not less than three veins. The length of the remnant of the wing is
3.13 mm.

Family Musidoromimidae Rohdendorf, 1962
Rohdendorf, 1962, p. 311

Description. — The wings are moderately elongated (approximately two-and-a-half times as long as wide), about 3 mm. The structure of the base of R is not known. The costal field is narrow, R runs nearly to its apical part parallel with C. SC is delicate, running into C beyond the middle of the wing. RS branches out from R in two places: in the vicinity of the base and approximately in the middle of the wing, forming a closed radial cell and three almost straight branches. There are some sturdy radiomedial veins. The medial system consists of a weak MP, at the end combined with CuA, and of three main branches, branching abundantly close to the border of the wing. CuA is very sturdy almost straight, with an abrupt branch near the apex directed forward and connected with MP. CuP is thin, strongly drawn together with CuA. A_1 thin, parallel with CuA and CuP. Apart from sturdy radiomedial veins, other transverse veins are weak and poorly marked.

Comparison and composition. — The parallel sturdy R, RS, M and CuA, a number of branches of RS, the branch at CuA and two branches of R — the chief divisions of RS — characterize well this family which is comparatively close to the Eopolyneuridae and is described on the basis of a single genus.

Genus *Musidoromima* Rohdendorf, 1962
Rohdendorf, 1962, p. 311

Type of genus: *M. crassinervis* Rohdendorf, 1962.

Description. — The anterior border is uniformly convex, the apex is short, poorly isolated. C is highly thickened, thinner behind the apex of the wing. SC thin, irregularly curved, in its main part drawn together with R, further located at the middle between R and C, at the end arch-like, bent forward and united with C. R is long, equal to nearly four-fifths of the wing, straight basally and irregularly bent apically. From the main division of R there branches out a sturdy basal branch RS which branches at the middle of the wing sending off a straight posterior branch to the apex and an anterior branch to the anterior edge. This anterior branch unites with R a sturdy, oblique, transverse vein and at this same point forms a fork — anterior and middle branches of RS. The anterior branch of RS is very thin; all four radial veins break up very weakly, nearly parallel. The posterior branch of RS constitutes an almost straight continuation of the common trunk of RS and occupies together with the anterior branch of M a central axial position in the blade of the wing. There are four sturdy radiomedial veins of which three are thicker. The place where M branches from Cu is unknown. The main trunk of M is quite straight and at the level of the common trunk of R sends off a thin but clearly visible branch which is the chief half of MP running parallel with CuA and at a sharp angle to M. At the level of the basal half of the common trunk of RS from the trunk of MA is sent off a second posterior branch running approximately parallel with MP and running in the posterior border of the wing. Further in front of the fork of RS the middle branch of MA branches out, diverging somewhat from the posterior branch and dichotomized twice before the edge of the wing. An anterior branch of MA forms an apical fork (reminiscent of such in the Sciaridae). Branches of M are combined with a few transverse veins. CuA is sturdy, irregularly curved, almost straight, with a sturdy offshoot near its end, directed forward and connected with MP; the distal section of CuA is more delicate, bent back, and running in the edge of the wing. CuP drawn together but not parallel with CuA, extending to the middle of the distal section of CuA. A_1 at the base apparently branches from CuP; it is just as thin as this vein and runs into the margin of the wing not far from CuA. A_2 is noticeable only partly in the form of a thin vein parallel with A_1.

Composition. – There is one species, the type of the genus.

Musidoromima crassinervis Rohdendorf, 1962 (fig. 50*C*)
Rohdendorf, 1962, p. 311, fig. 977

Holotype. – Positive impression of left wing (base of wing and part of its posterior margin not preserved). Coll. PIN No. 358/94 Issyk-kul, Upper Triassic. (Rhaetian?).

Description. – There are transverse veins between R and the anterior branch of RS (1), between the anterior and middle branches of RS (4), middle and posterior branches (4), between R and the main section of RS (3?), between the posterior branch of RS and the anterior of M (five weak veins not counting the sturdy real radiomedials), between the anterior and middle branches of M (3), middle and posterior (2). Besides transverse veins it is necessary to note the free posterior branch of vein R right after the branching of RS and of the vein RS not far from its branching from R and in front of the fork, between the two distal radiomedial transverse veins. The length of the remnant of the wing is 2.94 mm, of the whole wing about 3.2 mm.

Material. – The holotype.

Superfamily Tipulidea Latreille, 1802

One of the superfamilies of the infraorder surviving up to the present epoch. Brief characteristics are given above. In the Upper Triassic fauna they are represented by the single family Architipulidae.

Family Architipulidae Handlirsch, 1906
Handlirsch, 1906, p. 490. Rohdendorf, 1962, p. 312

Description and comparison. – Small or medium-sized Diptera, the wing length of which ranges from 3 to 6 mm. The main division of R is moderately thickened without bend, break or phragma. The wings are elongated two-and-a-half to three-and-three-quarter times as long as wide; anterior border moderately convex, sometimes in the greater part nearly straight with well-isolated apex. The costal field

is very narrow, parallel to the edge. RS long, always considerably longer than
half of the wing, going into C and sometimes, moreover, connecting at the end
with R. RS branches out from R in the main half of the wing; sometimes there
are additional free branches from R, lying distally or proximally from the real
RS, giving three distal branches. The medial system consists of a straight common
trunk which branches out in the base of the wing from CuA and which branches
only beyond the middle forming from three to six branches; always there is an
isolated intermedial cell, formed by the main fork of M and a transverse vein
between the middle and the last of its branches. Sometimes (genus *Diplarchitipula*)
there is a posterior branch of MP located parallel with MA.

CuA is sturdy and straight, bent back at the level of the forks of RS and MA;
CuP more delicate, adjacent and parallel with CuA. Anal area very wide, usually
with A_1 and A_2 long and A_3 shorter; sometimes A_1 forms a fork (some species
of the genus *Architipula*), occasionally this vein is double as a result of drawing
together with the thin A_2 (or branch of this vein?). Transverse veins are scarce:
the radiomedial vein is generally one, more rarely there are transverse veins between the posterior branch of RS and the anterior of M. Rarely transverse veins
are in the costal, subcostal, and anal fields (genus *Dictyotipula*). The type of the
family is the genus *Architipula* Handlirsch, 1906. The elongate common trunks of
RS and M, the long Cu, and the narrow costal and subcostal fields draw this
family near to the Limoniidae from which it is distinguished by smaller size,
wide base of wing, very broad anal region, and frequent presence of forks on A_1.

Composition and distribution. – In the first place this family was described
according to a Jurassic species from the Liassic of Germany and includes in the
Liassic fauna of Europe not less than 10 genera (see p. 212). In the reported Upper
Triassic fauna of central Asia are found species of the genera *Architipula* Handlirsch, *Diplarchitipula* g.n. and *Dictyotipula* g.n.

Key to the Upper Triassic genera

1 (2)   Costal and subcostal fields with weak transverse veins; there is
               an oblique curved transverse vein between the base of RS and the middle
               common trunk of M; behind A_1 there is in the form of a thin line – a
               branch of A_2 (or strictly A_2 branching from A_3?) . . . .
               . . . . *Dictyotipula* (p. 151)

2 (1)   Costal and subcostal fields without transverse veins; no transverse
               vein between the base of RS and M; vein A_1 simple, without a parallel
               vein behind it 3

3 (4)   Behind MA there is a delicate vein MP located parallel, reaching
               to the edge of the wing; costal and subcostal fields not especially narrow,
               RS distributed somewhat at a distance from R . . . .
               . . . . *Diplarchitipula* (p. 151)

4 (3)   MA without vein MP, simple; costal and subcostal fields narrow,
               RS sharply drawn together with R . . . .
               . . . . *Architipula* (p. 154)

Genus *Dictyotipula* Rohdendorf, 1962
Rohdendorf, 1962, p. 312

Type of genus: *D. densa* Rohdendorf, 1962.

Description. — Anterior border of wing for the most part quite straight. C
moderately thickened, SC thick and parallel with C, rather close to R which is
thin and not thickened at the base. Between C, SC and R there are straight, delicate transverse veins. R at the base unites with M by an oblique branch. RS branches
out from R in the main half of the wing and this vein apparently has several bases
of which one is clearest. From the base of RS there runs an oblique somewhat
S-shaped curved transverse vein. M is sturdy and straight, branching out from CuA;
the latter is connected with CuP which branches off from A_1. This latter sturdy
vein in its turn is closely connected with a thin vein — a branch of A_2 (?). The
anal field with thin veins or folds is hardly noticeable. The structure of the greater
part of the blade of the wing is unknown.

Comparison. — The unusualness of the costal and subcostal fields which bear
transverse veins and other features sharply distinguish this genus from all other
Architipulidae; it is very possible that investigation of more complete material
will show that in reality this form was a representative of a special family.

Composition. — A single species which is the type of the genus.

Dictyotipula densa Rohdendorf, 1962 (fig. 51*A*)
Rohdendorf, 1962, p. 312, fig. 980

Holotype. — Negative impression of the basal part of the right wing (base and
posterior border of blade not preserved). Coll. PIN No. 358/196 Issyk-kul, Upper
Triassic (Rhaetian?).

Description. — In the retained portion of the wing which comprises approximately two-fifths of the whole blade, in the costal field there are nine, in the
subcostal seven transverse veins. Side by side with the base of RS are three curved
transverse veins located proximally from RS (not represented in the illustration)
and one nearly straight transverse vein lying immediately distally from it. Length
of remnant is 1.13 mm; total length of wing was probably about 2.8 mm.

Material. — The holotype.

Genus *Diplarchitipula* Rohdendorf, 1962
Rohdendorf, 1962, p. 313

Type of genus: *D. multimedialis* Rohdendorf, 1962

Description. — The anterior border of the wing is straight in its greater part,
weakly and uniformly convex to the apex; wing elongate, three times as long as
wide with isolated large apex (*D. multimedialis*). C moderately thickened, SC
sturdy, parallel to C and R, situated approximately in the middle between the
two named veins or somewhat closer to R, apparently going into C. R is sturdy,
unthickened at the base, straight, reaching the last quarter of the wing (*D. multi-*

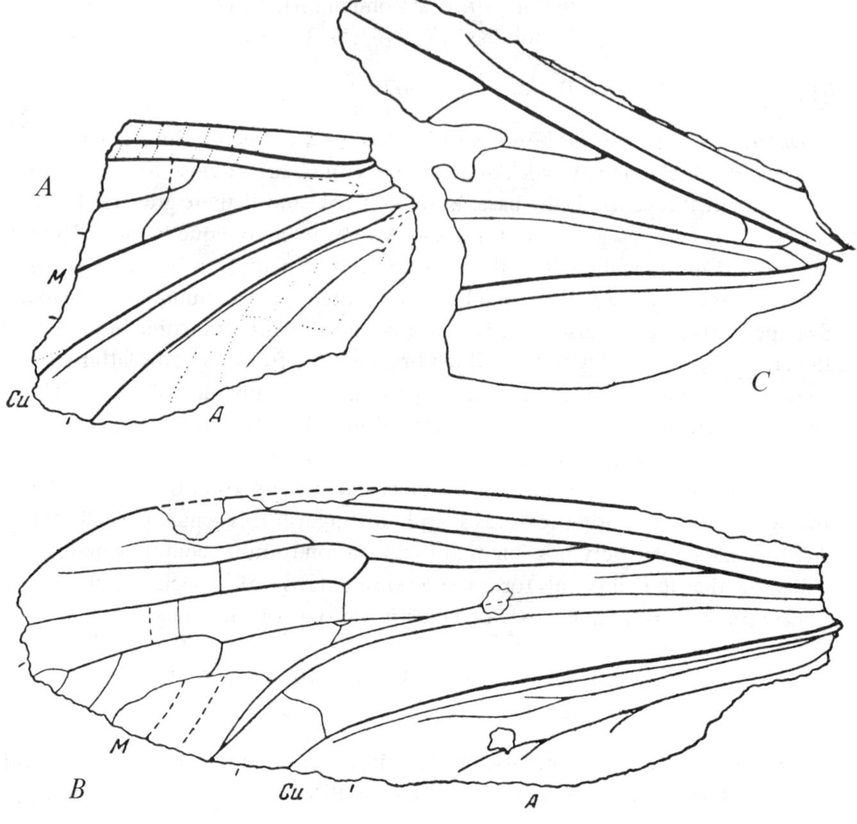

Fig. 51. Tipulidea of the Upper Triassic of central Asia. *A. Dictyotipula densa* Rohd. (Architipulidae).
Base of wing. Length of remnant 1.13 mm. *B. Diplarchitipula multimedialis* Rohd. (Architipulidae).
Wing. Length 4.45 mm. *C. D. destructa* Rohd. sp. n. Base of wing. Length of remnant 1.5 mm. Coll.
PIN No. 371/952. (*A. B.* according to Rohdendorf, 1962; *C.* original.)

medialis) or its middle (*D. destructa*). RS branches out from R at the end of the
first quarter of the wing and is divided at once beyond the middle of the wing,
sending off a single posterior branch, running to the point of the apex of the wing
and an anterior which soon again divides forming the anterior and middle branches
of RS. From R, beside the principal trunk of RS, there branch out some other
weak veins; sometimes one located proximally to RS and dividing soon into two
small free branches (*D. multimedialis*) or two proximal transverse veins (*D. destructa*)
Distally RS sometimes has a longitudinal branch running toward the apex of the
wing, almost to the level of the fork of RS (*D. multimedialis*). The three distal
branches of RS are parallel, thin at the end. The posterior branch of RS not far
from its base is connected by a sturdy straight transverse vein with M; furthermore there is a transverse vein between the middle and posterior branches of RS
and two vague transverse veins between the distal division of the posterior branch

of RS and the anterior branch of M. The medial system is abundantly branching;
the common trunk of M is very long, nearly straight and is accompanied by an
almost parallel MP that goes behind it. MA for the first time branches to the place
of connection with the radiomedial vein sending off to the posterior edge of the
wing a posterior branch which, bending in an arch, runs parallel with MP and
unites with it before the point where it goes into the border of the wing. At the
place of junction of MA with the radiomedial transverse vein from MA there
branches off backwards an oblique and sturdy transverse vein which unites with
the posterior branch of MA; a similar oblique transverse vein is arranged proximal
to the anterior fork of MA and unites the latter with MP. The anterior branch of
MA running to the apex of the wing forms a straight continuation of the main
trunk and forms some branches directed to the posterior border in a 'comb-like
sector'; in all there are no less than five (to eight?) distal branches of M. Arch-
like transverse veins isolate the intermedial cell. There is a large arch-like transverse
vein uniting the distal ends of CuA and MP. CuA is sturdy, considerably longer
than half of the wing, almost strictly straight, slightly bent at the apex; in front
of this vein, in the base of the wing, sometimes there is a weak parallel vein or
fold reaching only to the end of the first third (*D. multimedialis*). Behind, CuA
is closely accompanied as far as the distal bend by the vein CuP. The anal region
is small; sometimes there are two very weak branching veins (A_1 and A_2 *D. multi-*
medialis?), sometimes the anal veins are quite invisible (*D. destructa*). The base
of the wing is wide.

Comparison. – The location of SC, the presence of some branches of RS on R,
the well-expressed MP and the abundance of branches of M are very peculiar,
sharply distinguishing this form from all other representatives of the family: it is
possible that the genus *Diplarchitipula* comprises a special isolated group, a subfamily or maybe even a related family.

Composition. – The genus is described on the basis of two well distinguished
species.

Diplarchitipula multimedialis Rohdendorf, 1962 (fig. 51*B*)
Rohdendorf, 1962, p. 312, 313, fig. 979

Holotype. – Positive impression of left wing (base of wing, posterior border
and region of costal border with end of R and SC not preserved). Coll. PIN No.
358/645, Issyk-kul, Upper Triassic (Rhaetian?).

Description. – SC longer than half of wing, midway between C and R; anterior
branch of RS curved, drawn together with the middle branch in its proximal half;
transverse vein between middle and posterior branches of RS close to the centre
of the whole anterior fork of RS and located at the level of the fork of the anterior branch of M, parallel with the radiomedial transverse vein; two transverse
veins between the posterior branch of RS and the anterior of M are located distally from the level of the transverse vein between the branches of RS, parallel
with other transverse veins; in front of CuA there is a weak vein; anal veins weak
and inconspicuous. Length of remnant 3.45 mm, total length about 3.75 mm.

Material. – The holotype.

Diplarchitipula destructa Rohdendorf, sp. n. (fig. 51*C*)

Holotype. – Coll. PIN No. 371/952, Issyk-kul, Upper Triassic (Rhaetian?).

Description. – Wing apparently wider, SC shorter, about half the length of the wing; branches of R proximal to the base of RS in the form of weak transverse veins; in front of CuA there is no thin vein; anal field quite narrow without traces of veins. Length of remnant 1.50 mm, whole wing about 3 mm. This species differs quite highly from the preceding and more complete material would presumably show it to be a representative of another genus.

Material. – The holotype.

Genus *Architipula* Handlirsch, 1906
Handlirsch, 1906, p. 490

Type of genus: *Bibio seebachi* Geinitz, 1884 (Liassic of Germany).

Remarks. – The genus was reported for the first time on the basis of one form from Liassic deposits of northern Germany and according to the data of Handlirsch (1908, 1939) contains 24 species. In the fauna of the Upper Triassic of central Asia were discovered three species quite sharply distinguished from one another. The Liassic species of western Europe were described very superficially and presumably belonged in reality to a series of separate genera; the existing reports and descriptions of the wings are very inaccurate and do not allow the possibility of ascertaining the real system of these Diptera. The redescription of the types of Handlirsch is the only course for the determination of this question. The distinction of this genus from other Upper Triassic architipulids is indicated in the key; a more detailed description would logically be done after revision of Handlirsch's collections and special investigations of the wings of different groups of contemporary Limoniidae to which the genus *Architipula*, together with the genus *Protipula* Handlirsch and *Eotipula* Handlirsch are closest.

Key to the Upper Triassic species

1 (2)   SC before the end connects with R by an oblique transverse vein; base of M is not united with R; common trunk of RS less than half of common trunk of M; A_1 without fork . . . . *A. radiata* (p. 155)

2 (1)   SC not connected at the end with R; base of M either with short branch, directed forward; or right after the branching of this vein from CuA, M is connected by a short transverse vein with R; trunk of RS is always considerably more than half of common part of M from base to fork. A_1 with fork 3

3 (4)   Wing narrow, nearly four times as long as wide; general part of anterior and middle branches of RS more than twice the length of the part of RS from the fork to the radiomedial transverse vein; humeral transverse vein located proximal to the beginning of M; A_1 branches at the end . . . . *A. turanica* (p. 155)

4 (3)   Wing wider, only three times as long as wide; general part of
                    anterior and middle branches of RS almost equal to the part of RS
                    from the fork to transverse rm; humeral transverse vein is located at
                    the level of the beginning of M; A_1 branches at the middle . . . .
. . . . *A. asiatica* (p. 156)

Architipula radiata Rohdendorf, 1962 (fig. 52*A*)
Rohdendorf, 1962, p. 312, fig. 980

Holotype. – Almost complete impression of left wing (only the posterior edge
and part of the anterior border of the base not preserved). Coll. PIN No. 371/403,
Issyk-kul, Upper Triassic (Rhaetian?).

Description. – Wing wide, 2.8 times as long as the width, with comparatively
convex anterior edge; apex poorly separated, very large. R at the end has a fork
and three indstinct short branches, running to the edge; well-marked end of R
unites with an anterior branch of RS. End of R occupies a large pterostigmal spot.
The total length of the anterior and middle branches of RS is one-and-a-half times
greater than the length of RS from the fork to rm. The common trunk of RS is
arching back equal to 0.42 of the whole trunk of M which is almost straight. The
anterior and middle branches of RS gradually break up; the latter is almost parallel
with the posterior and at the end four branches of M meet with it; the intermedial
cell is pentagonal, sharply separated, an anterior branch of M gives a fork shorter
than its common trunk. The transverse mcu is clear although not especially thick.
Apparently A_2, stronger than A_1, does not reach to the margin of the wing; A_1
long and straight, tapering only at the very end and bent to the margin. Length
of wing is 5.9 mm, width 2.1 mm.

Material. – The holotype.

Architipula turanica Rohdendorf, sp. n. (fig. 52*B*)

Holotype. – Positive impression of left wing (venation of distal part is not clear).
Coll. PIN No. 371/1025, Issyk-kul, Upper Triassic (Rhaetian?).

Description. – Wing narrow, 3.8 times as long as wide; anterior border straight
to two-thirds of the extent with well-isolated apex. Structure of the end of R is
unknown. Pterostigmal spot is not developed. The total part of the anterior and
middle branches of RS more than three times greater than the part of RS from
the fork to the transverse rm. Common trunk of RS almost straight, it equals
0.63 of the common trunk of M which also is almost straight. All the branches
of RS are almost parallel. The veins of the system M are weak; the number of
distal branches of M is unknown, the proximal angle of the intermedial cell is
very acute. A transverse vein between the branches of M and mcu is vague. A_1
is weak and irregularly curved, forming a fork at its end. A_2 is more sturdy; it has
a well developed A_3 in the form of a short vein equal approximately to one-third
of A_2. Length of wing is 6 mm.

Material. – The holotype.

Architipula asiatica Rohdendorf, sp. n. (fig. 52C)

Holotype. – Positive impression of left wing (poorly preserved veining of distal part and greater part of anal blade). Coll. PIN No. 371/980, Issyk-kul, Upper Triassic (Rhaetian?).

Description. – Wing not particularly narrow, three times as long as wide with a straight anterior border and sharply isolated short apex. R quite straight, running to the edge of the wing, its branches are vague. Pterostigmal spot is not clear. General part of anterior and middle branches of RS only one-and-a-half times as long as the part from the fork of RS to the transverse vein rm. The common trunk of RS is irregularly convex backwards, equal to 0.53 of the common trunk of M which is noticeably convex forward. The structure of the branches of RS and M is poorly distinguished: apparently there are transverse veins between R and RS; there are apparently four branches of M. The intermedial cell has an acute proximal angle and distally is vague. The transverse mcu is almost straight. A_1 has a fork at its middle. A_2 is sturdy. Length of wing is 5.15 mm.

Material. – The holotype.

Superfamily Chironomidea Macquart, 1838
[n. transl. Malloch, 1917 (ex Chironomidea Macquart, 1838)]

It is characterized by the development of the phenomenon of costalization at the base of the already quite elongated wings. The only Triassic family is the Architendipedidae.

Family Architendipedidae Rohdendorf, 1962
Rohdendorf, 1962, p. 317

Description. – The elongated wings measure from 4.5 to 5 mm, approximately four times as long as wide, with nearly straight anterior border and characteristic venation which shows the special features of developing costalization. R at the base has a rudimentary phragma in the form of an outgrowth or transverse vein. The costal field is quite wide with straight transverse veins the number of which is variable. SC is thin and long, strongly drawn together with R, the latter is sturdy and reaches nearly four-fifths the length of the wing, with transverse veins or a short fork at the end. The branches RS and M to the border of the wing break up and become thinner; transverse veins are only in the middle part of the wing. Posterior edge has poor venation. Forks of RS are located in the middle of the wing and are not removed to its apex. M in the basal part reduced or very thin. The anal region is very large. CuA is sturdy and in its greater part is straight, running parallel with the thinner CuP. Besides transverse veins in the costal field, such may be present in the middle part of the wing between R and CuA. Type of family: genus *Architendipes* Rohdendorf, 1962.

Comparison and composition. – The general character of the structure of the wing is peculiar: its features combine archaic characteristics of different Dictyodipteridae, Dyspolyneuridae, Eopolyneuridae as well as traits of the younger

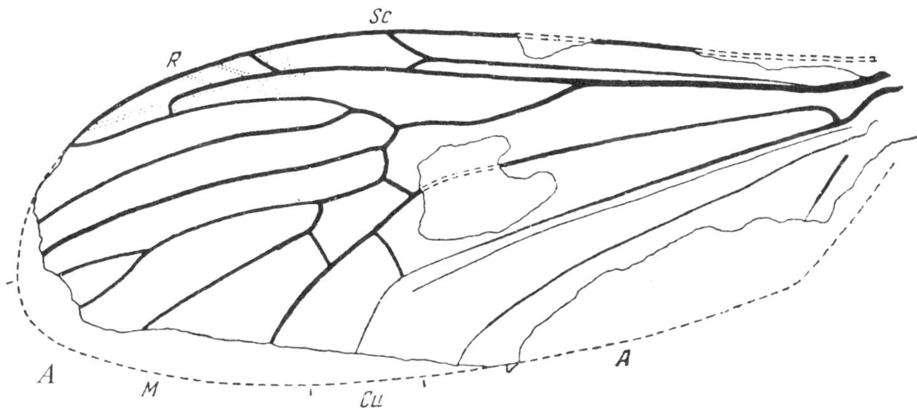

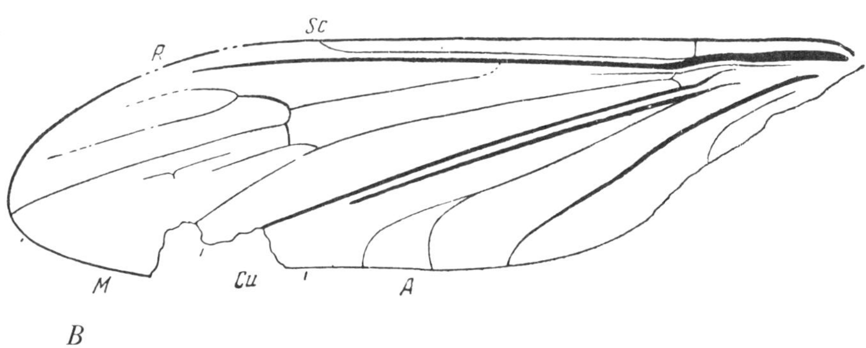

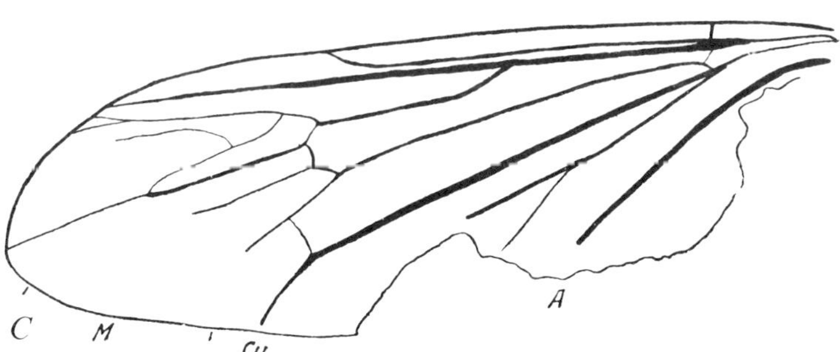

Fig. 52. Tipulidea of the Upper Triassic of central Asia. *A. Architipula radiata* Rohd., (Architipulidae). Wing. Length 5.9 mm. *B. A. turanica* Rohd. sp. n. Wing. Length 6 mm. Coll. PIN No. 371/1025 *C. A. asiatica* Rohd., sp. n. Wing. Length 5.15 mm. PIN No. 371/980. (*A.* according to Rohdendorf, 1962; *B. C.* original.)

progressive Chironomidae. This family is established by two genera — *Architendipes* Rohd. and *Palaeotendipes* Rohd., quite remote from one another and presumably representatives of separate families.

Key to the genera

1 (2)   Wing with numerous transverse veins in the costal and radial fields and between the branches of RS; there is a characteristic double pterostigmal spot of rectangular form; CuA and A_1 unite in front of the edge of the wing forming a closed sharp cubital cell; common trunk of M is in the form of a very thin vein between the sturdy RS and CuA .... *Architendipes* (p. 158)

2 (1)   Costal field with only one sturdy humeral transverse vein and short transverse veins near the end of SC; between the branches of RS and R transverse veins are almost lacking; pterostigma in the form of a small spot in the terminal fork of R; CuA and A_1 do not unite at the end and do not form a closed cell; the common trunk of M is reduced and from the medial system of veins there are only two well-preserved distal branches .... *Palaeotendipes* (p. 159)

Genus *Architendipes* Rohdendorf, 1962
Rohdendorf, 1962, p. 317

Type of genus: *A. tshernovskyi* Rohdendorf, 1962.

Description. — Wing is 3.8 times as long as wide with straight anterior border and an isolated small apex. SC very thin, lying close to R and noticeable only in the basal half of the wing, apparently united with R at the end. Costal field with transverse veins, gradually tapering to the end. R at the base connected with C and forms a break or rudimentary phragma. In the greater part of its extent R is straight, sturdy, gradually tapering, at the end free, ending in a double rectangular pterostigmal spot in which are placed three short transverse veins which connect the end of R with C. RS branches out from R at the beginning of the second quarter of the wing in the form of a sturdy vein, directed straight to the apex of the wing and just at the middle of the wing forks into anterior and posterior branches. The posterior branch constitutes a straight continuation of the common trunk of RS and runs rectilinearly to the point of the apex of the wing, producing at the end the main third branch forward, the continuation of which is not clear. The anterior branch is deflected forward, bent in an arch and with an indistinct fork before its end. Between the basal trunk of RS and R there are oblique transverse veins of which two are especially sturdy and well marked; besides transverse veins in this cell there is a weak, poorly-distinguished longitudinal vein or fold which branches from R distal to the branching of RS and which unites with the base of the anterior branch of RS. Between the branches of RS there are several delicate transverse veins. Before the fork of RS there is a sturdy radiomedial transverse vein bent in an arch and connecting with the last branch of M: besides this sturdy rm, distally and proximally from it, there are still two shorter transverse veins connecting with the weakly developed M which branches off from CuA somewhat proximally to the level of the branching of RS from R. The chief parts of the system of M lying proximally from the main radiomedial transverse vein in the form of a pair of weak longitudinal veins, partially form a vague and irregular network; distally to rm, M continues in the form of a simple weak vein to the last quarter of the wing where it divides into two diverging thin veins. The posterior branch of M which appears at the first glance as a continuation of the curved rm (in reality of the main remnant of the posterior branch of M!) is almost straight and gives in the vicinity of its end a short anterior little branch. From the base of the posterior branch of M there branches out an oblique sturdy transverse vein mcu. CuA is very sturdy with a sharply isolated basal division (basicubitus) after the place of connection with mcu bent in an arch to the border of the wing; before the end of this vein there forms a branch forward (similar to the same structure of the posterior branch of M!) which terminates freely in the membrane of the posterior medial cell. CuP is quite sturdy, its basal section is vague and with it is connected an anterior branch of A_1, the main trunk of which goes rectilinearly and unites with CuA before its end. The structure of the greater part of the anal blade is unknown.

Comparison and composition. — Sharply distinguished from the genus *Palaeotendipes* by the presence of numerous transverse veins in the costal and radial fields, sturdy posterior branch of M, closed cubital cell and other features. It is described from a single species, the type of the genus.

Architendipes tshernovskyi Rohdendorf, 1962 (fig. 53*A*) [9]
Rohdendorf, 1962, p. 317, fig. 999

Holotype. — Positive impression of left wing (a great part of the anal blade not preserved). Coll. PIN No. 358/197, Issyk-kul, Upper Triassic (Rhaetian?).

Description. — Costal field with five straight and two oblique transverse veins in the main part and four short, straight transverse veins in the pterostigmal zone; from the anterior branch of RS there arise seven transverse branches of which four unite with the posterior branch of RS and three with the middle. Length of wing 4.5 mm.

Material. — The holotype.

Genus *Palaeotendipes* Rohdendorf, 1962
Rohdendorf, 1961, p. 317

Type of genus: *P. alexii* Rohdendorf, 1962.

Description. — The ratio of the length of the wing to its width is unknown; anterior border irregularly curved with simple indentations in the base and at the level of the end of vein SC which is very delicate and running parallel with R and going into C; before its end SC connects with the edge of the wing by three short and oblique transverse veins. At the very base of the wing, proximal to the humeral

---

9. The species is named in memory of Alexi Alexeevich Tshernovsky who perished prematurely in Leningrad in 1942, the author of the first complete summary for the larvae of the Chironomidae.

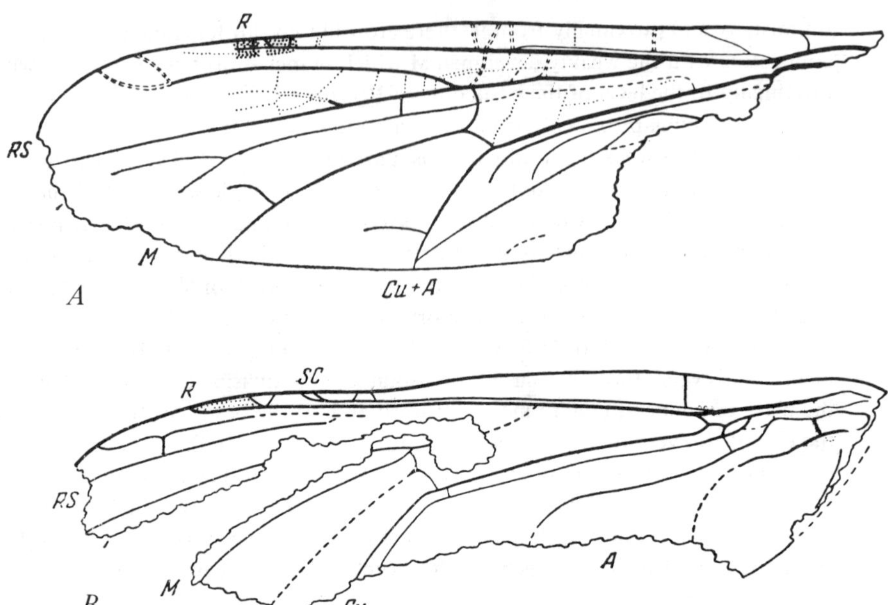

Fig. 53. Chironomidea of the Upper Triassic of central Asia. *A. Architendipes tshernovskyi* Rohd. (Architendipedidae). Wing. Length 4.5 mm. *B. Palaeotendipes alexii* Rohd. (Architendipedidae). Wing. Length 4.9 mm. (According to Rohdendorf, 1962.)

transverse vein, SC is concealed by a thick R. Costal field wide at the middle, narrowing toward the end and at the base; there is a sturdy and well-marked humeral transverse vein located at the end of the first quarter of the wing. R at the base is connected with C and is strongly thickened, forming a bend and phragma in the form of a transverse vein. For most of its length this vein is straight and is completed by a fork forming a pterostigmal spot. The place of branching of RS is not clear and apparently is located at the middle of the wing in the form of a weak vein which in the distal part of the wing forms three almost parallel branches. There is a sturdy transverse vein between the ends of the anterior and middle branches of RS. The common trunk of M is reduced and the definite medial vein is arranged in the form of branches of the posterior RS; the other, a very weak medial vein, branches out from the middle stout transverse vein which forms the base of the anterior branches of M and CuA. CuA is sturdy, located almost at the middle of the wing, nearly straight to the place of union with the transverse mcu and further bends obliquely to the border of the wing. Behind CuA is located a parallel thinner CuP which before the place of bending is united by a straight and thin transverse vein with CuA. The main division of the cubital and anal veins is complexly constructed and with difficulty yields to homologization. CuA is connected by a transverse (phragmal) vein with R and by an oblique irregularly thickened vein with CuP which, in its turn, is closer to the base of the wing and combines with CuA, forming the common trunk of Cu. A_1 is irregularly curving and connecting with CuP by a straight transverse vein. A_2 is sturdy, straight at the base and thin, arch-like at the middle of the anal division. A_3 is noticeable in the form of a transversely curved line.

Comparison and composition. – Well-distinguished from the genus *Architendipes* by features indicated in the key. This genus approaches the family Chironomidae owing to the greatly advanced costalization of the venation. It is described on the basis of a single species, the type of the genus.

Palaeotendipes alexii Rohdendorf, 1962 (fig. 53B)[10]
Rohdendorf, 1962, p. 317, fig. 1000

Holotype. – Positive impression of left wing (posterior border, apex and greater part of the medial field not preserved). Coll. PIN No. 358/195 Issyk-kul, Upper Triassic (Rhaetian?).

Description. – In the subcostal cell there is one thin transverse vein located near the anterior part of R; anterior and middle branches of RS approximate, posterior branch of RS more removed; anterior branch of M located at the middle between the last of RS and the posterior branch of M which, in its turn, lies exactly at the middle between anterior branches of M and CuA. Length of wing is 4.88 mm.

Material. – The holotype.

Superfamily Rhaetomyiidea Rohdendorf, 1962
Rohdendorf, 1962, p. 318.

Description. – Characterized by the nearly complete reduction of the anal veins, by a sturdy and clear phragma, comparatively moderately elongated wing and well-isolated basal cells. There is a single family, the Rhaetomyiidae.

Comparison. – This peculiar representative of the tipulomorphs is very isolated from other groups of the infraorder; the structure of the wings of *Rhaetomyia* was the most mechanically perfect of all the Upper Triassic Diptera.

Family Rhaetomyiidae Rohdendorf, 1962
Rohdendorf, 1962, p. 318

Description. – Moderately elongated wings, in size about 3.5 mm, the length of which is approximately 2.2 times the greatest width with very weakly convex anterior border and small apex. R with sturdy almost transverse phragma, united with CuA. Costal field wide with only one sturdy humeral transverse vein. SC is sturdy, drawn together with R and reaching only to the middle of the wing. R reaches the apex of the wing and before the middle of the wing there branches off RS, constructed typically for tipulomorphs. The main trunk of M is noticeably thinner than R and CuA. The transverse veins rm and mcu are sturdy and straight. The apical part of the branches of RS and M are noticeably diminished from their basal sections. The forks of RS and M and the transverse veins rm and mcu are located approximately at the middle of the wing; other transverse veins

---

10. The species is named in memory of Alexi Alexeevich Tshernovsky who perished prematurely in 1942.

are very weak and vague. CuA is straight and sturdy and accompanies the weak parallel CuP; it equals approximately half of the wing. The type of the family is the genus *Rhaetomyia* Rohdendorf, 1962.

Comparison. – General character of the wing structure is very peculiar, suggesting some other tipulomorphs, for example the Orphnephilidea or even the distant asilomorphs.

Genus *Rhaetomyia* Rohdendorf, 1962

Rohdendorf, 1962, p. 318

Type of genus: *R. necopinata* Rohdendorf, 1962.

Description. – Moderately elongated wing with weakly convex anterior border. SC sharply drawn together with R, which is divided at the end into two branches and which is connected by a weak transverse vein with R. The latter vein is straight in its greater part, weakly curved before the apex, with a fork at the end. RS branches off from R before the fork of SC and divides into anterior and posterior branches just distally to the end of SC. The common trunk of the anterior branch of RS is located parallel with R and forms two branches, anterior and middle, of which the latter is parallel with the posterior branch of RS and runs together with it to the apex of the wing. The anterior branch of RS is curved and goes into the anterior edge by means of some secondary forks. The posterior branch of RS is weakly curved and directed to the apex of the wing; with the base of this branch of RS is united a straight and sturdy rm, the other end of which is united with the base of the anterior branch and M. The thin main trunk of M branches out from CuA right after the union with the last phragma of R and divides into two branches approximately at the level of the fork of SC. The two forks of M are almost parallel, very weakly divergent. The anterior fork of M is divided into two weak branches; the posterior branch of M is simple, almost straight. The intermedial transverse vein is very weak. The transverse vein mcu is sturdy; at the point of connection with this transverse vein the posterior branch of M forms a sharp break. CuA is very sturdy and to the connection with mcu almost completely straight; its distal portion is directed obliquely to the posterior border, irregularly curved and gradually tapering to the end. CuP thin and parallel with CuA reaching to the middle of the distal section of the transverse vein. The anal veins have the appearance of weak, thin and short veins. A_1 is very short; A_2 is longer. The anal blade is wide, larger in area than both basal cells together. This is described on the basis of a single species, the type of the genus.

Rhaetomyia necopinata Rohdendorf, 1962 (fig. 54)

Rohdendorf, 1962, p. 318, fig. 1006

Holotype. – Positive impression of right wing of good preservation (only part of the anal blade not preserved). Coll. PIN No. 358/76, Issyk-kul, Upper Triassic (Rhaetian?).

Description. – Posterior branch of RS branches off just at the middle between

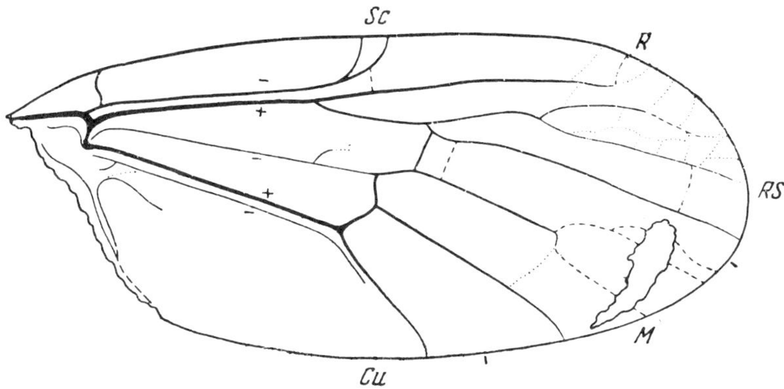

Fig. 54. Rhaetomyiidea of the Triassic of central Asia. *Rhaetomyia necopinata* Rohd., (Rhaetomyiidae). Wing. Length 3.44 mm. (According to Rohdendorf, 1962.)

the branching of RS from R and the anterior fork of RS. Before the end of R there are very thin transverse veins; between the end of R and the anterior branch of RS one, between the posterior branch of RS and the common trunk of the anterior fork of RS two, between RS and M distal to rm are three very thin transverse veins. Furthermore, there are poorly distinguished transverse veins which form almost a network of veins between the anterior and middle branches of RS. The anterior fork of the middle branch of M is divided into two thin branches. Length of wing is 3.44 mm, the width 1.57 mm.

Material. – The holotype.

Infraorder Bibionomorpha

Description. – Wings moderately elongated, usually with convex anterior border. RS forms a distinguishable number of branches which, in turn, generally do not branch and are always directed forward to the anterior border; sometimes the branches of RS are completely absent. The basial, as a rule sharply isolated, is usually without a phragma. Transverse veins are scarce and always absent from the anal blade and costal cell. The size of the wings in the Upper Triassic species are from 1.5 to 5 mm.

Comparison and composition. – A vast infraorder of Diptera including in the Upper Triassic no less than five superfamilies of which three are ancient, extinct groups and two, the Fungivoridea and Rhyphidea, are living until the present epoch. These Diptera are characterized by the process of reduction of the wings, more accurately by early development of the lifting flight which is expressed by a decrease of venation, expansion of the blade and the development of costalization. The extinct superfamilies of the infraorder are very distinct in their phylogenetic importance. The first superfamily, Pleciodictyidea, shows very primitive features and clearly connects all the bibionomorphs with the ancient dictyodipteromorphs. Two other extinct superfamilies, Protoligoneuridea and Phragmoligoneuridea, are apparently specialized forms as shown by the peculiar venation of the wings. The two main superfamilies, Fungivoridea and Rhyphidea, have the greatest importance; they turned out to be the original forms each for different younger groups of Diptera. Thus the superfamily Rhyphidea turned out to be the origin for all 'Brachycera', i.e., of the infraorder Asilomorpha and its derivatives. The phylogenetic relations of this infraorder are set forth in detail on p. 295.

Superfamily Pleciodictyidea Rohdendorf, 1962
Rohdendorf, 1962, p. 319

Description. – Known only from the wing of a moderately elongated form with very weak veining of the middle part; only R and CuA are sturdy. Between M and CuA there are weak transverse veins. The system of M is very weak but undoubtedly rich, with an isolated posterior branch ('M_4'). There are no phragmata.

Comparison. – Undoubtedly this group is the most archaic of the whole infraorder of bibionomorphs, close to the original forms of the dictyodipteromorphs. There is a single family, the Pleciodictyidae.

Family Pleciodictyidae Rohdendorf, 1962
Rohdendorf, 1962, p. 319

Description. – SC thin but reaching to the middle of the wing. R with break at the base. RS branches out from R at the end of the first third of the wing and sends a series of thin branches forward, the greater part of which have the appearance of transverse veins between RS and R. Medial vein very thin and consists of a series of branches, in succession and dichotomously divided. MP is present but short and rudimentary. M_4 (more correctly MA_4) unites by transverse veins with CuA and gives a branch backwards, not far from the end of CuA. CuP joins with CuA but does not reach the bend of the latter. The structure of the anal veins is unknown; if they were present then they were very thin and delicate.
Type of family: the genus *Pleciodictya* Rohdendorf, 1962.

Genus *Pleciodictya* Rohdendorf, 1962
Rohdendorf, 1962, p. 319

Type of genus: *P. modesta,* 1962.

Description. – Anterior border of wing straight to the middle; length of wing is 2.5 times as great as the width; posterior border moderately convex, apex large, poorly isolated. SC is weak and straight, not drawn together with R, at the end apparently with a fork. R very sturdy at the base; after a break it is straight, gradually tapering to the apex. RS branches out from R proximally to the middle of R and is located parallel with this vein; between them are found three branches of RS in the form of curved and branching transverse veins. From the distal half of RS two weakly developed branches go off. There are not less than three rm veins. The forks of M are very long although consisting of weak veins; before the fork

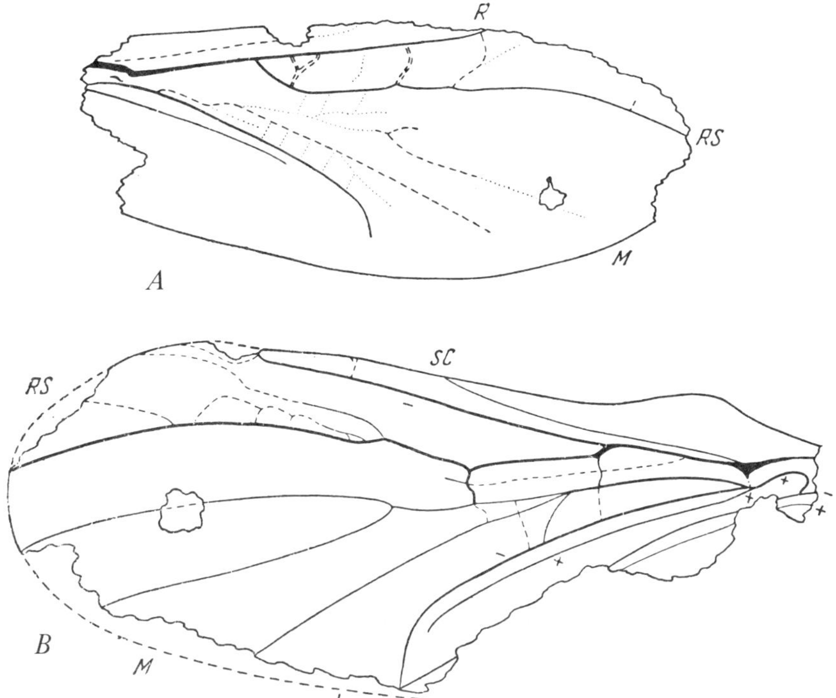

Fig. 55. Pleciodictyidea and Protoligoneuridea of the Triassic of central Asia. *A. Pleciodictya modesta* Rohd.
(Pleciodictyidae). Wing. Length 2.2 mm. *B. Protoligoneura fusicosta* Rohd. (Protoligoneuridae). Wing. Length
2.5 mm. (According to Rohdendorf, 1962.)

of M is a weak vein noticeable only in its basal portion with which two rm veins
unite. The posterior branch of M is nearly perfectly straight, connecting by many
transverse veins with CuA; the latter vein is curved, shorter than half of the wing,
at its apex bent to the posterior border. The anal blade is large but not convex
along the border. The anal veins are not clear.

Composition. — It is described on the basis of a single species, the type of the
genus.

Pleciodictya modesta Rohdendorf, 1962 (fig. 55A)
Rohdendorf, 1962, p. 319, fig. 1008

Holotype. — Positive and negative impressions of left wing (the greater part of
the anterior border, apex and base of the anal region not preserved). Coll. PIN No.
358/919, Issyk-kul, Upper Triassic (Rhaetian?).

Description. — Between M_4 and CuA there are not less than five transverse veins.
Length of the wing is 2.2 mm.

Superfamily Protoligoneuridea Rohdendorf, 1962
Rohdendorf, 1962, p. 319.

Description. — Only the wing is known. The moderately elongated wings have
a characteristic, quite sturdy venation: the anterior border has a sharp protuberance;
R has thin phragma and thicker part. There are two sturdy rm veins forming a peculiar cell parallel with the edge. The anal veins are sturdy and long, drawn close to
one another.

Comparison and composition. — This superfamily is related most closely to the
Fungivoridea and Bibionidea, well distinguished by the indicated peculiar features;
they are apparently a special trunk of the bibionomorphs in the representatives of
which there developed very early important mechanical improvements of the wing,
(phragma, costal protuberance, central cell). There is a single family, the Protoligoneuridae.

Family Protoligoneuridae Rohdendorf, 1962
Rohdendorf, 1962, p. 319

Description. — SC thin but sturdy, reaching to the middle of the anterior border.
R at the base with a swelling and thin phragma. RS branches out from R at the
end of the first quarter of the wing and sends off from its distal part numerous
and irregular oblique branches to the anterior border (but not to R). Medial veins
quite sturdy, to the number of three branches. CuA and CuP are parallel but moderately drawn together. A_1 and A_2 drawn together and almost parallel. A_1 before
the edge of the wing unites with CuA forming a closed cubital cell. Between RS,
M and Cu in the middle part of the wing there are numerous but quite thin transverse veins. It is described according to the single genus *Protoligoneura*, Rohdendorf, 1962, which is the type of the family.

Genus *Protoligoneura* Rohdendorf, 1962
Rohdendorf, 1962, p. 319

Type of genus: *P. fusicosta* Rohdendorf, 1962.

Description. — The protuberance of the anterior edge of the wing is equal
approximately to half of the length of SC. The length of the wing is 2.6 times as
great as the width; the apex is large, isolated. SC is thin, almost straight, going
into C at the middle of the edge of the wing, at the base drawn together with R
and towards the end gradually branching out to C. R at the base convex in front
not particularly thick, dilated only at the level of the phragma, further on almost
straight; to the place of branching of RS it is sturdy; before its end this vein is
sharply bent to the anterior border. RS at the base very sturdy and between both
transverse rm veins almost straight; distally the branchings of the latter rm are
quite sharply bent back to the anterior edge and separate their first, the clearest,
oblique branch; the place of division of the anterior branch is located approximately at the level of the middle between the ends of SC and R and also is marked
by the noticeable break of the chief trunk of RS. The anterior branch of M forms

the fork usual for bibionomorphs approximately at the level of the branching of
the first branch of RS; the common trunk of the fork (more distal than rm) is
many times shorter than the branches. The posterior branch of M branches out
from the common trunk of M between the two transverse rm veins; immediately
before the branching of the posterior branch of M, between the common trunks
of M and CuA, is located an oblique, quite sturdy transverse mcu besides which
there are also special straight mcu veins. Between the base of RS and the common
trunk of M there is a weak longitudinal vein (fold?), cutting the two transverse
rm veins in half and branching out from the base of R, like a rudimentary RS.
This is described according to a single species, the type of the genus.

Protoligoneura fusicosta Rohdendorf, 1962 (fig. 55B)
Rohdendorf, 1962, p. 319, fig. 1009

Holotype. – Negative impression of right wing (almost all of the anal blade,
part of the posterior edge and some portions of the anterior edge in the region
between the ends of R and RS are not preserved). Coll. PIN No. 358/164, Issyk-
kul, Upper Triassic (Rhaetian?).

Description. – Between the terminal sections of R and C, approximately at the
middle between the ends of SC and R there is a thin straight transverse vein; in the
field between RS and R there are comparatively numerous branches of RS, partly
reaching the edge of the wing, partially however ending freely in the membrane.
Both sturdy transverse rm veins curving; between them are irregular, vague transverse veins (not shown in the illustration). In the fork between the posterior branch
and the common trunk of M there are thin transverse veins. Length of the wing
is 2.5 mm.

Material. – The holotype.

Superfamily Fungivoridea Latreille, 1809
[n. transl. Rohdendorf 1962, (ex Fungivorae Latreille, 1809)]

One of the chief superfamilies of the infraorder, richly developed in the later
faunas including the contemporary. They are descendants of forms related to the
Pleciodictyidea. The characteristics are given above (see p. 65 and 136). In the
Upper Triassic fauna it is represented by two families, Pleciofungivoridae and
Palaeopleciidae.

Family Pleciofungivoridae Rohdendorf, 1946
Rohdendorf, 1946, p. 51

Remarks. – Representatives of this family were described for the first time
from the Middle Jurassic of Karatau. Reported below are numerous Upper Triassic
representatives of the family. Most of them differ from Karatau forms, on the one
hand in the poorer aerodynamic qualities of the wings (the presence of thin and
irregular transverse veins in species of the genera *Rhaetofungivora, Rhaetofungivorella, Archipleciofungivora, Protallactineura, Archihesperinus*), and on the other

hand in a series of traits which characterize other later groups (genus *Archiplecio-mima* reminding us, according to the form of the wing, of the Karatau Pleciomimi-dae; g. *Protallactoneura* approaches, according to venation, the Allactoneuridae). All this demonstrates the important meaning of this group of fungivorids, undoubtedly being the original family for later Pleciomimidae: Allactoneuridae and others. Altogether in the Upper Triassic fauna of central Asia there were discovered eight genera of this family including 27 species.

Key to the Triassic genera

1 (2)   Wing short and wide, its length not more than twice as great as the width; a single branch of RS has the form of a sturdy, almost straight transverse vein which connects with R at right angles. Other transverse veins besides the sturdy rm are absent. Dimensions small — length of wing about 1.9 mm (fig. 64*B*) . . . . *Archipleciomima* (p. 195)

2 (1)   Wing usually considerably longer: its length at least two-and-a-half times as great as width, usually even greater, rarely the wing is very short (*Rhaetofungivora curta*, sp. n.); sturdy branches of RS as a rule located obliquely and never connected with R at right angles. Some-times many transverse veins are present 3

3 (6)   SC at the end with branches or short transverse veins, between the last branch of M and CuA usually there is an oblique transverse vein 4

4 (5)   There is only one long, not particularly weak, curved branch of RS, reaching from C nearly to the end of R; distally to this branch within the cell between the end of RS and C; transverse or small branches of RS are absent. Length of wing upwards of 2.5 mm (fig. 63*E*) . . . . *Archihesperinus* (p. 192)

5 (4)   Number of branches of RS indefinite: they branch out from the chief trunk of RS through all of its extent in the form of curved, short transverse veins to the anterior border; length of wing less than 2 mm (fig. 62*A*) . . . . *Rhaetofungivorella* (p. 185)

6 (3)   SC at the end without branches or transverse veins; oblique transverse vein between the last branch of M and CuA is only in some species of the genus *Rhaetofungivora* 7

7 (8)   The end of SC goes into C considerably distally to the level of the transverse rm, reaching the base of the first branch of RS; between the chief trunks of R and M there is a weak longitudinal vein or fold; between the last branch of M and CuA there are oblique transverse veins (fig. 63) . . . . *Rhaetofungivorodes* (p. 189)

8 (7)   As a rule SC is much shorter, never reaches the level of the base of the anterior branch of RS and only seldom reaches somewhat behind the level of rm; between the main trunks of R and M there are no longi-

tudinal weak veins; oblique transverse veins between M and CuA are
only in some species of the genus *Rhaetofungivora* 9

9 (10)   Branches of RS very weak, sharply differing according to size
from the main vein RS, having the form of an irregular, poorly-distinguished network; the posterior branch of M is segregated from the main
trunk of M and is combined with CuA (fig. 64*A*)   . . . .
. . . . *Protallactoneura* (p. 193)

10 (9)   Branches of RS of diverse structure, often sturdy: if they are
vague then the posterior branch of M always branches out from the
main trunk of M 11

11 (12)   In the costal field there is a sturdy transverse vein located somewhat proximally to the place of branching of RS from R; two clear,
closely located rm veins; there is a single, quite short and sturdy branch
of RS, going into C distally to the end of R . . . .
. . . . *Archipleciofungivora* (p. 191)

12 (11)   In the costal field there are no transverse veins; the number of
branches is different; if there is a single ordinary sturdy branch of RS,
then it is disposed obliquely   13

13 (14)   Wing moderately elongated, its length is 2.5 times as great as
the width; venation consists only of clear veins; there is one sturdy
oblique branch of RS and a single thick transverse rm (fig. 64*C*) . . . .
. . . . *Palaeohesperinus* (p. 196)

14 (13)   Wing broader, its length at the most only 2.2 times as great as
the width; side by side with sturdy and clear veins, almost always there
are weak and thin ones; if there are almost no thin veins in the wing
then rm veins are two in number or the number of branches of RS is
always greater than one (fig. 56) . . . .
. . . . *Rhaetofungivora* (p. 169)

Genus *Rhaetofungivora* Rohdendorf, 1962

Rohdendorf, 1962, p. 320

Type of genus: *R. reticulata* Rohdendorf, 1962

Description. – The wing is irregular in form. The anterior edge of the wing is
straight in the main third or half and further uniformly convex to the apex. The
size of the wings are from 1.50 to 3.25 mm. SC thin but well marked, of variable
length, reaching the level of the branching of RS or reaching with its end beyond
the level of rm. R equal to two-thirds or three-quarters the length of the wing, of
variable form. Main section of R arched, sharply isolated from the distal part by
an abrupt break. The main trunk of RS more or less curved, equal in thickness
to R or somewhat thinner: the character of the branches of RS are very distinguishable in different species – sometimes there is only one rudimentary branch (*R.
simplex* sp. n.); more often there are two (many species) or even a whole network
of irregularly joined veins which form a similarity to a network (group *R. reticulata*,
sp. n. and related species). Transverse rm veins are usually two in number, some-

times three and more; more rarely there is only one rm but in this case there are
always present different thin transverse veins between the branches of R, M, Cu
or the thin branches of RS.

Comparison and composition. — This genus includes a relatively large number
of species and is the most archaic according to the structure of the wings. We are
convinced of this by the presence of numerous branches of RS in many species,
especially the occurrence of peculiar transverse veins between M and Cu, and
finally by the multiplicity of transverse rm veins. All this allows us to consider
the genus *Rhaetofungivora* as the original group for all other forms of the family
Pleciofungivoridae and presumably of some other related forms (for example, the
Pleciomimidae). It is still necessary to note the presence of distinct species groupings, apparently of subgeneric character. Such for example, are the species related to
R. reticulata, which are characterized by rich venation and the species close to *R.
major* that have a peculiar paired rm; finally there are forms with connected mechanized venation as, for example, *R. amasioides*. However, in the present investigation
no attempt is made to determine the intraspecific system of the *Rhaetofungivora*
as a result of little study of these insects; it is quite evident that up to now there
are still very few known species of this genus, the number of which was undoubtedly large.

Key to the species

1 (8)   There is only one sturdy and clear rm; sometimes transverse
                 veins are present between RS and M in the distal part of the wing . .
                 2

2 (3)   Between M_3 and CuA there are no transverse veins; there is
                 only one quite clear branch of RS; SC long, reaching the level of rm;
                 common trunk M_1+M_2 long (fig. 61*B*) . . . .
                 . . . . *R. amasioides* (p. 185)

3 (2)   Between M_3 and CuA there is at least one transverse mcu. RS
                 with numerous partly short branches 4

4 (5)   Between M and Cu there is only a single transverse mcu, which
                 looks like the base of M_3, while the true base of this vein is a sharply-
                 isolated transverse vein; one of the branches of RS is in the form of a
                 quite sturdy S-shaped curved vein (fig. 59*D*) . . . .
                 . . . . *R. magniradius* (p. 179)

5 (4)   Between M and Cu there are not less than four transverse veins
                 and M_3 branches out from the common trunk of M; the branches of RS
                 are thin and irregularly bent 6

6 (7)   Between M_2, M_3 and CuA there are many sturdy transverse
                 veins; M_3 is a direct continuation of the common trunk of M; between
                 the distal ends of RS and M_1 there are two transverse veins; the fork
                 of M_1+M_2 is sharply diverging (fig. 58*C*). . . . .
                 . . . . *R. perreticulata* (p. 177)

7 (6)   Between M_2 and M_3 there are no transverse veins; between M_3
                 and CuA at the level of the bend of the latter there is a pair of long

transverse veins; M_3 branching out from the common trunk of M by means of a very short single section, with which is connected an oblique transverse mcu; between RS and M_1 besides the single sturdy rm there are no other transverse veins; the forks of M_1+M_2 are in the form of parallel branches (fig. 57) . . . . *R. mediicubitalis* (p. 173)

8 (1)   Transverse radiomedial veins never less than two; sometimes there are three or more of them 9

9 (20)   In the fork of R-RS proximal to the level of the transverse vein there is one, more rarely two or three sharply isolated transverse veins which enclose a characteristic triangular cell 10

10 (11)   SC long, its end goes into C approximately at the middle of the anterior border of the wing, two straight, close rm veins considerably more distal; wing very broad, the length of it less than twice the width (fig. 59C) . . . .*R. curta* (p. 179)

11 (10)   SC always far short of the middle of the anterior border and the level of the transverse rm veins 12

12 (13)   Between R and RS there is a large number of irregularly situated mutually connecting quite thin veins, branches of RS or transverse veins; rm number three or two sturdy straight veins not especially distant one from another (fig. 56) . . . . *R. reticulata* (p. 173)

13 (12)   Between R and RS besides short transverse veins in the very fork there are two or three curved branches of RS; rm usually a large number, rarely only three veins which are strongly removed one from another 14

14 (15)   The first branch of RS strongly converges with the end of R, going into C immediately side by side with R; there are no anterior branches of R; the transverse fork of R+RS very sturdy; in all there are three branches of RS; rm veins are straight, located distally to the level of the end of SC (fig. 59A) . . . . *R. radialis* (p. 179)

15 (14)   The first branch of RS parallel with R more rarely barely joining with it; often there are anterior branches of R; two branches of RS 16

16 (17)   Besides a sturdy middle rm there is still only one rm near the base of RS (proximally to the level of the end of SC) and a third rm in the distal part of the wing; anterior ends of the branches of RS approach each other; R without anterior branches; M_3 and base of the anterior branches of M are weak (fig. 58B) . . . . *R. destructimedia* (p. 177)

17 (16)   Transverse rm veins drawn together; ends of the branches of RS not drawn together at the border of the wing; R near the end with short anterior little branches (transverse) 18

18 (19)   Between RS and M there are not less than six parallel rm veins of different size, moreover the majority of them are located in the middle part of RS; from R two thin anterior branches go off; wing more than 2.25 mm, comparatively elongate (fig. 58*A*) . . . . *R. radiimedialis* (p. 174)

19 (18)   There are four rm veins located at the level of the main part of RS; these veins are of different position, the greater part oblique (fig. 59*B*); from R there branches off only a short pre-apical transverse vein; wing less then 2 mm, relatively wide . . . . *R. quadrimedialis* (p. 179)

20 (9)   In the fork of R-RS there are no transverse veins and a triangular cell is not separated 21

21 (26)   Two or three transverse rm veins, closely drawn together; distance between them less than the length of a separate vein 22

22 (23)   SC short, reaching only to the level of the base of RS; between the common trunk M_{1+2} and M_3 there is a sturdy transverse vein; two adjacent rm veins; wing comparatively large, about 3 mm (fig. 59*E*) . . . . *R. maxima* (p. 181)

23 (22)   SC always longer, the end of it noticeably distal to the level of the base of RS; there are no transverse veins between the branches of M 24

24 (25)   Two or three transverse rm veins, drawn together, irregularly bent; end of SC does not reach the level of rm; length of wing 2.5 mm (fig. 60*A, B*) . . . . *R. magna* (p. 181)

25 (24)   Two transverse rm veins moderately drawn together, parallel; end of SC combines with C distal to the level of rm; length of wing 2.3 mm (fig. 60*C*) . . . . *R. subcostalis* (p. 183)

26 (21)   Transverse rm veins not drawn together: distance between them greater than the length of the individual rm veins 27

27 (28)   Size small, wing not greater than 1.8 mm; SC short, going into C approximately at the level of the branching of RS; three rm veins of which the distal is sturdy and straight and the two proximal weak and oblique; rm closest to the base of the wing branches out almost from the place of branching of RS from R; between the common trunk of M and CuA there is a thin but clear transverse vein (fig. 58*D, E*) . . . . *R. parva* (p. 177)

28 (27)   Size large, not less than 2.1 mm; SC long; two rm veins moderately drawn together; between the common trunk of M and CuA there are no transverse veins 29

29 (30)   The part of RS between the base and the first rm vein is twice as large as the part of this vein between the two rm veins; between the base of M_3 and CuA there are no transverse veins (fig. 60*D*) . . . . *R. simplex* (p. 183)

30 (29)   This part of RS only one-and-a-quarter times as long as the distance between the two rm veins; from the main trunk of M_3 going to
CuA is a thin transverse vein (fig. 61A) ....
.... $R.$ *major* (p. 184)

Rhaetofungivora reticulata Rohdendorf, 1962 (fig. 56A, B)
Rohdendorf, 1962, p. 320, fig. 1011

Holotype. — Right wing, Coll. PIN No. 358/123, Issyk-kul, Upper Triassic
(Rhaetian?).

Description. — SC short, not reaching the level of the distal rm; thin, going into
C or into R (specimen No. 371/998). In the fork of R+RS, is a sturdy transverse
vein; three clear rm veins; sometimes the proximal rm is removed to the base of
the wing and is located between the main trunks of R and M (No. 371/998); between R and RS is a continuous network of numerous weak branches, of RS and
of transverse veins; distally to sturdy rm veins in this same field there are weak
transverse veins and branches of RS and M_1; between M_{1+2} and M_3 there is a
quite clear straight transverse vein and weak curved veins (No. 358/123); in another
specimen in this field there is only a weak longitudinal vein; between M_3 and CuA
are two or three (No. 371/998) weak transverse veins and one of them is disposed
very obliquely, forming part of the base of vein M_3 (No. 371/998). The length of
the wing is 1.75 mm in No. 358/123 and 1.9 mm in No. 371/998.

Comparison. — Is described according to the remnants of two wings differing
markedly from one another on the basis of veining; the comparative resemblance
in general plan of structure compelled us temporarily to refrain from the determination of a second species. This species is closest to *R. mediicubitalis* sp. n.

Material. — In addition to the holotype there is a second wing, Coll. PIN No.
371/998, Issyk-kul, Upper Triassic (Rhaetian?).

Rhaetofungivora mediicubitalis Rohdendorf, sp. n. (fig. 57 A to D)

Holotype. — Wing, Coll. PIN No. 358/98, Issyk-kul, Upper Triassic (Rhaetian?).

Description. — SC of average length, reaching or not reaching to the level of
the single, quite sturdy rm, without branches; in the fork of R+RS there are no
sturdy transversely distributed veins; rm is straight, removed from the base of
RS; between R and RS is a complex network of irregularly disposed branches of
RS of which only a few are clearly distinguishable; sometimes there is a weak
anterior branch of R (No. 371/1017), and an anterior branch of the main trunk
of M (No. 371/1015) or M_1 and M_2 (No. 371/1017); there is a weak longitudinal
vein between M_{1+2} and M_3, rarely it is absent (No. 358/98); between M_3 and CuA
there are from three to five transverse veins of different size and position; vein M_3
branches out from M in the form of a direct continuation of the common trunk
of M or even by means of a short transverse vein (No. 358/98). Length of the wing
is from 2.25 to 2.75 mm.

Comparison. — Is described according to three remnants differing somewhat
from one another; the differences of these fossil insects are evaluated in the quality

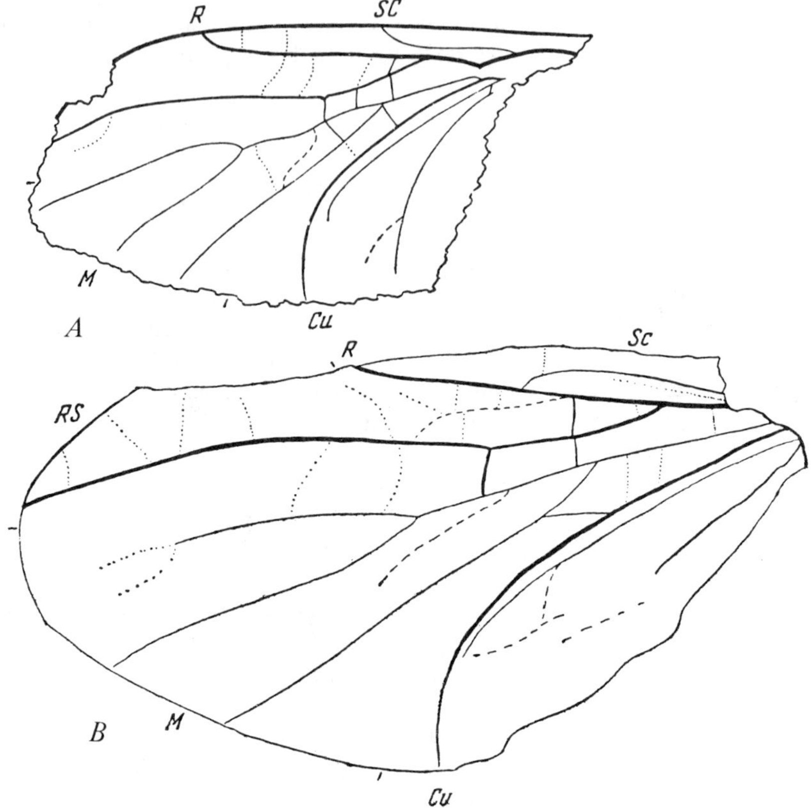

Fig. 56. Pleciofungivoridae of the Triassic of central Asia. *A. Rhaetofungivora reticulata* Rohd. Holotype. Length about 1.9 mm. *B.* The same, paratype, Coll. PIN No. 358/998. Length 1.9 mm. (*A.* according to Rohdendorf, 1962; *B.* original.)

of the individual variability of one form which is to a known extent a preliminary solution. The species is closest to the preceding one.

Material. − Besides the holotype there are two wings, No. 371/1015 and No. 371/1017, Issyk-kul, Upper Triassic (Rhaetian?).

Rhaetofungivora radiimedialis Rohdendorf, sp. n. (fig. 58*A*)

Holotype. − Wing, Coll. PIN No. 358/190, Issyk-kul, Upper Triassic (Rhaetian?).

Description. − SC of medium length, not reaching the level of rm; R with two weak anterior branches of which the distal has the form of a short transverse vein; the fork of R+RS is limited to a sturdy transverse vein besides which there are still two more weak transverse veins; there are two branches of RS of which the proximal is longer and curved; rm veins are not less than seven in number; two proximal rm veins are the most sturdy, parallel and highly drawn together; the structure of the medial and cubital systems are unknown. Length of the remnant

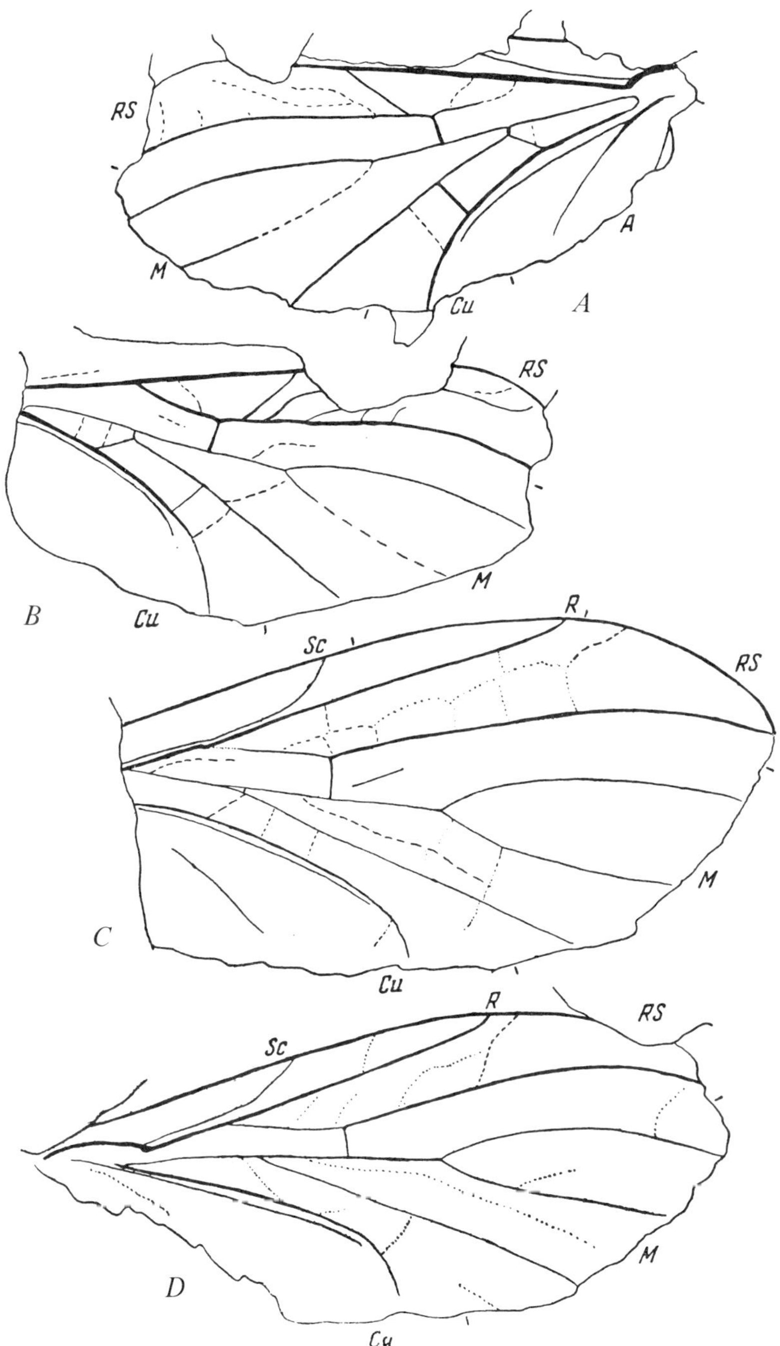

Fig. 57. *Rhaetofungivora mediicubitalis* Rohd. sp. n. (Pleciofungivoridae). Upper Triassic of central Asia, Variability of venation. *A*. Holotype, Coll. PIN No. 358/98. Wing. Length 2.75 mm. Negative impression. *B*. The same, positive impression. *C*. Paratype, Coll. PIN No. 371/1015. Length of impression 2.2 mm. *D*. Paratype, Coll. PIN No, 371/1017. Length of impression. 2.25 mm. (Original.)

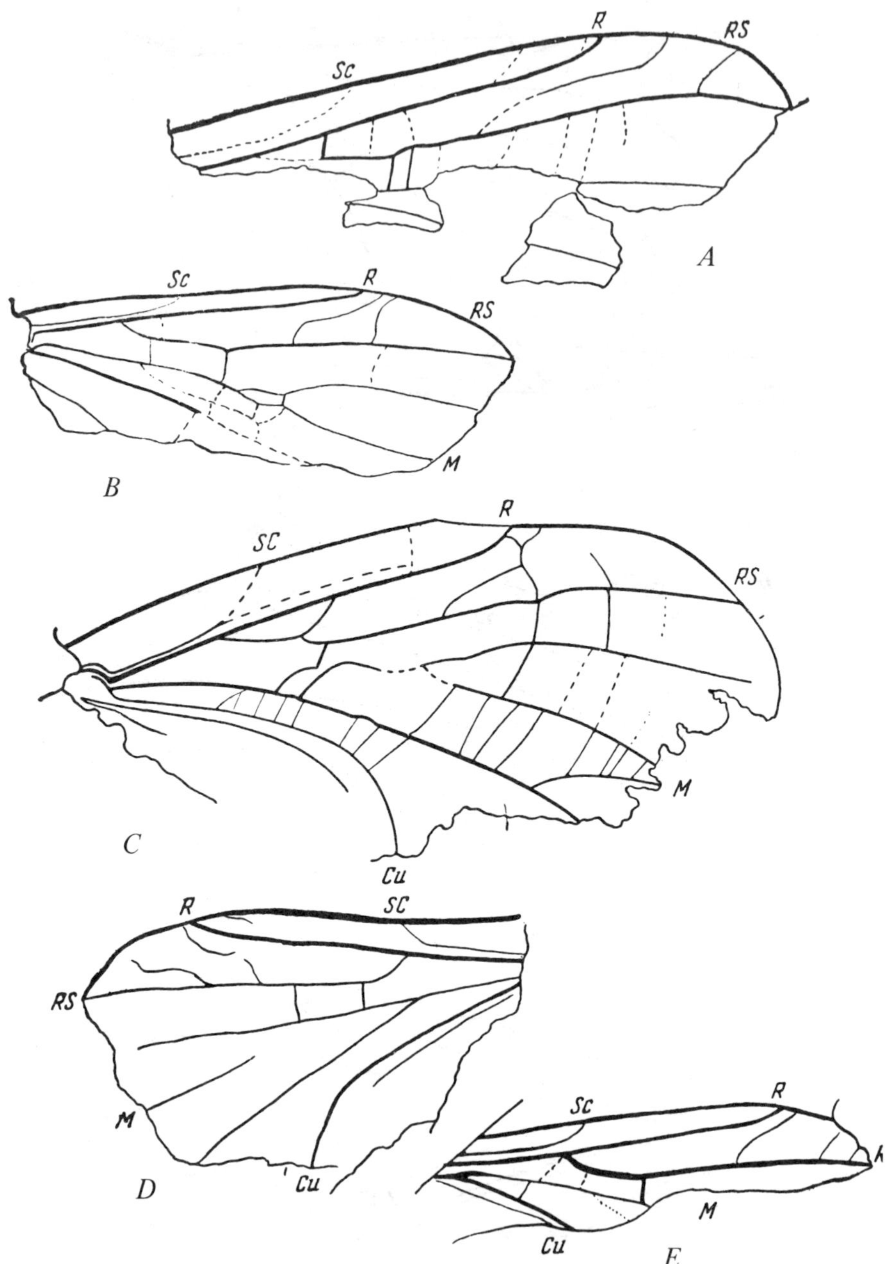

Fig. 58. Pleciofungivoridae of the Triassic of central Asia. *A. Rhaetofungivora radiimedialis* Rohd. sp. n. Holotyp Coll. PIN No. 358/190. Length of remnant 1.02 mm. *B. R. destructimedia* Rohd. sp. n. Holotype, Coll. PIN No. 371/190. Length of remnant 1.88 mm. *C. R. perreticulata* Rohd. sp. n. Holotype, Coll. PIN No. 371/136. Length of remnant 2.3 mm. *D. R. parva* Rohd. sp. n. Holotype, Coll. PIN No. 371/289. Length of remnant 1.5 mm. *E.* The same. Paratype, Coll. PIN No. 358/138. Length of remnant 1.57 mm. (Original.)

is 2.03 mm, of the whole wing about 2.3 mm.

Comparison. – This is described on the basis of one remnant. The form is closest to *R. destructimedia*, sp. n. and especially to *R. perreticulata*, sp. n.

Material. – The holotype.

Rhaetofungivora destructimedia Rohdendorf, sp. n. (fig. 58*B*)

Holotype. – Wing, Coll. PIN No. 371/910, Issyk-kul, Upper Triassic (Rhaetian?).

Description. – Costal field narrow; SC thin, not particularly short, without anterior branches; fork of R+RS limited to a weak transverse vein; there are two curved branches of RS, the ends of which clearly approach each other; there are two sturdy proximal and one weak distal rm veins and the first proximal rm is located at the level of the fork of R+RS and the second proximal considerably more distal to the level of the end of SC; all veins of the M system are displaced and compressed; M_3 in the form of a weak supplementary vein, connected by a few weak transverse veins with M_{1+2} and CuA. The length of the remnant of the wing is 1.88 mm, the whole wing probably about 2.0 mm. Is described on the basis of a single impression; quite an isolated form, closest to *R. radiimedialis*.

Rhaetofungivora perreticulata Rohdendorf, sp. n. (fig. 58*C*)

Holotype. – Positive impression of right wing, Coll. PIN No. 371/136 Issyk-kul, Upper Triassic (Rhaetian?).

Description. – Costal field wide, SC weak and very short, its end hardly going beyond the fork of R+RS; R with thin but clear anterior branch having the form of a straight transverse vein between C and R; fork of R+RS vaguely limited by a short oblique anterior branch of RS, besides which there are three thin and partial branches of RS, partly uniting one with another; there is one sturdy and short rm and two long and thinner distal rm veins; the fork of M_{1+2} large with widely diverging branches between which are indicated delicate transverse veins; between M_2 and M_3 not less than 10 and between M_3 and CuA eight thin and delicate straight transverse veins; M_3 is like a direct continuation of the common trunk of M. Length of remnant 2.3 mm, total length of the wing is presumably about 2.5 mm.

Comparison. – Quite a peculiar form, closest to *R. radiimedialis*.

Material. – The holotype.

Rhaetofungivora parva Rohdendorf, sp. n. (fig. 58*D, E*)

Holotype. – Negative impression of right wing, Coll. PIN No. 371/289, Issyk-kul, Upper Triassic (Rhaetian?).

Description. – Costal field moderately wide, SC thin, terminating at the level of the fork of R+RS; R without anterior branches; fork of R+RS not limited by a transverse vein; anterior branches of RS are only in the distal half of the vein and number two or three obliquely directed veins; rm veins number three and the most proximal vein is weak and obliquely disposed; all the rm veins are removed

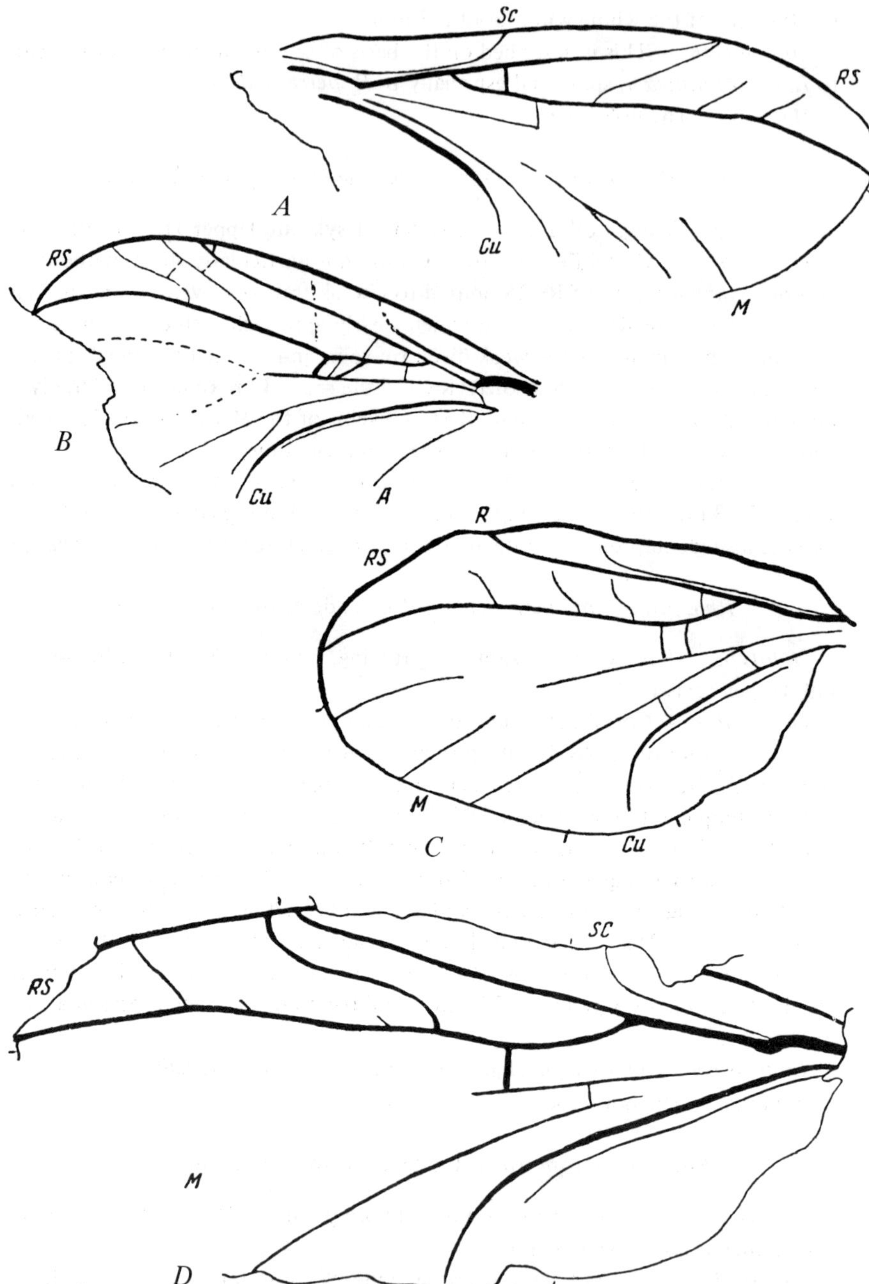

Fig. 59. Pleciofungivoridae of the Triassic of central Asia. *A. Rhaetofungivora radialis* Rohd. sp. n. Holotype. Coll. PIN No. 371/22. Length 2.3 mm. *B. R. quadrimedialis* Rohd. sp. n. Holotype. Coll. PIN No. 358/52. Length 1.94 mm. *C. R. curta* Rohd. sp. n. Holotype, Coll PIN No. 371/251. Length 2.0 mm. *D. R. magniradius* Rohd., sp. n. Holotype, Coll. PIN No. 371/234. Length 2.65 mm. (Original.)

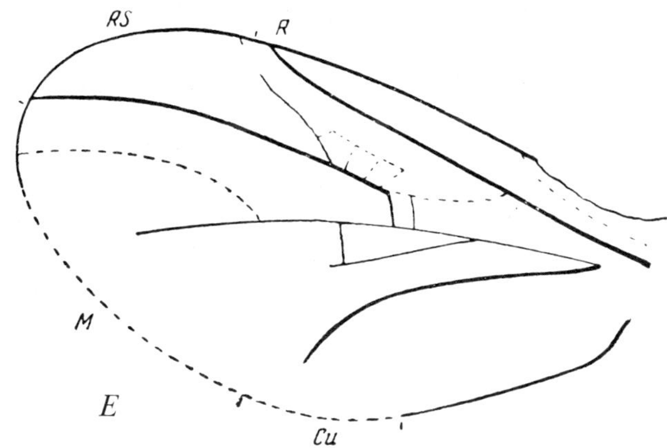

Fig. 59. (cont.). Pleciofungivoridae. *E. R. maxima* Rohd. sp. n. Holotype, Coll. PIN No. 371/1031. Length 2.85 mm. (Original.)

from one another; the distal rm veins are absent; M_3 is removed directly from the common trunk of M; sometimes there is a thin mcu (No. 358/138); the structure of medial and cubital systems of veins are poorly known. Length about 1.75 mm; No. 371/289, length of remnant 1.5 mm; No. 358/138, length of remnant 1.57 mm.

Comparison. — Is described from two fossils. The observed features in the venation of these two fossils may, with great probability, be accepted as individual differences. This species is closest to the group of large rhaetofungivorids (*R. major* – *R. magna*).

Material. — Besides the holotype there is a positive impression of part of the wing, Coll. PIN No. 358)138, Issyk-kul, Upper Triassic (Rhaetian?).

Rhaetofungivora radialis Rohdendorf, sp. n. (fig. 59*A*)

Holotype. — Negative impression of left wing, Coll. PIN No. 371/22, Issyk-kul, Upper Triassic (Rhaetian?).

Description. — Costal field moderately wide with well-expressed protuberance in the basal part; SC not especially short, drawn together with R; the latter vein without anterior branches; fork of R+RS sharply limited by a sturdy transverse vein; there are three anterior branches of RS of which the first is long and oblique, little curved and drawing together with R; it unites with C immediately behind its end; there are some weak rm veins; the structure of the medial vein is unknown; CuP sturdy, at the end going out from CuA. Length of the wing is 2.3 mm.

Comparison. — The species is close to *R. quadrimedialis* and in part to *R. radiimedialis.*

Material. — The holotype.

Rhaetofungivora quadrimedialis Rohdendorf, sp. n. (fig. 59*B*)

Holotype. – Positive impression of left wing, Coll. PIN No. 358/52, Issyk-kul, Upper Triassic (Rhaetian?).

Description. – Costal field quite wide, SC of moderate length, reaching the level of the middle of rm; R with short pre-apical transverse vein; fork of R+RS limited by two transverse veins; there are two curved and oblique branches of RS parallel with one another, going into C and joined by two weak transverse veins; there are four rm veins of which two are thin and two sturdy, oblique; there is one very short transverse vein between the bend of R and M (of its kind a rudiment of a phragma?); the fork of M_{1+2} very delicate, M_3 with a clear posterior branch running parallel with CuA. Length of wing is 1.94 mm.

Comparison. – This peculiar form is somewhat close to *R. radiimedialis* and *R. radialis.*

Material. – The holotype.

Rhaetofungivora curta Rohdendorf, sp. n. (fig. 59*C*)

Holotype. – Negative impression of right wing, Coll. PIN No. 371/251, Issyk-kul, Upper Triassic (Rhaetian?).

Description. – The costal field is moderately wide, SC is long, reaching almost to the middle of the wing, located near to R and curved only at the very top; R without anterior branches; the fork R+RS is limited by a fine transverse vein; it has not less than three curved anterior branches of RS; there are two straight highly adjacent rm veins; the fork M_{1+2} is of moderate size, only insignificantly longer than the common trunk of M_{1+2} (from rm to the fork); M_3 is a direct continuation of the common trunk of M and is connected by two straight, almost parallel transverse mcu veins with CuA. The length of the wing is 2.00 mm, the width is 1.1 mm.

Comparison. – A peculiar species which to some extent approaches *R. reticulata,* *R. mediicubitalis* on the one hand and *R. subcostalis* on the other.

Material. – The holotype.

Rhaetofungivora magniradius Rohdendorf, sp. n. (fig. 59*D*)

Holotype. – Positive impression of left wing, Coll. PIN No. 371/234, Issyk-kul, Upper Triassic (Rhaetian?).

Description. – Costal field is wide; SC is thin and short; R without anterior branches, with sharp break; fork of R+RS without transverse veins; there is a sturdy S-shaped curved anterior branch of RS besides which there are still three weaker, partly incomplete anterior branches; there is a single strong rm; the structure of the anterior medial veins is unknown – they are undoubtedly very fine; M_3 is thin but highly noticeable, branching out from CuA and connected with the common trunk of M by means of a short straight transverse vein. The length of the wing is 2.65 mm. The length of R from the fork of R+RS is 1.05 mm, of the main part of RS from the fork to rm is 0.35 mm, of the second part of RS from rm to the apex is 1.42 mm.

Comparison. – The species is characterized by a sturdy anterior branch of RS and by the peculiar branching of M_3 by which it is distinguished from all other forms of the genus.

Material. – The holotype.

Rhaetofungivora maxima Rohdendorf, sp. n. (fig. 59*E*)

Holotype. – Positive impression of left wing, Coll. PIN No. 371/1031, Issyk-kul, Upper Triassic (Rhaetian?).

Description. – The costal field is moderately wide; SC is short, situated far from R, going into C at the level of the base of RS; R straight, without anterior branches; the fork R+RS without transverse veins; anterior branches of RS are weak and poorly marked: there is not less than one long curved branch; there are two very contiguous rm veins; of which one is strong, the other weak; the structure of the medial veins is not clear; between the main trunk of M_{1+2} and M_3; distal to rm, is located a sturdy, straight transverse vein; M_3 is separated from the common trunk of M at a sharp angle and does not join with CuA by transverse veins. The length of the remnant is 2.85 mm, the total length of the wing is approximately 3.25 mm.

Comparison. – This species is characterized by a short SC, by a sturdy transverse vein between the medial veins and by other features. It is closest to *R. magna.*

Material. – The holotype.

Rhaetofungivora magna Rohdendorf, sp. n. (fig. 60*A, B*)

Holotype. – Positive impression of left wing, Coll. PIN No. 371/12, Issyk-kul, Upper Triassic (Rhaetian?).

Description. – Costal field is wide; SC quite long, almost reaching at its end the level of rm; R without anterior branches, with a sharp break at the base; the fork of R+RS without transverse veins; anterior branches of RS weakly developed and they are only at the middle of the vein where they form an irregularly curved, complex branch uniting with C immediately distal to the end of R; the distal part of RS deprived of branches or with numerous short blind branches (No. 371/911); two or three rm veins (No. 371/911) irregularly inclined, transverse; fork M_{1+2}, sometimes with delicate loop-shaped transverse veins (No. 371/12); transverse veins between M_{1+2}, M_3 and CuA are absent; vein M_3 branches out from the common trunk of M at a sharp angle. Length of wing in both specimens is 2.5 mm; length of veins in the type (No. 371/12): R from fork to C is 0.9 mm; RS from fork to rm 0.28 mm, RS from rm to C 1.3 mm, M from rm to fork of M_{1+2} is 0.45 mm, anterior branch of M_{1+2} is 0.9 mm.

Comparison. – Is closest to *R. maxima, R. simplex, R. major.*

Material. – In addition to the holotype there is still the impression of a wing, Coll. PIN No. 371/911, Issyk-kul, Upper Triassic (Rhaetian).

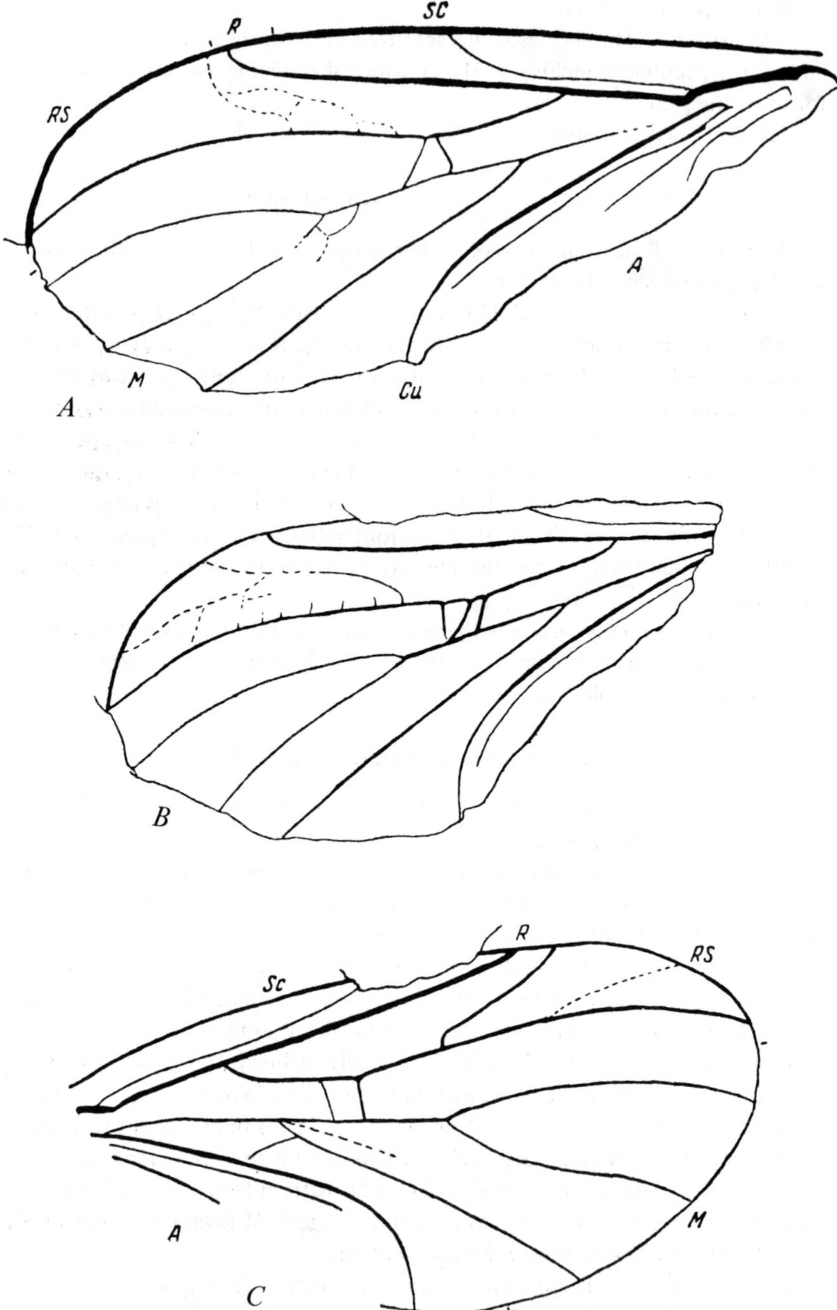

Fig. 60. Pleciofungivoridae of the Triassic of central Asia. *A. Rhaetofungivora magna* Rohd. sp. n. Holotype, Coll. PIN No. 371/12. Length 2.5 mm. *B.* The same. Paratype, Coll. PIN No. 371/911. Length 2.5 mm. *C. R. subcostalis* Rohd. sp. n. Holotype, Coll. PIN No. 358/74. Length 2.3 mm. (Original.)

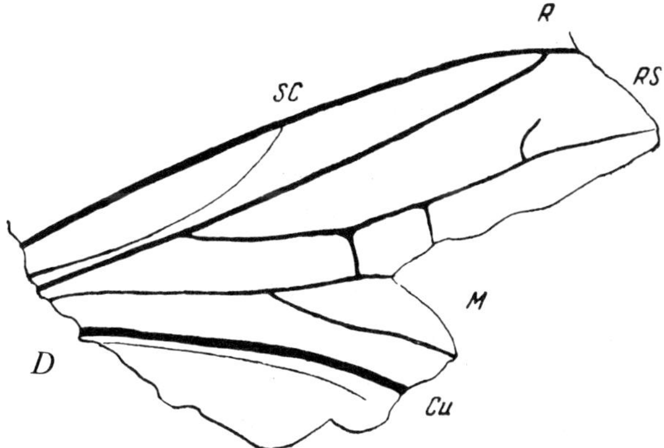

Fig. 60. (cont.). Pleciofungivoridae. *D. R. simplex* Rohd. sp. n. Holotype, Coll. PIN No. 371/19. Length of fragment 2.5 mm. (Original.)

Rhaetofungivora subcostalis Rohdendorf, sp. n. (fig. 60C)

Holotype. – Negative impression of left wing, Coll. PIN No. 358/74, Issyk-kul, Upper Triassic (Rhaetian?).

Description. – Costal field quite wide; SC long, reaching to the middle of the wing and connected to C distal to the level of rm, not particularly thin; R without anterior branches; fork of R+RS without transverse veins; two anterior branches of RS of which the distal is thin and poorly marked; there are two quite sturdy parallel straight rm veins, separated by a distance shorter than the length of one vein; branches of the fork M_{1+2} gradually diverge; M_3 branches off from the common trunk of M at an acute angle and unites by a fine oblique transverse vein with the middle of vein CuA; from the fork of M_{1+2} and M_3 there emerges a weak longitudinal vein or fold; CuA is sturdy and sharply bent at the end. The length of parts of the veins are: R from the fork of R+RS to C 0.93 mm; RS from the fork to the distal rm 0.38 mm; RS from the distal rm to C 1.13 mm; M_{1+2} from the distal rm to the fork of M_{1+2} 0.25 mm. M_1 from the fork to the edge of the wing 0.93 mm. Length of the wing is 2.3 mm.

Comparison. – A peculiar species which is characterized by a long SC and other features; closest to the large rhaetofungivorids – *R. magna* and related species, partly to *R. curta.*

Material. The holotype.

Rhaetofungivora simplex Rohdendorf, sp. n. (fig. 60D)

Holotype. – Fragmentary positive impression of right wing, Coll. PIN No. 371/19, Issyk-kul, Upper Triassic (Rhaetian?).

Description. – Costal field wide, SC relatively long, reaching level with the proximal rm vein, bent in an arch to C; R without anterior branches and transverse veins; fork of R+RS without transverse veins; anterior branches of RS poorly

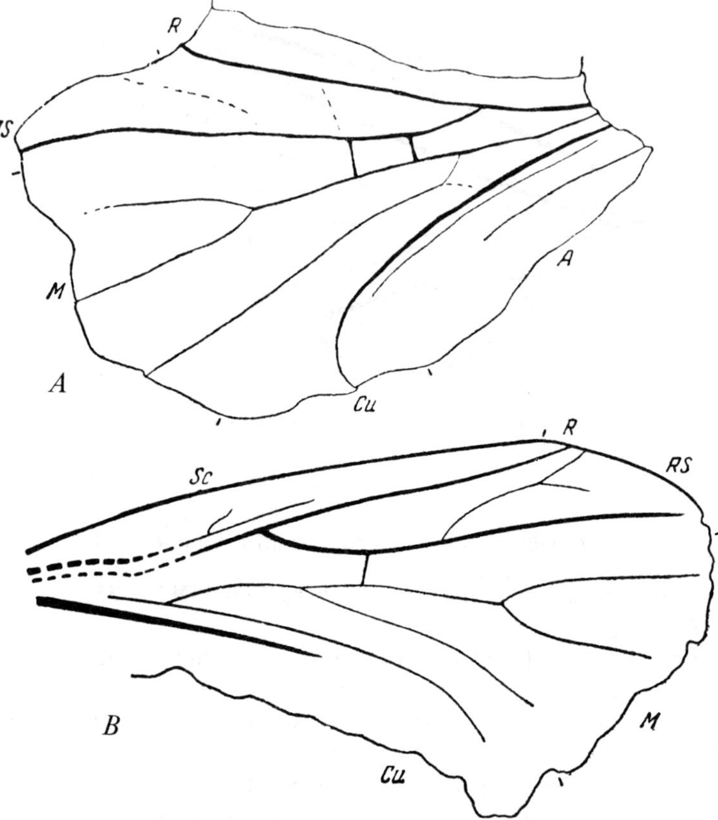

Fig. 61. Pleciofungivoridae of the Triassic of central Asia. *A. Rhaetofungivora major* Rohd. sp. n. Holotype, Coll. PIN No. 371/28. Length 2.3 mm. *B. R. amasioides* Rohd. sp. n. Holotype, Coll. PIN No. 371/1005. Length 1.85 mm. (Original.)

developed; noticeable is one incomplete branch at the middle of the vein, not reaching C; two sturdy straight parallel rm veins, separated by a distance somewhat greater than the length of one rm vein. M_3 branches off from the common trunk of M at an acute angle; transverse veins between M and CuA are absent; venation of the distal part of the wing is not known. The length of R from the fork to C is 1.0 mm; RS from the fork to the proximal rm vein is 0.38 mm; length of fragment of wing 1.9 mm; total length of the whole wing 2.5 mm.

Comparison. — Close to species of the group *R. magna.*

Material. — The holotype.

Rhaetofungivora major Rohdendorf, sp. n. (fig. 61*A*)

Holotype. — Incomplete negative impression of left wing, Coll. PIN No. 371/ 28, Issyk-kul, Upper Triassic (Rhaetian?).

Description. – Costal field wide; SC not clear in the impression; fork of R+RS without a transverse vein; anterior branches of RS are weak; there is a short delicate branch, almost a transverse vein at the level of the distal rm and a long curved branch situated more distally; two sturdy, straight rm veins, distant one from the other at a distance somewhat greater than the length of the distal rm; M_3 weak, branching out from M at an acute angle and connected near its base by a thin oblique transverse vein with the middle of vein CuA, which nearly in all its extent is straight and sharply bent backwards near its end. The length of R from the fork to C is about 1.15 mm; RS from the fork to the distal rm vein is 0.43 mm; M_{1+2} from the distal rm to the fork of M_{1+2} is 0.30 mm; length of the fragment of the wing is 2.3 mm; the whole wing approximately 3 mm.

Comparison. – The species apparently is close to *R. magna* and other large forms of this genus.

Material. – The holotype.

Rhaetofungivora amasioides Rohdendorf, sp. n. (fig. 61B)

Holotype. – Positive impression of right wing, Coll. PIN No. 371/1005, Issykkul, Upper Triassic (Rhaetian?).

Description. – Costal field wide; SC thin, adjacent to R and gradually moving away to C, quite long and entering at the end beyond the level of rm; on SC there is noticeable a vague branch; R without anterior branches or transverse veins, straight; fork of R+RS without transverse veins; there is one thin anterior branch of RS, going into C directly beyond the end of R; anterior branch of RS bears a short branch, directed to the apex of the wing; one sturdy rm; M_3 branches out from the common medial trunk at an acute angle, somewhat distally to the level of the fork of R+RS and not connected with CuA by transverse veins; fork M_{1+2} extending to the edge of the wing; its common trunk (=III part M) long; M_3 branches at the middle; CuA sturdy, straight in the main half and gradually bending back to the distal part. Length of wing is 1.85 mm.

Comparison. – A peculiar species which differs by the single, quite clear branch of RS and by the impoverished, mechanized remaining venation of the wing.

Material. – The holotype.

Genus *Rhaetofungivorella* Rohdendorf, 1962
Rohdendorf, 1962, p. 320

Type of genus: *R. subcosta* Rohdendorf, 1962

Description. – Wing moderately elongated, its length is two to two-and-a-half times greater than the width; anterior border weakly but uniformly convex, straight only in the extent of the first third, span of wing from 1.8 to 2.0 mm. SC from one-third to one-half of wing, thin but well distinguished, always branching, ending in two to four branches. R equal to two-thirds or three-quarters of the wing, straight, with anterior branches or transverse veins, rarely simple (*R. sectoralis*). The structure of the basal section of R is not known. RS curved, the same calibre

or somewhat thinner than R and bears numerous anterior branches which either
have the character of transverse veins and connect with R, or go freely into C;
the number of anterior branches or transverse veins varies from four to eight;
the anterior branches of RS do not form an irregular network. Transverse rm
veins are two or three in number, but only one rm vein is sturdy, the rest are thin
or indistinct.

Comparison and composition. — This genus is established on the basis of four
species which have been studied, and is well characterized by a series of primitive
features: branching SC, numerous branches of RS, the presence of transverse
veins in the system of M. This genus shows the greatest closeness to species of
Rhaetofungivora, namely to the group *R. reticulata* — *R. mediicubitalis*. There
is scarcely any point in investigating the features of the wing venation of insects
of these two genera with the aim of clarification of the greater or lesser nearness
of them to supposed original forms, presumably to peculiar pleciodictyids. The
inevitable irregularity in the development of different features and the scantiness
of material (one wing) makes such conclusions conditional and purely preliminary.
It is obvious only that representatives of this genus together with *Rhaetofungivora*
and *Rhaetofungivorodes* are one of the least perfect in aerodynamic relations of
the Pleciofungivoridae.

Key to the species

1 (2)   R does not have sturdy and distinct anterior branches; SC with
               four branches; two rm veins of which the proximal is very sturdy and
               located far from the fork of R+RS; the common trunk M_{1+2} is con-
               siderably larger than the short fork (fig. 62*A*) . . . .
               . . . . *R. subcosta* (p. 187)

2 (1)   R as a rule with well-developed anterior branches, rarely simple;
               SC with only two terminal branches; proximal rm vein thinner than the
               other; common trunk M_{1+2} always shorter than the fork 3

3 (4)   R without branches; fork of R+RS without transverse veins;
               proximal rm vein very much removed from the fork and situated at
               the level of the end of SC (fig. 62*B*) . . . .
               . . . . *R. sectoralis* (p. 187)

4 (3)   R always with anterior branches; fork of R+RS limited to a
               transverse vein 5

5 (6)   Proximal rm vein located not far from the fork of R+RS; R
               with only one branch close to the apex; in the distal part of RS there
               are only three anterior branches; between the distal section of M_3 and
               CuA there are no transverse veins (fig. 63*A*) . . . .
               . . . . *R. medialis* (p. 188)

6 (5)   Proximal rm vein situated at a considerable distance from the
               fork of R+RS; R with five branches; in the distal part of RS there are
               not less than four branches; between the distal portions of M_3 and
               CuA there are not less than three transverse veins (fig. 63*B*) . . . .
               . . . . *R. analis* (p. 189)

Rhaetofungivorella subcosta Rohdendorf, 1962 (fig. 62*A*)
Rohdendorf, 1962, p. 320

Holotype. – Positive impression of right wing, Coll. PIN No. 358/87, Issyk-kul, Upper Triassic (Rhaetian?).

Description. – SC reaches the middle of the anterior border and in the distal half is divided into four branches; after the end of SC, between C and R, is located a straight, thin transverse vein; R has no branches; the fork of R+RS has two transverse veins of which the distal is curved and is a continuation of rm; from the distal part of RS there arise four anterior branches, partly uniting with one another; two rm veins – a sturdy short proximal and a thin longer distal vein; both rm veins come together at a distance which is equal in length to the proximal rm; fork of M_{1+2} shorter than its common trunk; M_3 branches off from the base of M, of which the first part of M is very short, nearly imperceptible; between M_3 and the main trunk of M_{1+2} there is a short, clear transverse vein located at the level of the fork of R+RS; in the distal part of M_3 is indicated a weak anterior branch; between M_3 and CuA is a longitudinal fold or a weak transverse vein; the anal blade of the wing is quite large, the veins in it are not clear. The length of the fragment is 1.8 mm, of the whole wing about 1.9 mm.

Comparison. – A characteristic species readily distinguished from all related forms by the short fork of M, branching SC and other features.

Material. – The holotype.

Rhaetofungivorella sectoralis Rohdendorf, sp. n. (fig. 62*B*)

Holotype. – Positive impression of right wing, Coll. PIN No. 358/182, Issyk-kul, Upper Triassic (Rhaetian?).

Description. – SC does not reach the middle of the anterior border of the wing and is divided into two weak branches; R simple, without branches; fork of R+RS without transverse veins; RS in its distal half with four parallel anterior branches: in the radial field there are weak, poorly-marked longitudinal folds or delicate veins; there is a pair of straight, parallel rm veins of which the distal is considerably thicker than the proximal; apart from the clearly expressed rm veins, distal to the thick rm there are weak rudimentary veins in the form of three posterior branches to RS; the fork of M_{1+2} is more than twice as long as the common trunk of M_{1+2} (= III part M); M_3 is a direct continuation of the main trunk of M; between M_3 and the main trunk of M_{1+2} there are three transverse veins, located proximally to the level of the thick rm vein; between M_3 and CuA there are no transverse veins; the anal blade of the wing is of moderate size; CuA is curved; A in the form of a strong vein possessing a weak anterior branch. The length of the wing is 1.85 mm.

Comparison. – This species is close to *R. medialis* and together with it approaches some species of the genus *Rhaetofungivora* (*R. mediicubitalis, R. reticulata* and others).

Material. – The holotype.

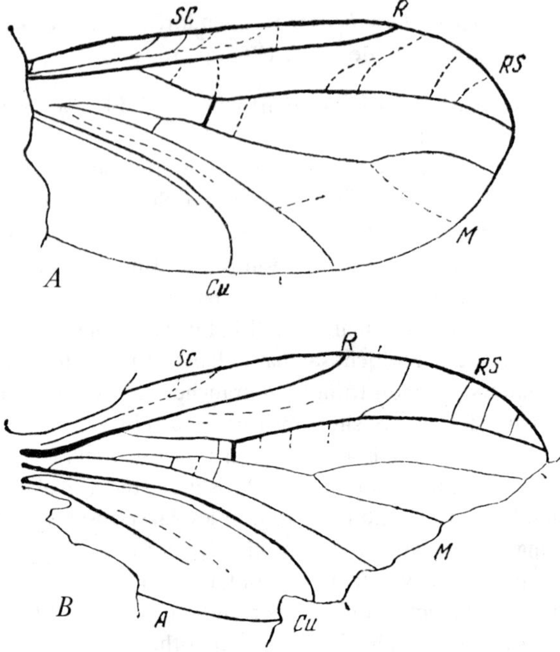

Fig. 62. Pleciofungivoridae of the Triassic of central Asia. *A. Rhaetofungivorella subcosta* Rohd. Holotype. Coll. PIN No. 358/87. Length approximately 2 mm. *B. R. sectoralis* Rohd. sp. n. Holotype. Coll. PIN No. 358/182. Length 1.81 mm. (*A.* according to Rohdendorf, 1962; *B.* original.)

Rhaetofungivorella medialis Rohdendorf, sp. n. (fig. 63*A*)

Holotype. — Impression of left wing, Coll. PIN No. 358/38, Issyk-kul, Upper Triassic (Rhaetian?).

Description. — SC not particularly long, at the end with a fork and with two weak transverse veins (not shown in the illustration); R at the end with a sturdy fork; fork of R+RS limited in the form of a triangle by the sturdy short transverse vein; in the distal part of RS there branch off three anterior branches of which the proximal joins with R, and the two distal branches terminate freely in the membrane; there are two quite sturdy rm veins widely separated from one another; the distal rm is thicker than the other; the structure of M_{1+2} is not clear, the fork apparently is large; M_3 is separated from the common trunk of M by a curve; between M_3 and M_{1+2} at the level of the distal rm and further toward the apex of the wing are located four oblique and thin transverse veins; from the place of branching of M_3 from M_{1+2} there branches out a thin oblique transverse vein to CuA; the most basal section of M joins with CuA by means of three short and clear transverse veins; CuA straight in the main half and is farther curved in an arch; the anal blade is quite large with the simple, quite thin A. The length of the wing is 1.8 mm.

Comparison. — This species is close to the preceding one and, at the same time,

suggests some forms of the genus *Rhaetofungivora.*
   Material. – The holotype.

Rhaetofungivorella analis Rohdendorf, 1962 (fig. 63*B*)
Rohdendorf, 1962, p. 321, fig. 1012

   Holotype. – Positive impression of right wing (base not preserved). Coll. PIN No. 358/822, Issyk-kul, Upper Triassic (Rhaetian?).
   Description. – SC quite short with a fork at the end; R straight and sturdy and bears four somewhat weaker anterior branches; the fork of R+RS is not preserved in the impression and only two transverse veins, which bound the impression, are visible; RS with numerous weak anterior branches; the distal half of the vein bears four almost parallel branches going into C; in the proximal part of the radial field are observed irregularly disposed weak transverse veins and one longitudinal vein (or fold) parallel with the main part of RS; there are three comparatively adjacent rm veins, of which the middle is the most delicate; the system of vein M is poorly distinguished and was possibly crumbled and distorted during burial; the fork of M_{1+2} is very large, several times as long as the common trunk of M_{1+2}; the place of branching of M_3 from M_{1+2} and the connections of the whole trunk of M with CuA are not clear; there are three well-distinguished weak transverse veins between the distal parts of M_3 and CuA. CuA is arched; the anal blade of the wing is very large, the veins in it are not clear. The length of the impression is 1.81, of the whole wing about 2.0 mm.
   Comparison. – Is close to the preceding species, well distinguished by the structure of R.
   Material. – The holotype.

Genus *Rhaetofungivorodes* Rohdendorf, 1962
Rohdendorf, 1962, p. 320

Type of genus: *R. defectivus* Rohdendorf, 1962.

   Description. – Total length of wing about 3 mm. SC long, reaching the level of the beginning of the anterior branch of RS, and therefore noticeably distal to the level of rm, without branches. R curved in the second part, simple, devoid of anterior branches; the structure of the basal section of R is unknown. RS of the same thickness as R with curved anterior branches, of which the proximal is clear and going parallel with R; the total number of branches of RS is not known, but is not less than three. There is a single sturdy rm, apart from which there is only a weak transverse vein, located proximally to rm in the vicinity of the fork of R+RS. Between the common trunk of R and M+Cu there is a thin longitudinal vein or fold running from the base of the wing and uniting with RS near the fork. M veins are thin but clear; the fork of M_{1+2} is very large, its common trunk short, a little longer than rm; M_3 is thin and branches off from M_{1+2}, approximately at the level of the middle of the main section of RS. The common trunk of M is relatively very large. CuA sturdy, curved in the distal half. Other features of the

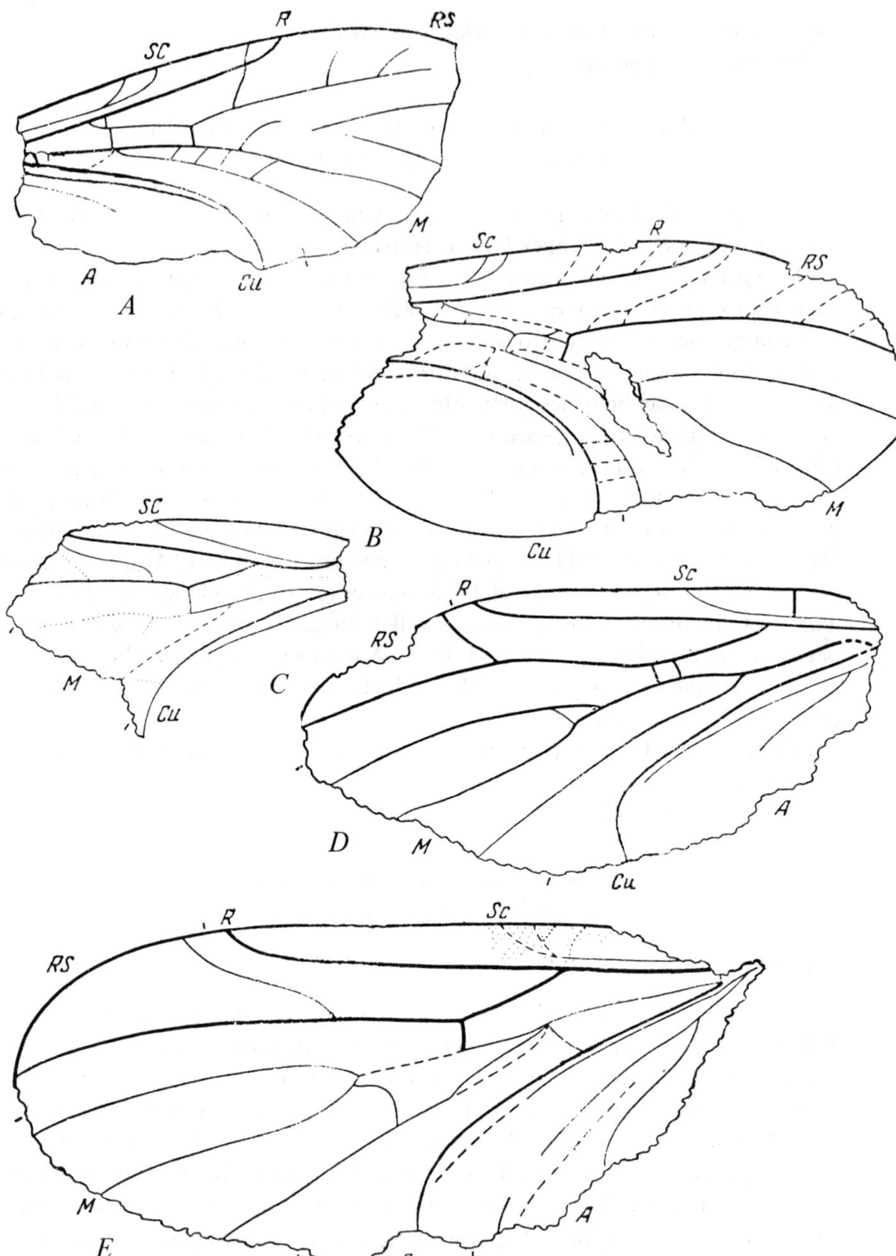

Fig. 63. Pleciofungivoridae of the Triassic of central Asia. *A. Rhaetofungivorella medialis* Rohd. sp. n. Holotype. Coll. PIN No. 358/38. Length about 2 mm. *B. R. analis* Rohd. Holotype. Coll. PIN No. 358/822. Length approximately 2.1 mm. *C. Rhaetofungivorodes defectivus* Rohd. Holotype. Coll. PIN No. 371/913. Length of remnant 1.56 mm. *D. Archipleciofungivora binerva* Rohd. Holotype. Coll. PIN No. 358/41. Length 2.1 mm. *E. Archihesperinus phryneoides* Rohd. Holotype. Coll. PIN No. 371/912. Length 2.62 mm. (*A. B.* original; *C. D. E.* according to Rohdendorf, 1962.)

structure of the wing are unknown because this genus is established on the basis of the investigation of the incomplete impression of a wing.

Comparison and composition. – The single species of the genus shows a similarity with species of the genera *Rhaetofungivora* and *Rhaetofungivorella* differing from them by the peculiar fold or thin vein between the main parts of R and M+ Cu_2 and by other features. The establishment of this genus is partly preliminary; further findings and study of the species of *Rhaetofungivorodes* undoubtedly will refine the description of the inter-relations of these forms. Already it is obvious that *Rhaetofungivorodes*, together with the genera mentioned above, is one of the forms of pleciofungivorids, which possess the least mechanically improved wings.

Rhaetofungivorodes defectivus Rohdendorf, 1962 (fig. 63C)
Rohdendorf, 1962, p. 320, fig. 1015

Holotype. – Incomplete positive impression of left wing (the upper part, the greater part of the terminal border and the anal blade are not preserved). Coll. PIN No. 371/913, Issyk-kul, Upper Triassic (Rhaetian?).

Description. – Between M_3 and CuA there are oblique, quite thin transverse veins; the chief part of RS from the fork to rm is more than three times as long as rm; the chief part of M_{1+2} only one-and-a-quarter times as long as rm and many times less than the fork of M_{1+2}; rm is located exactly in the middle between the branching of M_3 from M and the fork of M_{1+2}. The length of the remainder (to the vein R) is 1.56 mm; the total length of the wing is about 3.0 mm.

Material. – The holotype.

Genus *Archipleciofungivora* Rohdendorf, 1962
Rohdendorf, 1962, p. 321

Type of genus: *A. binerva* Rohdendorf, 1962.

Description. – Lenght of the wing is two-and-a-quarter times greater than its width; absolute sizes of wing are somewhat greater than 2 mm. SC equals approximately one-third of wing, the end of it does not reach the level of rm; between this vein and C there is a sharp thick transverse vein (axillary?). R nearly completely straight without anterior branches. RS the same thickness as R, branching out from it at the end of the first fifth of the wing. There is a single sturdy branch of RS going into C at right angles and outgoing from RS at the level of the end of R, at the middle between rm and the end of RS. Two sturdy rm veins come together at a distance, equal in length to each rm vein; both these veins in this way set limits to the cell of almost the proper square form. The fork of M_{1+2} is long, parallel, considerably longer than the common trunk of M_{1+2} from the fork to rm; there is an oblique transverse vein which borders the very fork of M_{1+2}. M_3 branches out from M at an acute angle; transverse mcu veins are absent. CuA is sturdy, almost straight, bent at the end. The anal area is quite large, in length equal to half of the wing. A is observed in the form of a short, thin vein, as a branch of CuP.

Comparison and composition. – This genus is closest to the *Pleciofungivora* Rohdendorf, described from the Jurassic of Karatau; some features of the new form suggest its closeness to the ancestral forms of the Karatau genus. There is a single species, the type of the genus.

Archipleciofungivora binerva Rohdendorf, 1962 (fig. 63*D*)
Rohdendorf, 1962, p. 321, fig. 1018

Holotype. – Positive impression of left wing in good preservation (only the posterior border and part of C between R and RS is not clear). Coll. PIN No. 358/ 41, Issyk-kul, Upper Triassic (Rhaetian?).

Description. – SC thin but clear, moderately drawn together with R: from the level of the beginning of RS this vein gradually diverges from R and goes into C at the level of the middle of the second part of M (from the base of M_3 to rm). R at the end is very slightly bent forward; the part of C between the ends of R and the anterior branches of RS is only half the length of the latter; the first part of RS is four times as long as rm; M_3 slightly convex in front. The length of the wing is 2.1 mm.

Material. – The holotype.

Genus *Archihesperinus* Rohdendorf, 1962
Rohdendorf, 1962, p. 321

Type of genus: *A. phryneoides* Rohdendorf, 1962.

Description. – The length of the wing is two-and-a-quarter times as great as the width; the absolute sizes are about 2.6 mm. SC is somewhat less than one-third of the wing; the end of it does not reach the level of rm, but branches and forms several short branches, poorly distinguishable as the result of an oval ptero-stigmal spot which envelops the whole terminal section of SC and is located at the level of the fork of R+RS. R straight, without anterior branches, about two-thirds of wing; the structure of the basal section is not known. RS of the same size as R, little curved and terminating in front of the apex of the wing. There is one thin, curved oblique branch of RS, which in the distal half is parallel with the end of R; apart from this branch there are weak rudimentary folds or trans-verse veins going out from RS between rm and the base of the anterior branch (not shown in the illustration). One sturdy rm vein located in front of the middle of the wing; save for it, in this field there are weak and irregular oblique trans-verse veins which unite among themselves; the chief part of RS is four times as great as rm. The veins of the system of M are weaker than R. The fork of M is long, more than three times as long as the common trunk of M_{1+2} (from the fork to rm). M_3 branches out from the common trunk of M approximately at the level of the fork R+RS. Between the fork of M_{1+2} and M_3 there is a thin, curved trans-verse vein which limits the peculiar "discoidal" cell. There is an oblique, thin transverse mcu. CuA arched, CuP in the distal part is poorly marked, very weak. A is well developed, in the form of a fork and a supplementary thin vein lying between its branches.

Comparison and composition. – The peculiar nature of this genus consists of the whole plan of venation of the moderately elongated wing, having a strenghened R and RS, a straight anterior border and reduced medial veins, features of mechanical improvement. At the same time there are other, different traits which bear an ancient character, and which show little mechanical improvement. Such is the condition of multiple venation which develops in the branching of SC, in the presence of transverse veins. The described genus shows a resemblance to the Jurassic genus *Prohesperinus* of Karatau, undoubtedly being its ancestor. There is a single species, the type of the genus.

Archihesperinus phryneoides Rohdendorf, 1962 (fig. *63E*)
Rohdendorf, 1962, p. 321, fig. 1017

Holotype. – Positive impression of left wing (anterior part of the basalar and posterior border of the whole wing not preserved, beginning with the apex). Coll. PIN No. 371/912, Issyk-kul, Upper Triassic (Rhaetian?).

Description. – Branches of the end of SC are peculiar: all branches go off from the upper division of this vein and are contained in the pterostigmal spot; the distal part of RS is more than four times as long as the proximal; weak transverse veins in both sides of rm have the appearance of oblique veins from which branch out short, arched longitudinal small veins or folds; the chief division of M_3 is complexly constructed and consists of a pair of veins, the anterior thicker and the posterior delicate and thin; the transverse mcu branches out just from the point of derivation of M_3; the transverse separating the "discoidal" cell joins the base of the branch M_2 with the middle of M_3. The length of the wing is 2.65 mm.

Material. – The holotype.

Genus *Protallactoneura* Rohdendorf, 1962
Rohdendorf, 1962, p. 321

Type of genus: *P. turanica* Rohdendorf, 1962.

Description. – Form of wing vague; total size about 2.3 mm. Anterior border is moderately convex, not straight. SC apparently about one-third of wing, thin, without branches and going into C. R is almost absolutely straight without branches. RS branches out from R nearly at right-angles with a straight distal part (after rm). Sturdy, clearly distinguishable anterior branches of RS are absent and instead of them there is a very thin indistinct network of irregular cells composed of weak and delicate longitudinal veins. One sturdy rm vein, little removed from the fork of R+RS. Medial veins almost of the same thickness as the radials; the structure of the whole system of M, however, because of poor preservation, is poorly known; M_3 branches off from CuA at the level of rm, the place of its junction with M_{1+2} is not clear. The common trunk of M is very long. CuA almost straight; CuP is comparatively remote from CuA.

Comparison and composition. – The genus is established on the basis of the investigation of a single remnant which is poorly preserved: as a result of this the composed description is undoubtedly incomplete and the relation of this form

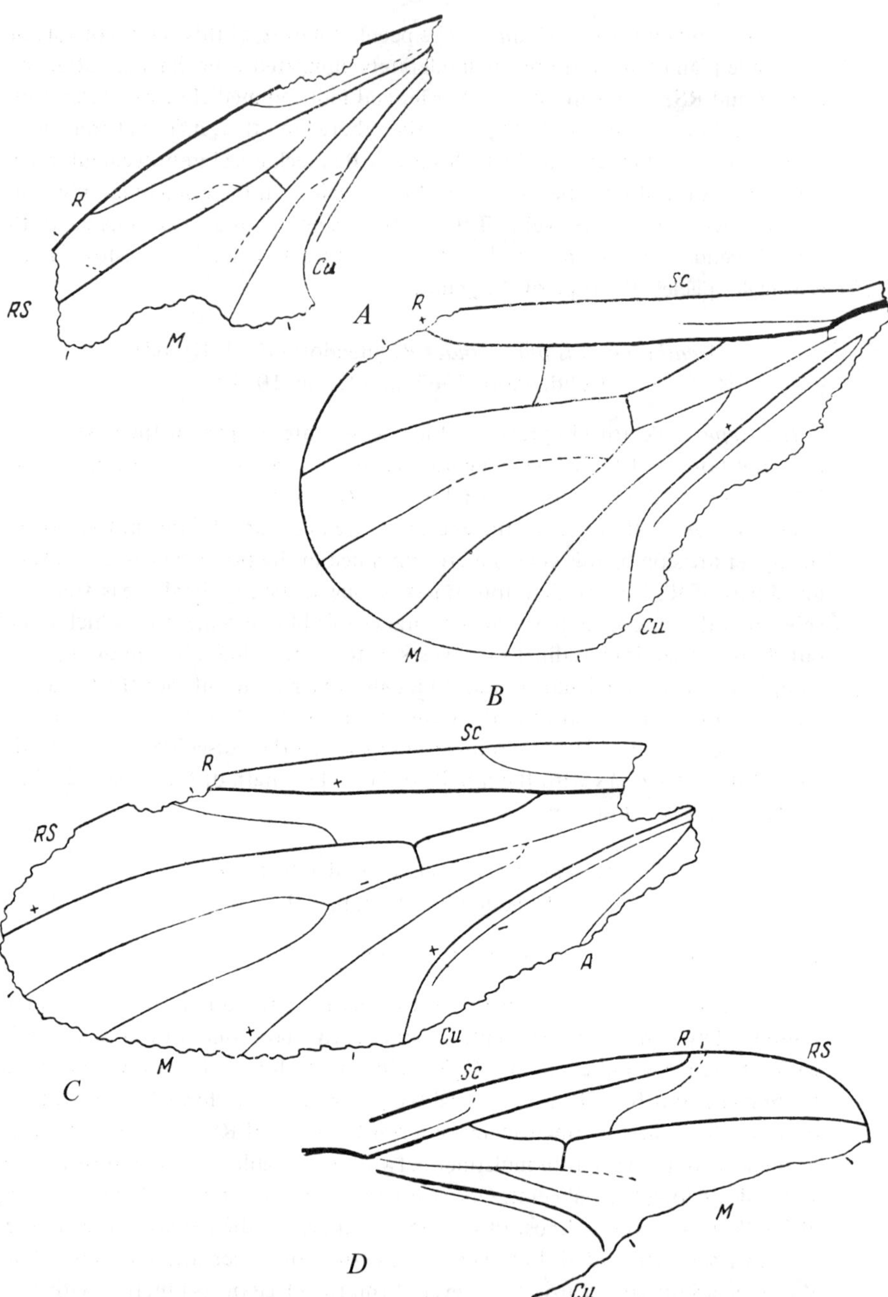

Fig. 64. Pleciofungivoridae of the Triassic of central Asia. *A. Protallactoneura turanica* Rohd. Holotype.
Coll. PIN No. 358/73. Length of remnant 2.0 mm. *B. Archipleciomima obtusipennis* Rohd. Holotype.
Coll. PIN No. 358/125. Length 1.9 mm. *C. Palaeohesperinus longipennis* Rohd. Holotype. Coll. PIN
No. 358/95. Length of remnant 3.13 mm. *D. P. minor* Rohd. sp. n. Holotype. Coll. PIN No. 358/96.
Length approximately 2.0 mm. (*A, B, C.* according to Rohdendorf, 1962; *D.* original.)

is still insufficiently clear. The unusual nature of the long and straight R and RS
and the curve of the base of the latter somewhat suggests representatives of the
family Allactoneuridae.

Protallactoneura turanica Rohdendorf, 1962 (fig. 64A)
Rohdendorf, 1962, p. 321, fig. 1016

Holotype. – Incomplete negative impression of right wing (apex of the wing,
greater part of the fork of M_{1+2} and the whole anal blade are not preserved).
Coll. PIN No. 358/73, Issyk-kul, Upper Triassic (Rhaetian?).

Description. – The distance from the place of branching of RS to the connection of the latter with rm is 2.9 times as great as the length of rm; the latter is
straight and sturdy; the place of junction of RS with R is delicate and indistinct;
apart from the weak veins in the radial field, there are no other delicate transverse
veins in the wing; the main part of M_3 which unites with CuA is arched and weak.
The length of the remnant is 2 mm; total length of the wing about 2.3 mm.

Material. – The holotype.

Genus *Archipleciomima* Rohdendorf, 1962
Rohdendorf, 1962, p. 321

Type of genus: *A. obtusipennis* Rohdendorf, 1962.

Description. – Wing short and wide; anterior border straight to the upper
quarter with sharply isolated, very wide apex; length of the wing is 1.8 times as
great as the width. The absolute size of the wing is about 1.9 mm. SC is weak,
straight, about one-third of wing, terminating freely in the membrane of the
costal cell, without branches. The basal part of R thick and convex in front,
separated by a break from the distal part, the latter almost straight, without
branches and equal to three-quarters of the wing. RS branches off at the end of
the first third of the wing, thinner than R, uniformly curved. There is one anterior
branch in the form of a straight, moderately thick transverse vein, which combines
with the middle of the distal part of R. RS unites with C in front of the apex of
the wing. One stout rm vein located just proximal to the middle of the wing; the
length of it is three times less then the first part of RS. The medial vein is somewhat thinner than RS and considerably thinner than R. The fork of M_{1+2} is very
large, its total trunk (from rm to the fork) is not less than six to seven times
smaller than M; the branches of the fork M_{1+2} are uniform but slightly divergent,
and the anterior branch of it, particularly in its chief half, is very thin and weak.
M_3 branches off from the common trunk of M at the level of the fork of R+RS,
is curved and not connected with CuA by transverse veins. There is a short but
noticeable transverse vein at the end of the basial which connects the curve of
R with M by a peculiar kind of rudimentary phragma. CuA is straight in the
greater part of its extent. The structure of the anal blade is unknown.

Comparison and composition. – This genus is characterized by a series of
features: shortening of the wing, reduction of SC, straight anterior border and

straightened R, structure of RS. All these traits of the wing of *Archipleciomima* indicate high mechanical properties: there is costalization and simultaneously the preservation of a great area. The structure of the wing of this form partly reminds us of some representatives of the younger family Pleciomimidae, known from the Jurassic of Karatau (such, in particular, as the reduction of branches of RS and the broad apex of the wing). More precisely, the relations of these forms are at present difficult to establish. Known from a single species, the type of the genus.

Archipleciomima obtusipennis Rohdendorf, 1962 (fig. 64*B*)
Rohdendorf, 1962, p. 321, fig. 1019

Holotype. — Positive impression of left wing of good preservation (only the greater part of the anal blade not preserved). Coll. PIN No. 358/125, Issyk-kul, Upper Triassic (Rhaetian?).

Description. — Costal field wide, is parallel with the outside; apex of R is quite sharply bent forward; basial (equals first part of RS) nearly straight, slightly concave in front; distal part curved; anterior branch of RS (a transverse vein) is located approximately at the middle between rm and the junction of R with C; the part of RS between rm and the anterior branch is equal to 0.85 of the section of RS between rm and the fork of R+RS; two basal sections of the common trunk of M (between the bend of R, branching of M_3 and rm) are approximately equal in size, the second one is somewhat shorter. The length of R from the break to C is 0.70 mm; of the main part of RS from the fork of R+RS to rm is 0.25 mm; of the second fork of RS from rm to C is 0.95 mm; the length of the whole wing is 1.875 mm.

Material. — The holotype.

Genus *Palaeohesperinus* Rohdendorf, 1962
Rohdendorf, 1962, p. 323

Type of genus: *P. longipennis* Rohdendorf, 1962.

Description. — Wing elongated; its length is approximately 2.6 times as great as the width; anterior border to distal third is straight; apex of wing narrow. The absolute sizes of the wing are 2 to 3 mm. SC thin but clear, not drawn together with R, going into C distally to the level of the fork of R+RS, without branches, the length of it less than one-third of the wing. The basal part of R convex in front, separated by a break, moderately thickened. The distal, large part of R almost straight, devoid of anterior branches, at the end uniformly curved and connected with C at quite a large angle; the total length of R is equal to three-quarters of the wing. RS branches off from R at the end of the first quarter of the wing; the main part of RS is equal to approximately one-third of the second part (from rm to C), which is moderately curved and terminates before the apex of the wing. There is one sturdy and curved anterior branch of RS, in the distal part nearly parallel with the end of R, which unites with C somewhat distally to

R; the part of RS from rm to the anterior branch is equal to 0.52 to 0.60 of the part of RS from rm to the fork of R+RS. There is one sturdy rm; the main part of RS is 3.8 to 5.4 times as long as rm. The veins of the system of M are somewhat thinner than RS. The branches of the fork M_{1+2} are almost parallel; the common trunk of the fork is equal to 0.32 of vein M_1. M_3 almost straight and branches out from the common trunk of M somewhat distally to the level of the fork R+RS; there is no transverse mcu. CuA is slightly arched, gradually bending back in its upper third. The anal blade is somewhat smaller than half of the wing; its venation is poorly known, only one simple A is noticeable.

Comparison and composition. – The described genus is closest to *Archihesperinus*, g.n., differing from it by the great development of features of mechanical improvement of the blade of the wing, by the reduction of rudimentary, weak transverse veins and by the strenghtening of the main trunks of R, RS and the single anterior branch of RS. This genus approaches more nearly to the Karatau *Prohesperinus*, being almost certainly its direct ancestor; this proposal is contradicted by very few features. It is described according to two very close species, *P. longipennis* Rohdendorf (type of the genus) and *P. minor*, sp.n.

Key to the species

1 (2)   Larger – length of wing more than 3 mm; SC goes in considerably distal to the level of the fork of R+RS; main part of RS more than five times as long as rm and 1.67 times as long as the second section (fig. 64*C*) . . . . *P. longipennis* (p. 197)

2 (1)   Smaller – length of wing about 2 mm; SC terminates at the level of the fork of R+RS; main part less than four times the length of rm and twice the length of the second section (fig. 64*D*) . . . . *P. minor* (p. 198)

Palaeohesperinus longipennis Rohdendorf, 1962 (fig. 64*C*)
Rohdendorf, 1962, p. 323, fig. 1023

Holotype. – Positive and negative impressions of left wing of good preservation (only the basial, a section of the anterior border of the wing in the position of the end of R and the anterior branch of RS and the posterior border of the whole wing, especially the anal blade not preserved). Coll. PIN No. 358/95, Issyk-kul, Upper Triassic (Rhaetian?).

Description. – Costal field is wide; SC thin, well distinguished in all of its extent, with its distal end going beyond the level of the forks of R+RS and M_{1+2} +M_3. RS with elongated main section which is five times as long as rm and 1.67 times as long as the second part; the last, biggest section of RS is noticeably curved; the anterior branch of RS is disposed obliquely; M_3 in all its extent nowhere approaches closer to CuA than to M_{1+2}. The length of the wing is 3.13 mm.

Material. – The holotype.

Palaeohesperinus minor Rohdendorf, sp. n. (fig. 64D)

Holotype. – Positive and negative impressions of right wing (greater part of
the medial field and the anal region are not preserved). Coll. PIN No. 358/96,
Issyk-kul, Upper Triassic (Rhaetian?).

Description. – Costal field not particularly wide; SC thin, at the end very
weak, poorly distinguishable, only insignificantly going beyond the level of the
fork of R+RS with its distal end; main part of RS only 3.8 times as long as rm
and twice as long as the second section; the last part of RS almost straight; anterior branch of RS more perpendicular; middle part of vein M_3 drawn nearer to
CuA than to M_{1+2}. Length of the wing is 2.03 mm.

Material. – The holotype.

Family Palaeopleciidae Rohdendorf, 1962
Rohdendorf, 1962, p. 319

Description. – Basal part of R thickened and passes into the distal part without break or bend. Wings elongated, more than three times as long as wide. RS
forms two branches: a middle running parallel with R and an upper in the form
of a narrow fork of the main trunk of RS. Between the end of R, C and the
middle branch of RS there are transverse veins. The transverse rm removed considerably from the fork of R+RS. The branches M_1 and M_2 with forks at the
ends, M_3 simple; medial veins weaker than the radial. M_3 strongly connected
with CuA but not united with it by transverse veins. Anal blade long with a sturdy
A. The type of the family is the genus *Palaeoplecia* Rohdendorf, 1962.

Comparison and composition. – It differs sharply from others by the presence
of transverse veins between R and the branch RS, by the elongate wings, by
branches at the distal parts of the veins RS and M and finally by the characteristic
union of M_3, CuA and CuP which are located near each other. The absence of
data on the structure of the body does not permit us to explain the true relations
of this peculiar form; we can only point to the completely isolated position of
the genus *Palaeoplecia* in the systematics of fungivorids and the very ancient
character of its association with them. Among the Cenozoic Diptera we do not
find any groups which show similar connections with the Palaeopleciidae. Only
some Jurassic Diptera may be descendants of this group; such, for example, as
the peculiar Tipulopleciidae and Sinemediidae, which are characterized by reduction of the venation and by the loss of veins of the medial system (see p. 252-255).

Genus *Palaeoplecia* Rohdendorf, 1962
Rohdendorf, 1962, p. 319

Type of genus: *P. rhaetica* Rohdendorf, 1962.

Description. – Costal field is moderately wide proximally and narrower distally where it is almost parallel to the edge. SC less than one-third of the wing,
thinner at the end, gradually converging and joining with C. Before the end of R
in the costal field are located two straight, thin transverse veins. R almost straight,

bending slightly before the end to the anterior border; R equals three-quarters of the length of the wing. RS branches out from R at the boundary of the first and second quarters of the wing, insignificantly less thick than R; at the level of the third and fourth fifths of the length of the wing RS divides into two branches of identical thickness. The anterior branch of RS is directed to the anterior border, it is parallel with the end of R and unites with C distal to the end of R; between this branch and R there are three transverse veins, of which the most proximal is thinner and more oblique. The posterior branch of RS (or its main trunk) is directed straight to the apex of the wing, bending somewhat back; at its middle the posterior branch forms a thinner anterior little branch running almost parallel with the main branch and which does not reach to the border of the wing; this little branch has the appearance of a fold. The medial veins, as also the transverse rm, are weaker than RS; only the posterior medial vein M_3 is comparatively sturdy. The common trunk is far away from R and RS and close to CuA, whereas M_3 is a direct continuation of it. The anterior trunk of M branches out from the common trunk of M at the level of the end of SC and forms poorly-distinguishable, irregularly-branching veins and cells. The fork of M_{1+2} is narrow and longer than the fork of RS; its branches gradually diverge and each of them in its distal part bears a thin oblique anterior branch directed to the apex of the wing and terminates freely in the membrane. There are weak transverse veins in the fork between M_1 and M_2 and between M_2 and the end of M_3. CuA is approximately equal to half of the wing, strongly thickened in the basal part, moderately thick for the greater part of its extent, and gradually becoming thinner in the distal part, which is bent back; CuP thin. A is thick at the base. There is one species, the type of the genus.

Palaeoplecia rhaetica Rohdendorf, 1962 (fig. 65*A*, 65*B*)
Rohdendorf, 1962, p. 319, fig. 1010

Holotype. — Positive and negative impressions of left wing in good preservation (posterior border and poorly-distinguished medial area not preserved). Coll. PIN No. 358/106, Issyk-kul, Upper Triassic (Rhaetian?).

Description. — The length of the wing is 3.4 times as great as its width; the part of RS from the fork of R+RS to rm is nearly twice as great as the section from rm to the branching of the anterior branch of RS; the distance between transverse veins in the costal field before the end of R is greater than the length of each transverse vein; the space between the straight transverse veins which occur between R and the branch of RS is still greater, almost twice as long as each; an oblique transverse vein in this field lies almost at the level of the first transverse vein in the costal field; between the branches of M_{1+2} are two parallel transverse veins. The length of the wing is 3.94 mm.

Material. — The holotype.

Superfamily Phragmoligoneuridea Rohdendorf, 1962
Rohdendorf, 1962, p. 332

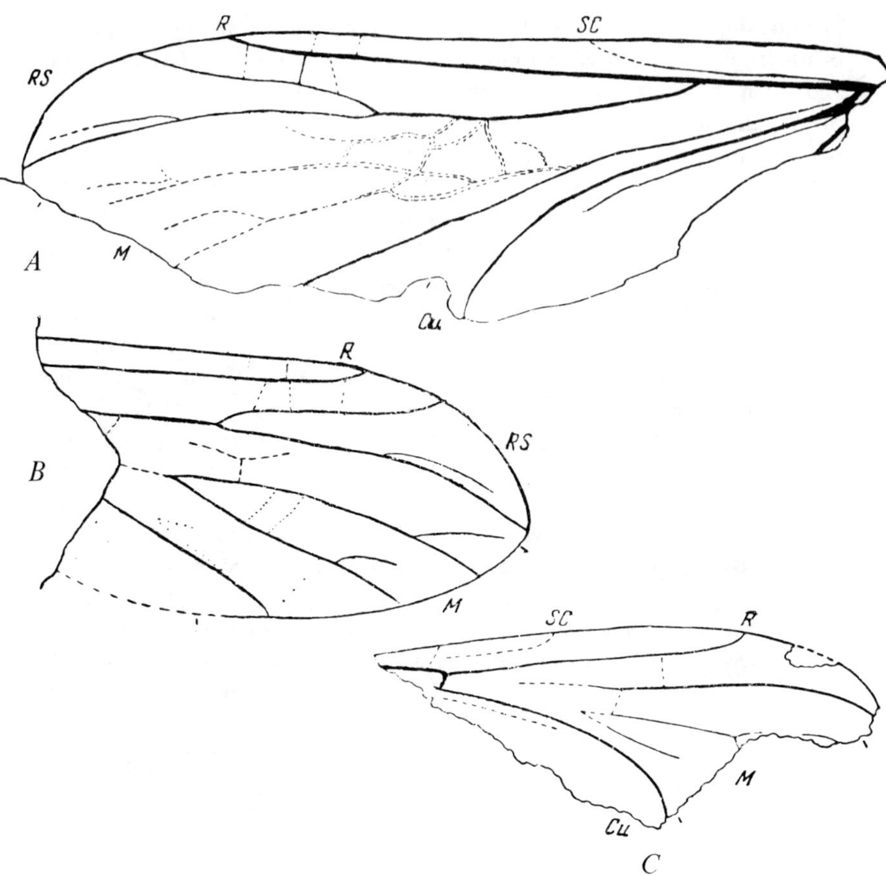

Fig. 65. Palaeopleciidae and Phragmoligoneuridae of the Triassic of central Asia. *A. Palaeoplecia rhaetica* Rohd. Holotype. Coll. PIN No. 358/106. Positive impression. Length about 4 mm. *B.* The same, negative impression. *C. Phragmoligoneura incerta* Rohd. Holotype. Coll. PIN No. 417/8. Length 2.4 mm. (According to Rohdendorf, 1962.)

Description. — R forms a well-expressed sturdy phragma, connecting the end of R with CuA. The main trunks of M and RS are reduced. RS is simple, without branches, only with a transverse vein. The veins of M do not form a closed cell. The structure of the body is unknown.

Comparison. — This superfamily is definitely a representative of the infraorder Bibionomorpha, besides being one of the most specialized. The development of the phragma, reduction of the main sections of M and RS, the poverty of branches of M — all these features indicate the high mechanical qualities of these wings, perhaps among the most perfect for bibionomorphs of the Triassic fauna. There is a single family, the Phragmoligoneuridae.

Family Phragmoligoneuridae Rohdendorf, 1962

Rohdendorf, 1962, p. 332

Description. – The wing is more than two-and-a-half times as long as wide. The basal part of R is not thickened. R is greater than three-quarters the length of the wing. RS is long, convex in front, with a straight, weak transverse vein. The transverse rm is located at the middle of the wing and far from the proximal end of RS which lies freely in the membrane. The vein M with three branches is noticeably thinner than RS and CuA, and is not connected with the last vein. The structure of the anal region is not clear. The type of the family and the only genus is *Phragmoligoneura*, Rohdendorf, 1962.

Genus *Phragmoligoneura* Rohdendorf, 1962
Rohdendorf, 1962, p. 332

Type of genus: *P. incerta* Rohdendorf, 1962.

Description. – The costal field is of moderate width. SC is very delicate and thin and reaches more than one-third of the wing, located at the middle between C and R and going into C; the base of SC is vague and this vein seems to be coming out from the middle (!) of a long oblique axillary transverse vein. R at the place of the phragma has a break in front, the greater part of this vein is weak and irregularly curved. RS begins as a very weak free vein, the proximal end of which lies at the level of the end of SC. Only after rm, closer to the apex, RS becomes a sturdy vein convex in front; the transverse vein between R and RS is located more distally than rm at a distance which exceeds one-and-a-half times the length of the same transverse vein. The part of R from the phragma to the transverse RS is 2.6 times as great as the section of R from the transverse RS to C. M_3 branches out from the common trunk of M at a distance of the length of rm; this vein is close to CuA, thin and terminating freely in the membrane, not reaching to the level of the end of R and the fork of M_{1+2}. The common trunk of M_{1+2} is long, apparently only insignificantly shorter than the anterior branch (M_1). In the wing, apart from the weak rm veins, anterior branches of RS and the axillary, there are no other transverse veins. CuA of uniform thickness as thick as R, uniformly convex. CuP is very thin and reaches only to the middle of CuA. There is a single species, the type of the genus.

Phragmoligoneura incerta Rohdendorf, 1962 (fig. 65C)
Rohdendorf, 1962, p. 332, fig. 1074

Holotype. – Positive impression of right wing (anal blade and medial veins not preserved). Coll. PIN No. 417/8, Issyk-kul, Upper Triassic (Rhaetian?).

Description. – Length of the wing is 2.39 mm. End of SC approximately at the middle between the phragma and the anterior branch (transverse) of RS; width of costal field at the level of rm is equal to the width of the radial field.

Material. – The holotype.

Superfamily Rhyphidea Newman, 1934
[n. transl. Rohdendorf, 1962 (ex Rhyphidae Newman, 1934)]

Description. – Medial veins to the number of four sturdy branches which form by their main divisions a closed cell. There is a rudimentary phragma in the form of a fold. RS with one or two sturdy and long anterior branches besides which there are sometimes short distal branches.

Comparison and composition. – A similar description of the superfamily is given elsewhere in this book (p. 72). It is represented in the Upper Triassic fauna by three families: Oligophryneidae, Protorhyphidae and Protolbiogastridae. This superfamily has an important phylogenetic significance, being the original group for the whole diverse complex of brachycerous Diptera, of the infraorders Asilomorpha, Myiomorpha and others. The Rhyphidea are represented in the faunas of the Mesozoic and Cenozoic; in the present-day fauna this group has a sharply relict character, including a few species of two families.

Family Oligophryneidae Rohdendorf, 1962

Rohdendorf, 1962, p. 332

Description. – Wing only a little more than twice as long as wide. The structure of the basial region is unknown. SC weak and short. To RS there are anterior branches: one is a long one branching out proximally from rm and short distal veins in the vicinity of the apex. The intermedial cell is very small and is located proximally to the middle of the wing in its basal half. There is a transverse vein between the branches of M_1 and M_2. The anal veins are not clear. The type of the family is the genus *Oligophryne*, Rohdendorf, 1962.

Comparison and composition. – This family is sharply distinguished from the others and approaches somewhat to the Bibionidea. There is a single genus, the type of the family.

Genus *Oligophryne* Rohdendorf, 1962

Rohdendorf, 1962, p. 332

Type of genus: *O. fungivoroides* Rohdendorf, 1962.

Description. – The costal field is wide. SC is observed in the form of a short line in the basal quarter of the wing, terminating freely in the membrane. R is straight, bent at the very apex. RS is irregularly curved, branching out from R in the basal quarter of the wing. A long anterior branch of RS separates from RS proximally to the level of the intermedial cell, uniformly arch-like. There is one sturdy, short rm vein, located proximally to the middle of the intermedial cell: this latter has a blunt proximal end and straight posterior edge, not forming a projecting angle by combining with the posterior transverse vein going to M_4. M_1 sturdy, combining with the edge of the wing exactly at its apex. M_2 issues from one point together with M_1; between these veins there is a thin straight transverse vein lying somewhat more proximal to the vertical branch of RS, parallel with it. M_2 and M_3 gradually diverging, curved. M_4 almost straight, going out from the common trunk of M and connected by the above-mentioned short transverse vein with the intermedial cell and by a long, oblique transverse vein with the curved CuA, uniformly bent backwards. CuP gradually tapers to the end. The anal veins are not clear. There is a single species, the type of the genus.

Oligophryne fungivoroides Rohdendorf, 1962 (fig. 66*A*)
Rohdendorf, 1962, p. 332, fig. 1068

Holotype. – Positive impression of left wing (main part of wing not preserved). Coll. PIN No. 358/120, Issyk-kul, Upper Triassic (Rhaetian?).

Description. – Upper part of RS has a straight perpendicular branch going into C; more distal to this branch there is a still weaker oblique branch, the end of which is not clear; M_1 terminates exactly at the apex of the wing. The length of the remnant is 1.65 mm; the length of the whole wing is about 2 mm.

Material. – The holotype.

Family Protorhyphidae Handlirsch, 1906
Handlirsch, 1906, p. 487; Rohdendorf, 1962, p. 332

Description. – SC sturdy, less than half the length of the wing. Two long anterior branches of RS branch out on both sides from rm and combine with C. The intermedial cell is small and is located at the middle of the wing. Between the branches of M_1 and M_2 there are no transverse veins. The structure of the basial is unknown. The type of the family is the genus *Protorhyphus* Handlirsch, 1906 (Liassic of Germany).

Comparison. – This family was described for the first time from the Liassic deposits of Germany, from whence are known three species of the related genera *Protorhyphus* Handlirsch and *Archirhyphus* Handlirsch. It is sharply distinguished from all known Rhyphidea by the presence of two long branches of RS; this feature at the same time points to the closeness of these insects to the original forms of asilomorphs.

Genus *Protorhyphus* Handlirsch, 1906
Handlirsch, 1906, p. 487; Rohdendorf, 1962, p. 332

Type of genus: *Bibio simplex* Geinitz, 1887.

Comparison. – It is distinguished from the other known genus of the family, *Archirhyphus*, by the absence of the common trunk of veins M_1 and M_2, which branch out directly from the intermedial cell. (Liassic and Upper Triassic.)

Protorhyphus turanicus Rohdendorf, 1962 (fig. 66*B*)
Rohdendorf, 1962, p. 331, fig. 1073

Holotype. – Wing, Coll. PIN No. 371/1, Issyk-kul, Upper Triassic (Rhaetian?).

Description. – Common trunk of M before the intermedial cell is very weak, indistinct; the distal anterior branch of RS has a fork at the end; the end of SC is at the level of the middle of the first part of RS (from the base to the branching of the proximal anterior branch); both branches of RS leave at an equal distance

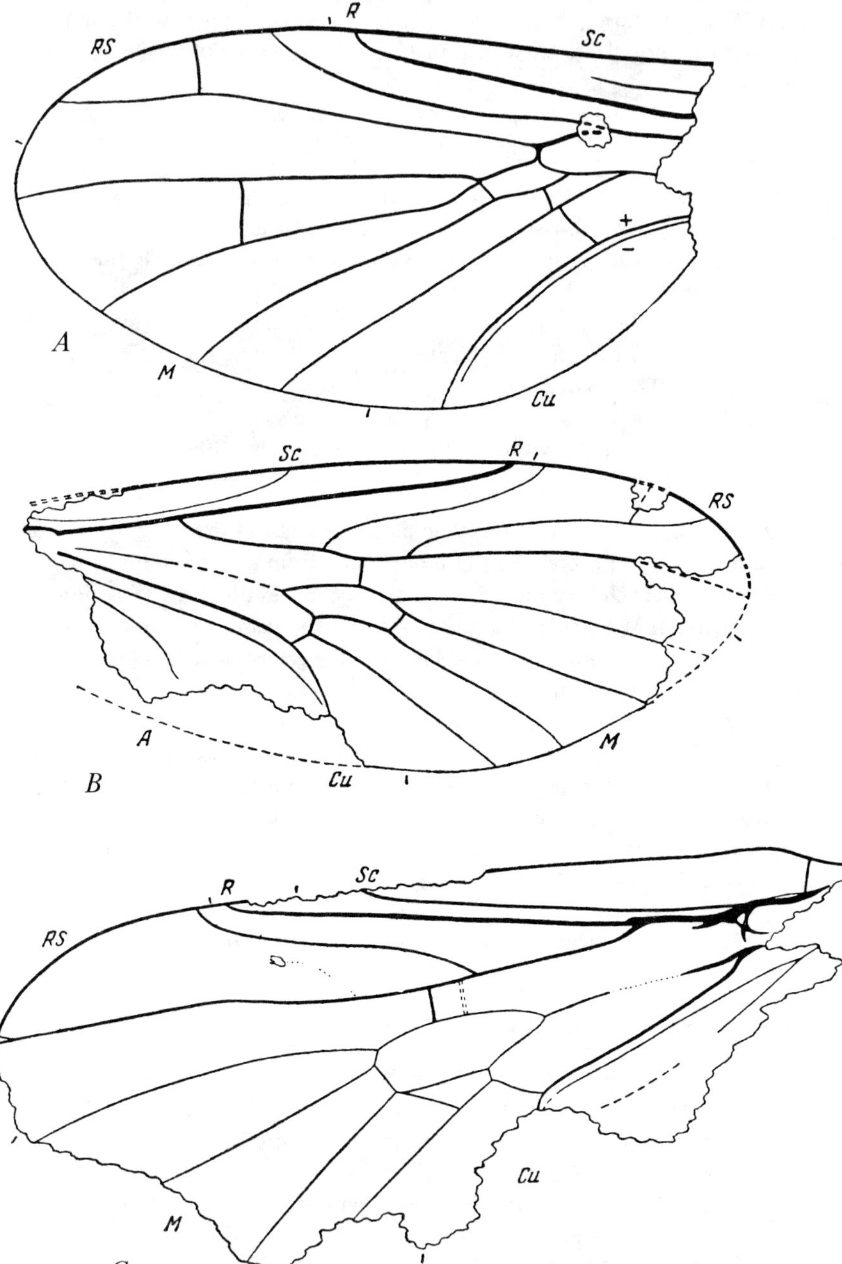

Fig. 66. Rhyphidea of the Triassic of central Asia. *A. Oligophryne fungivoroides* Rohd., (Oligophryneidae). Holotype. Coll. PIN No. 358/120. Length 1.65 mm. *B. Protorhyphus turanicus* Rohd., (Protorhyphidae). Holotype. Coll. PIN No. 371/1. Length 2.7 mm. *C. Protolbiogaster rhaetica* Rohd., (Protolbiogastridae). Holotype. Coll. PIN No. 371/114. Length 4.7 mm. (According to Rohdendorf, 1962.)

from rm; the distal branch of RS begins scarcely more distal than the fork of M_{1+2}; the chief section of M_2 passes uniformly into the distal part of the vein; between RS and M_1 there are no transverse veins; the chief section of M_4 has the appearance of a transverse vein only insignificantly shorter than mcu; CuP is sturdy and well distinguished. The length of the remnant is 2.7 mm; the length of the whole wing is about 3.0 mm.

Comparison. – It differs from the other two known species of the Liassic by the presence of a branch on the distal branch of R, by the thin common trunk of M and a series of other features of venation.

Material. – The holotype.

Family Protolbiogastridae, Rohdendorf, 1962

Description. – SC long, noticeably exceeding half of the wing. There is one sturdy anterior branch of RS, located proximally to rm. The intermedial cell, not particularly small, is located at the middle of the wing. Between the branches of M_1 and M_2 there are no transverse veins. There is a sturdy phragma; the chief trunk of M is sharply reduced, the costal field is wide. The type of the family is the genus *Protolbiogaster* Rohdendorf, 1962.

Comparison. – It is closest to the family Rhyphidae, differing from it in features of the venation (see p. 72).

Genus *Protolbiogaster* Rohdendorf, 1962
Rohdendorf, 1962, p. 332

Type of genus: *P. rhaetica* Rohdendorf, 1962.

Description. – Costal field wide with convexity right after the humeral transverse vein. SC weakly bent ahead, without branches and transverse veins. R almost straight, equal to 0.7 of the wing; basal part of R moderately thickened. The common trunk of R distal to the phragma is thick, with a short posterior branch, terminating freely in the membrane. RS branches from R at the boundary of the first and second quarters of the wing, almost the same calibre as R; anterior branch of RS branches off proximally to the middle of the wing and almost parallel with R, goes into C distally to the end of the latter vein. More distal from rm there branch off from RS indistinct anterior branches or folds; the chief trunk of RS terminates in front of the apex of the wing; rm sturdy and straight, located exactly in the middle of the wing; more proximal to rm is an indistinct supplementary transverse vein. The main trunk of M branches out from CuA in the form of a sturdy, thick vein which soon (at the level of the fork of R+RS) narrows down very strongly, almost disappears, and only in the second half of this part becomes noticeable again in the form of a thin vein. The sturdy rm is located more distal to the middle of the intermedial cell. M_1, M_2 and M_3 branch out from the distal end of the intermedial cell and the branching of the two main veins are sharply drawn together and located approximately at the level of the end of R. The chief section of M_4 has the appearance of a short transverse vein. From the base of M_3 to vein M_4 there runs an oblique transverse vein, which separates an incomplete, quadrangular irregular cell. The transverse long, oblique vein is considerably longer than the chief part of M_4. All veins of the system of M (apart from the base of the main trunk) are noticeably thinner than radial veins. CuA is sturdy, almost straight in its chief part, bending sharply in the distal third. CuP is thin, in the base far removed from CuA. A is apparently thin and short. There is a single species, the type of the genus.

Protolbiogaster rhaetica Rohdendorf, 1962 (fig. 66*C*)
Rohdendorf, 1962, p. 332, fig. 1069

Holotype. — Negative impression of right wing of comparatively good preservation (posterior part of wing, especially the anal blade, the end of vein Cu and the basial are not preserved). Coll. PIN No. 371/114, Issyk-kul, Upper Triassic (Rhaetian?).

Description. — Between the ends of R and the anterior branch of RS there are two thin transverse veins located at an angle one to another; more distally a sturdy anterior branch of RS has two or three weak veins or folds, which branch from RS and terminate freely in the membrane; the end of RS with a narrow and short fork; more proximal to the sturdy rm, at a distance somewhat less than its length is located a parallel but thinner second rm besides which, still closer to the base of the wing between RS and the anterior part of the intermedial cell, are weak oblique transverse veins or folds (not shown in the illustration). The length of the wing is 4.7 mm.

Material. — The holotype.

General character of the Upper Triassic fauna

The dipterous fauna of the Upper Triassic, which is described for the first time, consists not only of numerous species, genera and families, but is also in great measure peculiar, differing sharply from all other younger faunas. As we see, the distinctions of the Upper Triassic Diptera from even the Lower Jurassic fauna of western Europe, which is near according to time, are quite substantial and deep-seated.

In the first place our attention is attracted by the presence of representatives of two quite peculiar infraorders not represented in later faunas. Such are six unusual forms, the structure of which shows that they belong to up-to-now unknown taxa of high rank — the infraorders Dictyodipteromorpha and Diplopolyneuromorpha. These Diptera, known only by the remnants of one wing, and possessing such a peculiar structure of venation, compel us to separate them into special infraorders, very important for our understanding of the paths of the phylogenesis of the order. The representatives of the dictyodipteromorphs have a special significance: they gave a beginning to the original forms of the Bibionomorpha and probably the Tipulomorpha. The second Upper Triassic infraorder of Diptera, the diplopolyneuromorphs, apparently was isolated, and did not give rise to any descendants of the group.

The Dictyodipteromorpha include three families, the unusualness of which indicates their isolation and justifies their superfamily rank. The more important

superfamily, the Dictyodipteridea (with a family of the same name), contains
three species of three genera; this group apparently is closest to the original
forms of bibionomorphs, namely to the superfamily Pleciodictyidea. Another
superfamily, Dyspolyneuridea, is known through a single species and shows a
similarity with the tipulomorphs. This similarity, however, is noticeably less
deep-seated than in the preceding case and does not allow us unconditionally to
consider the dyspolyneurids as ancestral forms of the tipulomorphs. These traits
of resemblance, presumably, are purely convergent. A third superfamily of the
dictyodipteromorphs, Hyperpolyneuridea, known through a single species, is
an isolated group. Its connections with younger Diptera are not clear.

The rest of the 47 species are distributed among the already-known and surviving infraorders Bibionomorpha and Tipulomorpha. But the composition of
these infraorders in Upper Triassic time was essentially different from that in
the Jurassic and more so from that in the Cenozoic.

Thirteen species of tipulomorphs of the Upper Triassic deposits are known,
distributed among five superfamilies, of which only two, the Tipulidea and
Chironomidea, live until the present day. The three other superfamilies appeared
only for the first time and are absent already from the Jurassic fauna. They are
known from a single species of the peculiar Tipulodictyidea, which links the
tipulomorphs with the dictyodipteromorphs, the still more peculiar Eopolyneuridea
(known from three species of two families) and, finally, the Rhaetomyiidea
(described from a single species). This last group of Upper Triassic tipulomorphs
attracts attention by the high degree of mechanical improvement of the venation
of the wing which reminds us of many younger Diptera, and in the series of
features shows traits which one can observe only in more complete Cenozoic
groups (for example, in representatives of the superfamily Orphnephilidea). It
is very probable that the rhaetomyiids were represented also in younger faunas.
The recent finding in the contemporary fauna of the unusual relict genus *Perissomma,* which shows a resemblance to the genus *Rhaetomyia,* is very interesting
(see p. 60).

The bibionomorphs are the most diverse and abundantly represented group
of Upper Triassic Diptera: up to now 34 species, 15 genera and eight families are
described. All these diverse bibionomorphs are distributed among five superfamilies,
of which only two, Fungivoridea and Rhyphidea, are known from younger faunas.
The remaining three superfamilies (known from single species) apparently are
purely Triassic groups; of them, the Pleciodictyidea undoubtedly the most ancient
group, is characterized by a slight degree of mechanized venation; two other
superfamilies, Protoligoneuridea and Phragmoligoneuridea, on the contrary, possess special features of great mechanical improvement of the venation of the wing.

The superfamily Fungivoridea, richest in species, is known for 28 species of
nine genera of two families; it, in fact, is represented by one big family, the
Pleciofungivoridae, to which belong 27 species of eight genera. This family, described for the first time from Jurassic deposits of Karatau, as already explained
by me earlier (Rohdendorf, 1946), is ancestral to many Cenozoic groups of fungivorids and probably also to the whole superfamily Bibionidea, still not represented

in the Upper Triassic fauna. The second family of fungivorids, Palaeopleciidae, is described on the basis of a single peculiar genus, the connection of which with recent Diptera is insufficiently clear.

In the last superfamily, Rhyphidea, three species were discovered of three families of one known before (Protorhyphidae) and of two new ones (Oligophryneidae and Protolbiogastridae). The first new family of rhyphids is very peculiar and indicates a connection of this important group of Diptera with the bibionids; the second, on the other hand, is already quite specialized.

This brief and concise characteristic of the Upper Triassic Diptera described for the first time shows the unusualness of the composition of this most ancient fauna. In fact, the resemblance to the recent fauna, even to the Liassic fauna of Europe, the nearest according to time, is very small. Of common species there are none; of common genera two (*Architipula* and *Protorhyphus*), of common families three (Architipulidae, Pleciofungivoridae and Protorhyphidae). Even of the 13 superfamilies only four are represented in the Jurassic fauna (Tipulidea, Chironomidea, Bibionidea, and Rhyphidea). All the remaining groups are peculiar and are located only in this fauna. Such sharp distinctions of this fauna indicate with clarity that attachment of it to more ancient time; this gives me the basis on which I consider it as Upper Triassic.

It is very tempting to attempt to throw light upon the ecological features of the described Upper Triassic fauna of Diptera to clarify the habitat of these diverse forms. However, the character of the preservation of this fauna (known from wings only), and its little resemblance to the contemporary (a single common family) very greatly hamper the solution of the problem. It is possible on the one hand to draw conclusions concerning the character of the flight of fossil forms and, on the other, to make assumptions about the way of life of representatives of those or other families and superfamilies which are known at present. On the basis of such data it is certainly very difficult to judge what the way of life and conditions of existance might have been.

The structure of the wing of the Dictyodipteridae is very peculiar. The small sizes (not exceeding 3 mm), broad base, absence of costalization or very weak development of it, their elongate form with no separation of apex and trailing edge — all this bears witness in the first place to the small velocity of the stroke, and simultaneously indicates the low speed of flight and the insignificance of the strenght of thrust. We may assume for dictyodipterids that flight was of small importance: apparently these insects flew little and their elongated wings to a great extent fulfilled the function of integumentary organs covering the body. The structure of the posterior pair of wings, unknown up till now, is of very great interest; there is every reason to suppose that in representatives of this group of Diptera we may find the first forms of the production of buzzing organs, probably not yet altered into the present quickly-moving, club-shaped halteres, but representing reduced wing plates. When considering the possible flying properties of the wings of dictyodipterids there comes to mind the partly similar structure of wings of the quite different equal-winged insects of the Derbidae, for example, of the genus *Muiria* (Rohdendorf, 1949, p. 65, fig. 386);

the resemblance lies only in the form of the wing, also strongly elongate and devoid of an isolated apex and termen. Of course this likeness is small and sharply disturbed by the costalized venation and greater absolute size of the Derbidae. Nor can we exclude the possibility of the presence of hairs or scales on the wings of the Dictyodipteridae or of the feather-winged type in conformity with the small sizes of the wing. However the relative clearness of the venation and the absence of any traces of scales or hairs (even in the form of their places of attachment, their sockets) on the impressions does not allow us to affirm this; we may, however, recall the similar form of the wing of some of the smallest feathered-winged Copeognatha, for example the genus *Perientomum* (Rohdendorf, 1949, p. 69, fig. 46*A*), not to mention thrips.

Concluding this brief consideration of the possible functional properties of the wings of the Dictyodipteridae and related forms we can with very great difficulty express an opinion concerning the way of life of these insects. The small size and the weak flight allow us to assume very great contact with the substrate on which these insects could dwell, but concerning the character of the substrate and the medium of the habitat it is almost impossible to say anything definite; the dictyodipterids could live in overgrowths of low-growing vegetation, on the ground of shady stations, on exposed soils of shores of reservoirs, or, finally, on the trunks of trees. Only with evidence can we affirm the impossibility of the habitat of these insects in dry, highly-isolated open stations devoid of coverings: the impossibility of rapid flight did not allow them to avoid overheating.

Other dictyodipteromorphs, namely the Hyperpolyneuridae and the Dyspolyneuridae, are quite sharply distinguished from the Dictyodipteridae and the characteristics of the functional features of their wings are entirely different. The Hyperpolyneuridae possessed quite wide (length-width ratio 2.4) and small (2.8 mm) wings with moderately expressed costalization, strenghtened complex basial and rich but weak venation. These wings undoubtedly could accomplish quite rapid strokes and possessed weak but noticeable predominance in lifting qualities.

Finally, the last group of dictyodipteromorphs, the Dyspolyneuridae, are characterized by fairly small (more than 3 mm) wings of a strongly elongate shape (length-width ratio 3.7), moderately costalized with large narrow apex. All these features indicate the well-expressed thrusting character of the wings which are not devoid of some lifting features. Unintentionally we are obliged to compare the dyspolyneurids with different chironomids, particularly with representatives of the Architendipedidae, from which they differ by the smaller development of costalization, consequently giving them wings of larger thrust. The speed of the stroke was not especially great but the flight was rapid. These insects possessed a comparatively good flight which permitted them to frequent relatively remote substrates.

As we see, the different representatives of the main infraorders of Triassic Diptera are quite diverse according to their flying features, including the poor flyers, the dictyodipterids, the peculiar hyperpolyneurids and the rapidly-flying dyspolyneurids.

The wings of representatives of the succeeding Upper Triassic infraorder of

Diptera, the Diplopolyneuridae, are very peculiar. In these insects, were produced very early peculiar mechanical improvements of the blade of the wing, which consist of the formation of stable longitudinal folds in the basal two-thirds and isolation of a large apex which was separated from the basal part by a series of transverse veins, i.e., a kind of transverse joint. Such a structure bears witness to the very great stability of the blade of the wing to transverse bending and at the same time to the known mobility of the apex. Taking into consideration the average sizes of the wing (about 5 mm) and its great elongation (more than three times) it is necessary to assume the presence of powerful muscles and the development of comparatively large thrust. The development of so sudden a folding (corrugation) of the blade of the wing in the diplopolyneurids is quite unusual for the Diptera and bears witness to the unusualness of the whole construction of the flying apparatus of these insects; to me it seems that in these insects we see the substitution of the complete basial of other Diptera by strengthening of the basial part of the blade. The wings of these insects are extremely peculiar and cannot be compared with any others. One can notice some similarity with the wings of the youngest forms of the Tipulidae, in which the apex becomes distinct by a series of transverse veins and the production of known corrugation of the blade takes place. However, in the Tipulidae is developed a peculiar tubular basalar and at the same time the apex and corrugation of the blade do not approach that development which we see in the Diplopolyneuridae.

The way of life of these Upper Triassic Diptera can be understood only in the most general terms. The diplopolyneurids possessed ability for relatively rapid flight which, presumably, played a substantial role in their activity. The small size of the body, the comparatively mechanically perfect wing — all this bears witness to the mobility of the diplopolyneurids which could frequent different remote substrates, including open, isolated surfaces. The connections of the flight of these insects with the functions of feeding, reproduction and dispersal are impossible to tell.

The features of the wing structure of Upper Triassic tipulomorphs is diverse and subjected partially to evaluation in the descriptive part. The least mechanically improved wings we observe are in representatives of the superfamily Eopolyneuridea. These are characterized by little elongation, by weak costalization and, in some cases, by a strong expansion of the costal field. All this indicates the absence of thrust properties and simultaneously weak development of features promoting the accomplishment of a rapid stroke. This does not exclude the possibility of integumentary qualities (wide wings with a broad costal field). It is necessary, therefore, to assume the small importance of the flying function in the activity of the eopolyneurids. It is otherwise concerning the representatives of the other three superfamilies of the infraorder, the Tipulodictyidea, Tipulidea and Chironomidea. The first two groups, on the basis of mechanical features of the wings, are close to one another and approximately match up with the present-day tipulids with thrust wings of a different degree of development. The two peculiar families Architendipedidae and Palaeotendipedidae are close to the contemporary Chironomidae. Differences in the flying features of Triassic and present-day forms consist of the greater development of costalization in the latter and consequently in the small lifting qualities of the wings of representatives of the Triassic groups.

Estimating the flying abilities of the Triassic tipulids and chironomids and determining their nearness to contemporary forms tempts us to assume also a similarity in their way of life — the habitat of tipulids in growths of vegetation and the closeness of these and other Diptera to reservoirs in which their larvae now live. Making such an assumption it is necessary to remember the fragmentary nature of the material and the complete absence of data concerning the structure of the body and transformation; an understanding of these parts of the organization may clarify and perhaps totally change our ideas on the biological characteristic of these Diptera. In particular we must mention the qualities of flight in representatives of the peculiar group Rhaetomyiidae. As already mentioned above, the wing of this dipterous form is perhaps one of the most mechanically perfect among all the Upper Triassic fauna.

The wings of rhaetomyiids undoubtedly were able to perform very rapid strokes, and according to all their structure they must be referred to the peculiar variation of thrust-lift (tabanoid) type. The flight of these Diptera presumably was quite controlled and they possessed rather large lifting power. To speculate upon the way of life of the rhaetomyiids is quite impossible owing to the extreme isolation of these Diptera from other tipulomorphs.

Passing to an appraisal of the ecological features of the Upper Triassic bibionomorphs it is necessary in the first place to note the great similarity in the mechanical features of the wings of representatives of the Fungivoridea (namely of the Pleciofungivoridae) with Cenozoic forms. The Upper Triassic representatives of this superfamily, in fact, are distinguished little according to the features of the structure of their wings, not only from Mesozoic (Jurassic) forms but also from the known diverse present-day forms. There is no doubt that the nature of the flying function in representatives of the fungivorids changed little in their long history. The ecological features of these ancient Diptera as a whole presumably were close to those which we observe at present. Aggregations of plant fragments, thick growths of forest plants, abundant flora of fungi — this ecological setting undoubtedly was the niche in which might be preserved Diptera having the character of a substantially relict group. Changes in the historical development of the Fungivoridea took place in a completely different direction (see p. 65). Everything described is also substantially true for other superfamilies of the infraorder, in the first place of the Rhyphidea, now clearly expressed relict forms. Concerning the ancient extinct superfamilies Pleciodictyidea, Protoligoneuridea and Phragmoligoneuridea, it is possible to note only the small mechanical improvement of the wings of representatives of the first superfamily and the relatively great mechanical improvement in the aerodynamic qualities of the wings of the Protoligoneuridea and especially of the Phragmoligoneuridea.

The Lower Jurassic Diptera

As already indicated, Lower Jurassic Diptera are known chiefly from western
Europe: in the U.S.S.R. at one time were recorded only the Upper Liassic Diptera
from the Ust-Baleya site on the Angara River.

Dipterous insects from the Lower Jurassic deposits of western Europe are
described by Handlirsch (1906-08, 1937-39) from the Upper Liassic of Dobertin
(Mecklenburg, Germany) and by Tillyard (1933) from the Liassic of England,
and are known to me only from data in the literature. Consequently a review of
the composition of this fauna and an appraisal of the relations of the forms
described has merely a preliminary character. I was obliged to examine only the
description of the insects cited in the works of the above-named authors and
many of these illustrations were definitely not very precise. The reports composed
by Handlirsch and Tillyard are insufficiently complete and often did not illuminate
important features which I had to look for in the figures.

Altogether up to now have been described 53 species of Lower Jurassic Diptera
which are distributed among 17 genera, 11 families (three are established for the
first time), seven superfamilies (one is new) and three infraorders.

List of the Lower Jurassic Diptera of Western Europe

Infraorder Tipulomorpha
  Superfamily Tipulidea
    Family Eolimnobiidae Rohdendorf, 1962
      Genus *Eolimnobia* Handlirsch, 1906
        E. geinitzi Handlirsch, 1906

    Family Architipulidae Handlirsch, 1906
      Genus *Architipula* Handlirsch, 1906; (first group of species)
        A. seebachi (Geinitz, 1884) (type of the genus!)
        A. seebachiana Handlirsch, 1906
        A. elegans Handlirsch, 1906
        A. simplex Handlirsch, 1939
        A. intermedia Handlirsch, 1936
        A. major Handlirsch, 1939
        A. conspicua Handlirsch, 1939
        A. clara Handlirsch, 1939
        A. dubia Handlirsch, 1939; (second group of species)
        A. latipennis Handlirsch, 1906
        A. pusilla Handlirsch, 1939
        A. geinitzi Handlirsch, 1939
        A. vicina Handlirsch, 1939; (third group of species)
        A. stigmatica Handlirsch, 1906; (fourth group of species)
        A. nana Handlirsch, 1939
        A. parva Handlirsch, 1939

A. pulla Handlirsch, 1939; (isolated species)
A. minuta Handlirsch, 1939
A. debilis Handlirsch, 1939
A. obliqua Handlirsch, 1939
A. areolata Handlirsch, 1939; (species of obscure position)
A. arculifera Bode, 1907
A. bodei Handlirsch, 1939
A. brunsvicensis Handlirsch, 1939

Genus *Mesotipula* Handlirsch, 1939
    M. brachyptera Handlirsch, 1939
    M. curvata Handlirsch, 1939

Genus *Protipula* Handlirsch, 1906
    P. crassa Handlirsch, 1906

Genus *Eotipula* Handlirsch, 1906
    E. parva Handlirsch, 1906 (type of the genus)
    E. lapidaria Handlirsch, 1906
    E. longa Handlirsch, 1939
    E. mortua Handlirsch, 1939
    E. defuncta Handlirsch, 1939
    E. coarctata Handlirsch, 1939

Genus *Liassotipula* Tillyard, 1933
    L. anglicana Tillyard, 1933

Superfamily Eoptychopteridea Handlirsch, 1906 n, transl. Rohdendorf, hic
(ex Eoptychopteridae Handlirsch, 1906)
  Family Eoptychopteridae Handlirsch, 1906
    Genus *Eoptychoptera* Handlirsch, 1906
      E. simplex (Geinitz, 1887)

    Genus *Proptychoptera* Handlirsch, 1906
      P. liassina Handlirsch, 1906
      P. maculata Handlirsch, 1939
      P. similis Handlirsch, 1939
      P. megapolitana Handlirsch, 1939

Infraorder Bibionomorpha
  Superfamily Fungivoridea
    Family Pleciomimidae Rohdendorf, 1946
      Genus *Archibio* Handlirsch, 1939
        A. mycetophilinus Handlirsch, 1939

Family Pleciofungivoridae Rohdendorf, 1946
   Genus?   Species?   Tillyard, 1933

Superfamily Bibionidea
  Family Eopleciidae Rohdendorf, 1946
    Genus *Eoplecia* Handlirsch, 1925
      E. primitiva Handlirsch, 1925

  Family Protopleciidae Rohdendorf, 1946
    Genus *Protoplecia* Handlirsch, 1906
      P. liassina Geinitz, 1884

Superfamily Rhyphidea
  Family Protorhyphidae Handlirsch, 1906
    Genus *Protorhyphus* Handlirsch, 1906
      P. simplex (Geinitz, 1887)
      P. stigmaticus Handlirsch, 1939

    Genus *Archirhyphus* Handlirsch, 1939
      A. geinitzi Handlirsch, 1939

  Family Olbiogastridae Hennig, 1948
    Genus *Mesorhyphus* Handlirsch, 1939
      M. nanus Handlirsch, 1939
      M. areolatus Handlirsch, 1939
      M. anomalus Handlirsch, 1939

Infraorder Asilomorpha
  Superfamily Stratiomyiidea
    Family Protobrachycerontidae Rohdendorf, 1962
      Genus *Protobrachyceron* Handlirsch, 1939
        P. liasinum Handlirsch, 1939

Diptera of the Upper Liassic Deposits of Ust-Baleya (Shore of the Angara River of the Irkutsk Region)

Infraorder Tipulomorpha
  Superfamily Psychodidea
    Family Psychodidae Newman, 1834
      Genus *Mesopsychoda* Brauer, Redtenbacher, Ganglbauer, 1887
        M. dasyptera B., R., G., 1887

Description of the Lower Jurassic Diptera

The compilation of keys for the recognition of representatives of the Lower Jurassic fauna of Diptera at present is premature as a result of their slight investigation. It is necessary only to consider some forms, the relations of which are clearest.

Infraorder Tipulomorpha

Superfamily Tipulidea Latreille, 1802
Family Eolimnobiidae Rohdendorf, 1962
Rohdendorf, 1962, p. 313

Description. – Closed intermedial (discoidal) cells are absent; the transverse rm is located at the middle of the wing; SC long, reaches the upper third of the wing; two clear A veins.

Comparison and composition. – Apparently is closest to the present-day Ptychopteridae, being the ancestral group for the latter. There is a single monotypical genus, *Eolimnobia* Handlirsch, 1906.

Family Architipulidae Handlirsch, 1906
Handlirsch, 1906, p. 490

Remarks. – As already noted above (p. 150), this family is close to the Cenozoic families of Trichoceridae, Limoniidae and others. It is necessary to add that numerous species described by Handlirsch are quite diverse and presumably belong to a considerably greater number of genera than the five indicated above; the final decision possibly will stand only after a revision of the types of Handlirsch or the obtaining of new material on the Liassic Diptera. An examination of the figures and descriptions of Handlirsch allows us only to indicate the classification of species of the genus *Architipula* without any firm and definitive judgement.

Superfamily Eoptychopteridea Handlirsch, 1906
[nom. trans. Rohdendorf, 1962 (ex Eoptychopteridae, Handlirsch, 1906)]
Family Eoptychopteridae Handlirsch, 1906
Handlirsch, 1906, p. 488

Remarks. – This group, another one separated by Handlirsch, includes forms of small size; the length of their wings does not exceed 6 mm. Especially interesting is the structure of the medial system of veins, four branches of which form a closed and very big medial cell which distinguishes these insects sharply from all other representatives of the tipulomorphs. Another characteristic of the venation of the eoptychopterids is the presence of transverse veins between R and the anterior branch of RS and the frequent presence of a second rm vein between the distal sections of veins RS and M. All these features indicate the sharply isolated position of the Eoptychopteridae in the system of tipulomorphs, being definitely representatives of a peculiar superfamily. The phylogenetic relationships of this group are analyzed below (p. 290).

Infraorder Bibionomorpha

Superfamily Fungivoridea Latreille, 1809
Family Pleciomimidae Rohdendorf, 1946
Rohdendorf, 1946, p. 61

Remarks. – This family, well known on the basis of the Middle Jurassic fauna of Karatau, is recorded for the first time in the Liassic. The genus *Archibio* was described by Handlirsch in 1939 without an exact indication concerning attachment to the family. In reality this genus is a definite representative of the pleciomimids; examining the relation of *Archibio* to known Karatau genera and subfamilies it is possible to record with clearness its features, but it is not possible to include it in any subfamily and still less in a definite genus. The greatest closeness is shown with the subfamily Antefungivorinae; in fact, however, the genus *Archibio* forms a special very ancient and presumably primary subfamily of pleciomimids (Archibioninae sbf. n.); there is no point in describing it on the basis of literature materials.

Family Pleciofungivoridae Rohdendorf, 1946
Rohdendorf, 1946, p. 51

Remarks. – So far the Liassic faunas of western Europe have not yielded one fossil which could possibly be referred to the pleciofungivorids, a family that was certainly richly developed in Upper Triassic and Jurassic times. Tillyard described only one wing, which he referred erroneously to the Protorhyphidae, but which almost certainly is a representative of the Pleciofungivoridae. Appraising the description of the venation of this form (no figures are presented by Tillyard!) one can come to a conclusion concerning the nearness of it to the Triassic genera *Archihesperinus* and *Palaeohesperinus* on the one hand and to the Upper Jurassic Karatau genera of *Eohesperinus* and *Prohesperinus* on the other. [11] One can affirm with very great probability that the absence of this family from the Liassic faunas of Europe is clearly by chance, reflecting the incompleteness of the investigations. Further work on these faunas will necessarily uncover great numbers of diverse forms of Pleciofungivoridae.

Superfamily Bibionidea Newman, 1834
Family Eopleciidae Rohdendorf, 1946
Rohdendorf, 1946, p. 43

Remarks. – Known only from one monotypical genus from the Upper Liassic of Germany. This family is characterized by the presence of two sturdy anterior branches of RS, of which the proximal branches off from the main trunk of RS closer to the base of the wing than rm; SC long, greater than half of the wing.

11. This fossil, described by Tillyard (1933, pp. 73-74) according to the specimen in the British Museum No. 1-10458, belongs to a special genus and species of Pleciofungivoridae.

Family Protopleciidae Rohdendorf, 1946
Rohdendorf, 1946, p. 42

Remarks. — There are one Liassic and two Upper Jurassic genera (see p. 255).
It differs from the preceding family by the absence of the proximal anterior
branch of RS and the shorter SC, less than half of the wing.

Superfamily Rhyphidea Newman, 1834
Family Protorhyphidae Handlirsch, 1906
Handlirsch, 1906, p. 488

Remarks. — This family is described above (p. 203) during the analysis of the
Upper Triassic fauna in which is represented one of the Liassic genera, the type
of the family.

Family Olbiogastridae Hennig, 1948
Hennig, 1948, p. 48
Rohdendorf, 1962, p. 332

Remarks. — Is characterized by a short, SC, by one branch of RS, by the sturdy
common trunk of M. There are two genera in the contemporary and two genera
in the Jurassic fauna.

It is known from the Upper Liassic fauna of bibionomorphs; therefore, in its
composition it includes representatives of three superfamilies, six families, seven
genera and only 10 species.

Infraorder Asilomorpha

Superfamily Stratiomyiidea Latreille, 1802
Family Protobrachycerontidae Rohdendorf, 1962
Rohdendorf, 1962, p. 334
(Protobrachyceridae, emend. n.)

Description, comparison and composition. — To this family belong Diptera
described by Handlirsch on the basis of a single impression of a wing of very
small size (4.5 mm) from the Liassic of Dobertin (Handlirsch, 1937-39). This
form is closest to representatives of the present-day family Solvidae, being distinguished by an open cell M_3 owing to the fusion of vein M_4 with the edge of
the wing but not with M_3. These peculiar traits (i.e. small size and absence of
the closed cell) allow us to regard the genus *Protobrachyceron* Handlirsch as a
representative of a special family, presumably, immediately antecedent to the
Solvidae. This form is the most ancient representative of the infraorder Asilomorpha, the first 'fly' — Diptera with an integrated body.

Common Character of the Liassic Fauna

The Liassic fauna of Diptera described up to now, although containing upwards of 50 species, is yet characterized by considerably smaller variety in comparison with both the most ancient Upper Triassic and with the last Upper Jurassic of Karatau. As a whole the Liassic Diptera show evident nearness on the basis of systematic composition to the Middle Jurassic, Karatau, differing considerably from the Upper Triassic fauna.

Of the three infraorders represented in the Liassic fauna the most numerous in species is the Tipulomorpha, to which belong two superfamilies, three families, eight genera and 41 species: in fact, four-fifths of the species composition of the Liassic fauna belong to the tipulomorphs. Examining the composition of this chief Liassic infraorder of Diptera it is necessary to note on the one hand the sharp predominance of tipulids represented by the big family, Architipulidae (including not less than five genera and 34 species) and the special Liassic family Eolimnobiidae, and on the other hand the presence of the peculiar Liassic superfamily Eoptychopteridea (two genera with five species).

The second, comparatively scarce infraorder, Bibionomorpha, is represented by 10 species. However it is relatively very diversified: it consists in the Liassic fauna of three superfamilies, six families, and seven genera. The most diverse superfamily, Rhyphidea, includes two families with three genera and six species. Characteristic among the bibionomorphs of the Liassic are forms of the superfamily Bibionidea which is represented by two families; one is purely Liassic (Eopleciidae); the other developed also in the Upper Jurassic. Another characteristic of this fauna is the appearance of the first representative of the Pleciomimidae (genus *Archibio*), a family well developed in the fauna of Karatau.

A particularly important trait of the Liassic fauna of Diptera is the presence of a representative of the infraorder Asilomorpha. This peculiarity of the fauna is very interesting and shows its great differences from the Upper Triassic faunistic complex.

Presenting this appraisal of the composition of the Liassic fauna it is necessary to have in view certain different ecological features of forms recorded up to now in comparison with the Upper Triassic Diptera of central Asia. In first place is the conspicuous absence or very slight development of some groups of Diptera discovered in the Upper Triassic fauna and well developed in the Upper Jurassic and partly also in the Cenozoic. Such are numerous Pleciofungivoridae which constitute in the Upper Triassic the main background of the whole faunistic complex, but in the Liassic fauna are discovered only in the form of single species. There is also the strange absence from the Liassic of representatives of the Chironomidea which have been discovered in the Upper Triassic and which are richly developed in all more recent faunas except the Liassic of western Europe. Apparently these differences reflect on the one hand the rarity of burial in these locations as a consequence of distinct natural laws; on the other hand they bear witness to the different ecological features of these faunas. Thus the abundance of small bibionomorphs and other minute Diptera ranging from 1.5 to 3 mm in

the Triassic faunistic complex sharply distinguish it from the Liassic fauna of
Dobertin in which insects are noticeably larger and rarely less than 3 mm; usually
their sizes range from 4 to 7 mm. The ecological characteristics of both faunas
may be concisely determined by the presence in the Triassic complex of Issyk-
kul of a large number of terrestrial species of insects not connected with reservoirs
(Bibionomorpha as a whole) and by the presence, on the contrary, in the Liassic
fauna of Dobertin of a large number of insects, the larvae of which are connected
with water.

The study of the Liassic fauna of western Europe which was led by Handlirsch
alone has fallen behind the present-day level of knowledge and, in fact, has now
stopped completely. This is in spite of the importance of the Liassic fauna for
an understanding of the history of the order of Diptera.

Middle Jurassic Diptera of Karatau

The Diptera of the Middle and Upper Jurassic are known from southern
Kazachstan (the Karatau fauna) and from some points in Germany and England.
The most thoroughly studied are the Karatau Diptera; European Jurassic faunas
are known only on the basis of single descriptions of isolated forms, published
partly by Handlirsch in his classical summary, partly also by some authors of
the nineteenth centry. The description of European Upper Jurassic Diptera in
most cases does not allow for family determination (Rohdendorf, 1946, p. 38),
and compels us to hope for better. The basis of our knowledge of the Jurassic
fauna is the data on the Karatau Diptera first discovered by Martynov (1925)
and described by me (Rohdendorf, 1938, 1946, 1947a). The age of the Karatau
fauna was synchronized by Martynov (1938) with upper parts of the Dogger of
Europe; this determination to me originally (1938, 1946) did not seem completely
reliable and I attempted to consider this fauna as Upper Jurassic, near to Solen-
hofen on the basis of age. At present it is necessary to reject this assumption and
to accept the first determination of age by Martynov — to consider the Karatau
fauna as Middle Jurassic.

The composition of the fauna of Diptera of the Jurassic of Karatau is still far
from completely known, although already it amounts to 64 species. In all there
have been reported (and are described in this work) 64 species, belonging to 55
genera, 25 families, 12 superfamilies and four infraorders.

List of the Middle Jurassic Diptera of Karatau

Infraorder Tipulomorpha
    Superfamily Tanyderophryneidea
        Family Tanyderophryneidae Rohdendorf, 1962
            Genus *Tanyderophryne* Rohdendorf, 1962
                T. multinervis Rohdendorf, 1962

Superfamily Tipulidea
   Family Architipulidae Handlirsch, 1906
     Genus *Architipula* Handlirsch, 1906
         A. longipes Rohdendorf, sp. n.
         A. protipuloides Rohdendorf, sp. n.

Superfamily Dixidea
   Family Dixamimidae Rohdendorf, 1962
     Genus *Dixamima* Rohdendorf, 1962
         D. villosa Rohdendorf, 1962

Superfamily Chironomidea
   Family Protendipedidae Rohdendorf, 1962
     Genus *Protendipes* Rohdendorf, 1962
         P. dasypterus Rohdendorf, 1962

   Family ?
     Genus *Eopodonomus* Rohdendorf, g.n.
         E. nymphalis Rohdendorf, sp. n.

Superfamily ?
   Family ?
     Genus *Pachyuronympha* Rohdendorf, g.n.
         P. karatauensis Rohdendorf, sp. n.

Superfamily Mesophantasmatidea
   Family Mesophantasmatidae Rohdendorf, 1962
     Genus *Mesophantasma* Rohdendorf, 1962
         M, tipuliforme Rohdendorf, 1962

Infraorder Bibionomorpha
   Superfamily Fungivoridea
   Family Archizelmiridae Rohdendorf, 1962
     Genus *Archizelmira* Rohdendorf, 1962
         A. kazachstanica Rohdendorf, 1962

   Family Pleciofungivoridae Rohdendorf, 1946
     Genus *Polyneurisca* Rohdendorf, 1946
         P. atavina Rohdendorf 1946

     Genus *Transversiplecia* Rohdendorf, 1946
         T. transversinervis Rohdendorf, 1946

     Genus *Pleciofungivorella* Rohdendorf, 1946
         P. binerva Rohdendorf, 1946
         P. brevisubcosta Rohdendorf, sp. n.
         P. proxima Rohdendorf, sp. n.

Genus *Prohesperinus* Rohdendorf, 1946
    P. abdominalis Rohdendorf, 1946
    P. pedalis Rohdendorf, sp. n.

Genus *Eopachyneura* Rohdendorf, 1946
    E. trisectoralis Rohdendorf, 1946

Genus *Pleciofungivora* Rohdendorf, 1938
    P. latipennis Rohdendorf, 1938
    P. major Rohdendorf, 1946

Genus *Eohesperinus* Rohdendorf, 1946
    E. martynovi Rohdendorf, 1946
    E. weberi Rohdendorf, sp. n.

Genus *Allactoneurites* Rohdendorf, 1938
    A. jurassicus Rohdendorf, 1938

Family Pleciomimidae Rohdendorf, 1946
  Genus *Lycoriomima* Rohdendorf, 1946
    L. ventralis Rohdendorf, 1946

  Genus *Paralycoriomima* Rohdendorf, 1946
    P. sororcula Rohdendorf, 1946

  Genus *Lycorioplecia* Rohdendorf, 1946
    L. elongata Rohdendorf, 1946

  Genus *Archilycoria* Rohdendorf, 1962
    A. magna Rohdendorf, 1962

  Genus *Lycoriomimodes* Rohdendorf, 1946
    L. deformatus Rohdendorf, 1946

  Genus *Lycoriomimella* Rohdendorf, 1946
    L. minor Rohdendorf, 1946

  Genus *Megalycoriomima* Rohdendorf, 1962
    M. magnipennis Rohdendorf, 1962

  Genus *Pleciomima* Rohdendorf, 1938
    P. sepulta Rohdendorf, 1938
    P. secunda Rohdendorf, 1946

  Genus *Pleciomimella* Rohdendorf, 1946
    P. karatavica Rohdendorf, 1946

  Genus *Mimallactoneura* Rohdendorf, 1946
    M. vetusta Rohdendorf, 1946

Genus *Antefungivora* Rohdendorf, 1938
    A. prima Rohdendorf, 1938

Genus *Antiquamedia* Rohdendorf, 1938
    A. tenuipes Rohdendorf, 1938

Genus *Paritonida* Rohdendorf, 1946
    P. brachyptera Rohdendorf, 1946

Family Fungivoritidae Rohdendorf, n. n.
  Genus *Mesosciophila* Rohdendorf, 1946
    M. venosa Rohdendorf, 1946

  Genus *Mesosciophilodes* Rohdendorf, 1946
    M. angustipennis Rohdendorf, 1946
    M. similis Rohdendorf, sp. n.

  Genus *Mimalycoria* Rohdendorf, 1946
    M. allactoneuroides Rohdendorf, 1946

  Genus *Eoboletina* Rohdendorf, 1946
    E. gracilis Rohdendorf, 1946

  Genus *Allactoneurisca* Rohdendorf, 1946
    A. indistincta Rohdendorf, 1946

  Genus *Fungivorites* Rohdendorf, 1938
    F. latimedius Rohdendorf, 1938

Family Tipulopleciidae Rohdendorf, 1962
  Genus *Tipuloplecia* Rohdendorf, 1962
    T. breviventris Rohdendorf, 1962

Family Sinemediidae Rohdendorf, 1962
  Genus *Sinemedia* Rohdendorf, 1962
    S. angustipennis Rohdendorf, 1962

Superfamily Bibionidea
Family Protopleciidae Rohdendorf, 1946
  Genus *Mesoplecia* Rohdendorf, 1938
    M. jurassica Rohdendorf, 1938
    M. stigma Rohdendorf, sp. n.

  Genus *Mesopleciella* Rohdendorf, 1946
    M. minor Rohdendorf, 1946

Family Paraxymyiidae Rohdendorf, 1946
  Genus *Paraxymyia* Rohdendorf, 1946
    P. quadriradialis Rohdendorf, 1946

Family Protobibionidae Rohdendorf, 1946
  Genus *Protobibio* Rohdendorf, 1946
    P. jurassicus Rohdendorf, 1946

Superfamily Scatopsidea
  Family Protoscatopsidae Rohdendorf, 1946
    Genus *Protoscatopse* Rohdendorf, 1946
      P. jurassica Rohdendorf, 1946

Superfamily Rhyphidea
  Family Protorhyphidae Handlirsch, 1906
    Genus *Archirhyphus* Handlirsch, 1939
      A. asiaticus Rohdendorf, sp. n.

Infraorder Asilomorpha
  Superfamily Tabanidea
    Family Eostratiomyiidae Rohdendorf, 1962
      Genus *Eostratiomyia*, 1962
        E. avia, 1962

    Family Rhagionempididae Rohdendorf, 1962
      Genus *Rhagionempis* Rohdendorf, 1938
        R. tabanicornis Rohdendorf, 1938

    Family Rhagionidae Latreille, 1802
      Genus *Archirhagio* Rohdendorf, 1938
        A. obscurus Rohdendorf, 1938

      Genus *Protorhagio* Rohdendorf, 1938
        P. capitatus Rohdendorf, 1938

      Genus *Rhagiophryne* Rohdendorf, 1962
        R. bianalis Rohdendorf, 1962

  Superfamily Stratiomyiidea
    Family Archisargidae Rohdendorf, 1962
      Genus *Archisargus* Rohdendorf, 1938
        A. pulcher Rohdendorf, 1938

    Family Palaeostratiomyiidae Rohdendorf, 1951
      Genus *Palaeostratiomyia* Rohdendorf, 1938
        P. pygmaea Rohdendorf, 1938

Family Eomyiidae Rohdendorf, 1962
Genus *Eomyia* Rohdendorf, 1962
E. veterrima Rohdendorf, 1962

Family Protocyrtidae Rohdendorf, 1938
Genus *Protocyrtus* Rohdendorf, 1938
P. jurassicus Rohdendorf, 1938

Superfamily Asilidea
Family Protomphralidae Rohdendorf, fam. nov.
Genus *Protomphrale* Rohdendorf, 1938
P. martynovi Rohdendorf, 1938

Infraorder?
Family Palaeophoridae Rohdendorf, 1951
Genus *Palaeophora* Rohdendorf, 1951
P. ancestrix Rohdendorf, 1938

Key to the infraorders, superfamilies and families of the Middle Jurassic Diptera of Karatau

1 (10)   Radial sector with three or even four parallel branches, directed to the apex of the wing; wings generally elongated with straight anterior border ... Infraorder Tipulomorpha 2

2 (3)   Radial sector with five branches forming two long parallel forks and an unpaired posterior branch; medial system without a closed cell; legs moderately elongated, running; head large, thoracic section small, little inflated (fig. 67, 68*A*) .... Superfamily Tanyderophryneidea, Family Tanyderophryneidae (p. 228)

3 (2)   Radial sector of two or three branches; medial system usually with closed cells 4

4 (5)   Wings narrow and delicate, often with hairs and generally with thin and obscure venation; thoracic section very strongly dilated and hangs over the small head; legs thin (fig. 70) .... Superfamily Chironomidea (p. 235)

5 (4)   Wings with clear and well distinguishable venation; thoracic section moderately dilated 6

6 (7)   Forks of the anterior section of the medial system and radio-medial transverse shifted into the distal half of the wing, usually disposed at one level, approximately at the boundary of the second and the last third of the wing; legs of well-expressed thin type; head small (fig. 68*B*) .... Superfamily Tipulidea, Family Architipulidae (p. 231)

7 (6)   Radial transverse and forks of M not shifted into the distal part of the wing; legs running; head large 8

8 (9)   Wing very narrow and long, four times as long as wide; middle branch of RS short and sharply bent to the anterior border; base of wing narrow, without isolated and vannal lobes, or tegula (fig. 72) ... . Super-family Mesophantasmatidea, Family Mesophantasmatidae (p. 239)

9 (8)   Wing wide, the length not more than twice the width, both posterior branches of RS long, parallel; base of the wing with clearly expressed anal and vannal lobes (fig. 69) . . . . Superfamily Dixidea, Family Dixamimidae (p. 232)

10 (1)   Radial sector of a main posterior trunk and going out from it two or three (rarely more) anterior branches, never parallel and going out from it at great angles into the anterior border; form of wings diverse, often wide 11

11 (30)   Medial veins with three branches, of which the two anterior form a characteristic fork, often indistinct; intermedial cell absent; anal veins poorly distinguished, much thinner than anterior cubital; antennae with homonomous segments, usually long and thin; abdomen always longer than the head and thorax taken together . . . Infraorder Bibionomorpha (in part) 12

12 (15)   Venation of wings sharply costalized – displaced to the anterior border, quite sturdy; greater part of wing devoid of veins, distal parts of medial veins vague 13

13 (14)   Subcostal vein thin, nearly reaching the middle of the wing; medial vein greatly removed from RS; anterior branch of RS in the form of a transverse vein between R and RS . . . . Superfamily Bibionidea, Family Protobibionidae

14 (13)   Subcostal vein absent; medial vein in the main part united with RS, which bears two adjacent branches in the form of transverse veins between C and RS . . . . Superfamily Scatopsidea, Family Protoscatopsidae

15 (12)   Venation not costalized, uniformly distributed over the membrane of the wing, medial veins more or less definite. 16

16 (21)   RS simple, always with free branches going into C: but only occasionally has branches disposed in the form of transverse veins between R and RS . . . Superfamily Fungivoridea (in part) 17

17 (18)   R long, not less than three-quarters of wing; medial veins sturdy, highly noticeable . . . . Family Fungivoritidae (p. 252)

18 (17)   R short, not more than three-quarters of wing; medial veins weak, often indistinguishable 19

19 (20)   SC indiscernible; wing narrow; a transverse vein between the common trunk of M_{1+2} and M_3 (fig. 77F) . . . . Family Sinemediidae (p. 254)

20 (19)   SC present, sometimes weak; wing wide; no transverse veins between the branches of M (fig. 77A) . . . . Family Pleciomimidae (p. 249)

21 (16)   RS always bears well-distinguished anterior branches, part of which always unite with C 22

22 (25)   RS highly approximated to R, but some branches of RS have the appearance of transverse veins between R and RS; R long, not less than three-quarters of wing . . . Superfamily Fungivoridea (in part). 23

23 (24)   Base of RS reduced and sometimes obscure; has only one transverse vein between R and RS and a single anterior branch of RS which joins with C (fig. 73) . . . . Family Archizelmiridae (p. 241)

24 (23)   Trunk of RS sturdy throughout; there are some transverse veins between R and RS and one or two anterior branches of RS joining with C (fig. 72) . . . . Family Tipulopleciidae (p. 252)

25 (22)   Field between R and RS wide and usually does not include "transverse" veins (anterior branches of RS), R always shorter, not greater than two-thirds of the wing 26

26 (27)   Basal part of RS from the place of branching from R to the level of the transverse rm short, at the most four times as long as rm; middle part of RS from the level of rm to the base of the anterior branch of RS almost always very large and considerably larger than the anterior branch. Small forms: length of wing 1.5 to 3.5 mm . . . . Superfamily Fungivoridea, Family Pleciofungivoridae (p. 243)

27 (26)   Basal part of RS long, generally more than four times as long as rm, rarely shorter; middle part of RS short, always shorter than the anterior branch of RS. Larger: length of wing 2.2 to 8 mm . . . Superfamily Bibionidea (in part) 28

28 (29)   There are two long branches of RS located on both sides from the level of rm; the anterior part of RS as a result of the presence of the basal branch is divided into two. Sizes small: wing somewhat more than 2 mm, body more than 3 mm . . . . Family Paraxymyiidae

29 (28)   One long branch of RS situated distal to the level of rm. Sizes large: wing 3.75 to 8 mm, body 5 to 10 mm (fig. 78) . . . . Family Protopleciidae (p. 255)

30 (11)   The number of medial veins variable, from two to four; often there is a closed intermedial cell; antennae as a rule short, consisting of a few differently constructed segments 31

31 (34)   Antennae thin, consisting of homonomous cylindrical segments; head considerably smaller than the thoracic section; there is a small, clearly isolated discal cell; the anal and cubital veins do not join or come together . . . Superfamily Rhydidea, Infraorder Bibionomorpha 32

32 (33)   Two anterior branches of RS located on both sides of rm (fig. 79) . . . .

. . . . Family Protorhyphidae (p. 256)

33 (32)   Only one branch of RS, branching from RS proximal to rm. .

. . . . Family Olbiogastridae

34 (31)   Antennae short, being made up of a few differently constructed segments (usually two basal and one distal, bearing jointed arista); head large, usually the same width as the thoracic section; venation very characteristic; if there is a discal cell then it is always large; anal and cubital veins generally join or even fuse at the ends 35

35 (36)   Thoracic section sharply dilated in the form of a hump covering the comparatively small head; venation costalized with shortened and thickened R and anterior branches of RS; there are two large cells in the middle of the wing; posterior medial cubital and anal veins considerably thinner than the radials. . . . .

. . . . Family Palaeophoridae (Infraorder?)

36 (35)   Thoracic section not dilated and not covering the head; venation of different structure 37

37 (46)   Wings with little costalized venation, located on the whole blade of the wing; there is a well-formed discal cell 38

38 (39)   Body shortened, abdomen four-segmented, truncated; wings blunt; two posterior branches of RS parallel, both coming out to the apex of the wing. Sizes small: wing 2.5 mm. . . . .

. . Superfamily Stratiomyiidea, Family Palaeostratiomyiidae (p. 264)

39 (38)   Body moderately elongated, abdomen tapering to the end, consisting of 5 to 7 segments; both posterior branches of RS always more or less diverge. Sizes large: wing 3.5 to 13 mm   . . . Superfamily Tabanidea (in part) 40

40 (41)   There are several transverse rm veins and there is one transverse between the branches of RS and another between RS and R at the level of the end of SC; abdomen gradually tapering to the end, more or less conical (fig. 80A) . . . .

. . . . Family Eostratiomyiidae (p. 259)

41 (40)   One transverse rm 42

42 (43)   Wings narrow and long without vane and scales, with sturdy and clear venation; abdomen parallel at the edges, consisting of seven well-separated segments; antennae short, considerably smaller than the head; thorax narrow. Sizes large: wing 13 mm. . . . .

. . . . Family Rhagionidae, Subfamily Vermileoninae (p. 261)

43 (42)   Wings quite wide with well-developed basal expansions; abdomen more or less conical 44

44 (45)   Antennae protruding ahead with a third segment that is expanded in the form of a blade; head narrow and long; posterior two branches of RS form a sharply diverging fork. Sizes small: wing 3.5 mm . . . .

. . . . Family Rhagionempididae (p. 261)

45 (44)   Antennae short; head wide; posterior branches of RS long, slightly diverging, almost parallel. Sizes larger: 4.25 to 5 mm . . . .

. . . . Family Rhagionidae, Subfamily Protorhagioninae (p. 261)

46 (37)   Venation of wings considerably costalized, posterior half of the
wing blade usually without veins; closed cells usually indistinct . . 47

47 (48)   Veins of wing are moderately shifted to the front margin; RS
forms dichotomous branches, sending off veins to the front margin.
Sizes small: wing about 2 mm . . . .
. . . . Superfamily Asilidea, Family Protomphralidae (p. 266)

48 (47)   Venation sharply costalized . . . Superfamily Stratiomyiidea
(partly) 49

49 (50)   Wings narrow and long (about 16 mm) with well-developed radial
system, R and anterior branch of RS considerably stronger than posterior
branches of RS; abdomen narrow and long with clearly separated segments numbering not less than five; head wide, only slightly narrower
than thorax . . . .
. . . . .Family Archisargidae (p. 263)

50 (49)   Smaller insects; wings (1.5 to 3.5 mm) short and wide; venation
otherwise 51

51 (52)   Venation not clear; third segment of antennae rounded, with a
terminal short arista; abdomen consists of seven distinct segments; length
of wing 1.5 mm (fig. 80*B*) . . . .
. . . . Family Eomyiidae (p. 265)

52 (51)   Venation clear, span of wing greater: 2.5 to 3.5 mm 53

53 (54)   Venation complete; there is a closed medial cell; segments of
abdomen distinct; length of wing 2.5 mm . . . .
. . . . Family Palaeostratiomyiidae (p. 264)

54 (53)   Venation simplified, consisting of simple veins: R, RS, M and
Cu; third segment of antennae elongated; segmentation of abdomen
indiscernible; sizes of wing 3.5 mm. . . . .
. . . . Family Protocyrtidae

Infraorder Tipulomorpha

Superfamily Tanyderophryneidea Rohdendorf, 1962
Rohdendorf, 1962, p. 314

Description and composition. — This peculiar group of tipulomorphs is characterized by a wing which shows numerous veins of very little costalization, narrowed
at the base and devoid of an isolated basial. Medial veins five, radials of six branches;
anal blade almost absent. There is a single family, the Tanyderophryneidae.

Family Tanyderophryneidae Rohdendorf, 1962
Rohdendorf, 1962, p. 314

Description. — Wing constricted at the base and wide in the distal half. SC
less than half of the wing, at the end with a fork connecting it with C and R;
forks of middle branches of RS located at a different level, medial veins form an

anterior fork and a posterior contiguous group of veins: common trunk of M is reduced; transverse rm located somewhat proximal to the middle of the wing. A is not clear, CuA is short. Legs are of the running type, quite thin: ends of tarsi with spurs. The thorax is slightly arched with a well-expressed transverse joint at the back edge. Abdomen considerably longer than the head and thorax taken together, eight-segmented. Head large, antennae apparently quite short, little longer than the head. There is a single genus, the type of the family.

Genus Tanyderophryne Rohdendorf, 1962
Rohdendorf, 1962, p. 314

Type of genus: T. multinervis Rohdendorf, 1962.

Description. — The wing is 2.5 times as long as wide, with a short apex. Anterior fork of RS is divided proximally to the posterior at the level of the middle section of RS_{3+4} between the branching of RS_{1+2} and the fork of RS_{3+4}; vein RS_2 terminates at the anterior edge of the apex, sharply diverging from RS_3; veins RS_3 and RS_4 are parallel and directed to the apex of the wing; RS_5 for the greater part of its extent parallel to RS_4 and only in its last third diverging from the latter to the posterior border; transverse rm at the level of the end of SC; transverse mcu has the form of a curved basal section of MP (i.e. three posterior branches of M). CuA is shorter than half of the wing. Femora twice as wide as the tarsi and slightly shorter; the middle and posterior feet are somewhat shorter than the corresponding tarsi; first tarsomeres are the largest.

Composition. — There is a single species, the type of the genus.

Tanyderophryne multinervis Rohdendorf, 1962 (fig 67, 68A)
Rohdendorf, 1962, p. 314, fig. 989

Holotype. — Female, lying on the left side of the body with folded wings and legs (the antennae and, in part, the legs were not preserved). Coll. PIN No. 2452/301, Halkino, Middle Jurassic, A.V. Martynov, 1924.

Description. — SC is situated centrally between C and R, weaker than both of these veins; R weakly arched bent forward, RS separated from R somewhat proximally to the level of the end of SC; two anterior branches of M form the fork of veins diverging approximately at an angle of 25 degrees, the common trunk of which is shorter than the anterior branches; medial veins are thicker than the others. Legs sturdy; length of anterior shank 1.38 of posterior 1.78; of anterior leg 1.15; of middle 1.25 and of posterior 1.15; of anterior metatarsus 0.58, of middle 0.7 and of posterior 0.73 mm. Length of wing 3.55 mm. Total sizes of the body cannot be measured accurately as a result of the distortion of the abdomen (which was inflated owing to decomposition): the length of the abdomen of the fossil is equal to 4.5 and the length of the whole insect 5.5, which in life did not exceed 3.1 and 4.2 mm. The length of the thoracic section is 1.1, the depth of it is 1.05 mm. The length of the head is 0.35 mm.

Material. — The holotype.

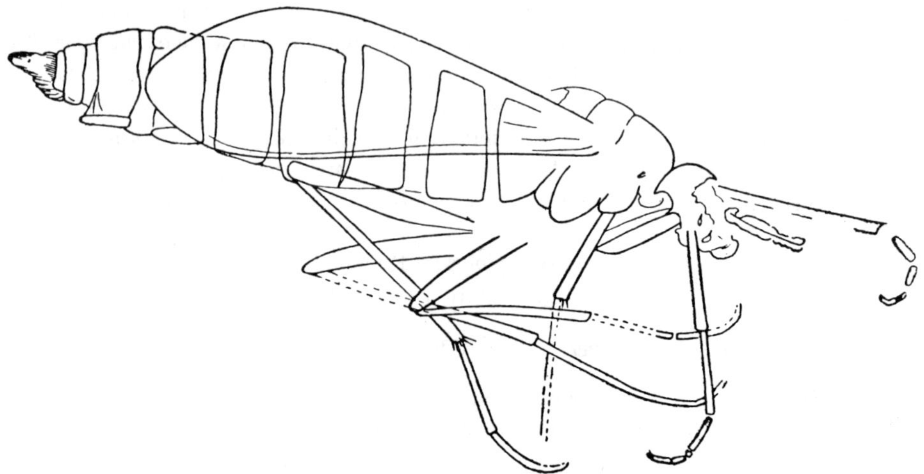

Fig. 67. *Tanyderophryne multinervis* Rohd., (Tanyderophryneidae). Female, general view of the holotype. Coll. PIN No. 2452/301. Middle part of the Jurassic of Karatau. Length 5.5 mm. (Original.)

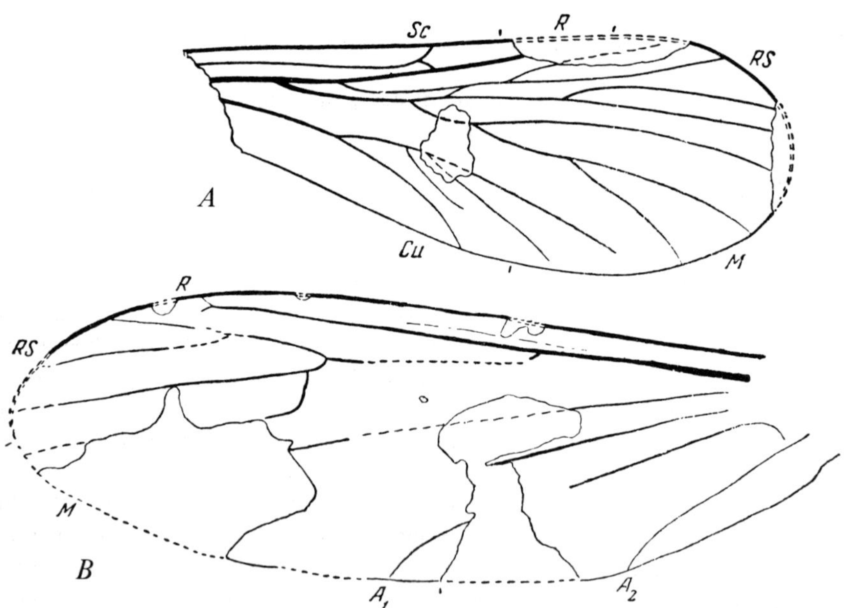

Fig. 68. Tipulomorphs of the Middle Jurassic of Karatau. *A. Tanyderophryne multinervis* Rohd., (Tanydero-phryneidae). Wing. *B. Architipula protipuloides* Rohd., sp. n. Holotype. Coll. PIN No. 2452/303H. Length of wing 4.5 mm. (*A.* according to Rohdendorf, 1962; *B.* original.).

Superfamily Tipulidea Latreille, 1802
Family Architipulidae Handlirsch, 1906
Handlirsch, 1906, p. 490. Rohdendorf, 1962, p. 312

Remarks. — For description see above (p. 149). Until recently the materials on the architipulids of the Karatau fauna remained unstudied on account of their imperfect preservation. The character of the burial of these Diptera in the rocks of Karatau was very peculiar: during the comparatively good preservation of remnants of the body and legs, that of the wings does not allow us to analyze their venation, the basis of the distinction of genera. Altogether in the collection of the Plaeontological Institute of the Academy of Sciences, U.S.S.R. there are only 30 remnants of the tipulids of Karatau, most belonging apparently to the family Architipulidae. Below are described two species of the genus *Architipula*, but one of them on the basis of the remnants of the body; in the absence of the venation of the wings the generic attachment of this form is uncertain.

Architipula Handlirsch, 1906
Handlirsch, 1906, p. 490

Type of genus: *Bibio seebachi* Geinitz, 1884, (Liassic of Germany)

Remarks. — The description of the genus given by Handlirsch requires complete revision together with working over and a more accurate illumination of the systematics of the whole family (see p. 149).

Architipula protipuloides Rohdendorf, sp. n. (Fig. 68*B*)

Holotype. — Wing. Coll. PIN No. 2452/303H, Halkino, Middle Jurassic of Karatau, A. V. Martynov, 1924.

Description. — The wing is moderately elongate, 2.9 times as long as wide, straight for two-thirds of anterior border with large apex not particularly clearly separated. SC is weak, noticeable only in places drawn together with R. R at the end with a vague fork, without anterior branches. RS branches out from R somewhat proximally to the middle of the wing; main part of RS from R to branching approximately equal to the anterior branch of RS_{1+2} which in its main third branches, sending off a posterior branch (RS_2), parallel with RS_3, to the apex of the wing; RS_3 is longest, parallel with RS_2 and M_1. Transverse rm veins in number are two, of which one (proximal) joins the common trunk of RS with M, and the other, (distal) joins the posterior branch of RS with M. The medial veins are weak, they have been poorly retained in the fossil: there is a weak common trunk of M, diverging at small angles with R (about 11 degrees) and with Cu (about 7 degrees). Forks and number of branches of M are unknown. CuA sturdy and long, terminating considerably distal to the middle of the posterior edge of the wing. A_1 sturdy, nearly straight, at the end weakly bent, reaching approximately the middle of the posterior border of the wing. A_2 well developed, almost straight, greatly removed from A_1. Both anal veins go to the border of the wing. Length of the wing is 4.5, width 1.55 mm.

Material. – The holotype. The fossil was in the horizon of paper shales together
with numerous remnants of the pupae of chironomids (genera *Eopodonomus* and
Pachyuronympha (see p. 236-238).

Architipula (?) longipes Rohdendorf, sp. n.
Architipula sp.: Rohdendorf, 1951, p. 24, fig. 13

Holotype. – Remnant of insect (veining is not clear). Coll. PIN No. 2452/286,
Halkino, Middle Jurassic of Karatau, A.V. Martynov, 1924.

Description. – Thoracic section strongly enlarged; prothorax protrudes between
the head and the mesothorax. The legs are of a sharply expressed thin type. Coxae
small, the joints are comparatively large. Femora – the thickest sections of the
legs; anterior femora are parallel to the distal end, equal together with the joint
to 2.94 mm; middle femora barely widening toward the apex, somewhat longer
than the anterior, equal together with the joint to 3.06 mm. Posterior femora
nearly equal to the middle, noticeably thickened towards the apex, 3.0 mm together with the joint. Tarsi parallel distally, almost twice as thin as the femora
but always longer than them; anterior tarsus equal to 3.5, middle 3.31 and posterior
3.75 mm. Legs thinner than tarsi, of different lengths: anterior legs longest, equal
to 3.75 mm and almost two-thirds of their length is composed of the first segment,
the metatarsus; middle legs the shortest, equal to 2.5 mm, and the first segment
composes somewhat more than half of the leg; posterior legs of medium length,
equal to 3.0 mm and are distinguished by the short first segment, approximately
equal only to one-third of the whole leg. The wings are equal to 5.9 mm; their
venation is unknown. Length of the body is 4.9 mm.

Material. – The holotype.

Superfamily Dixidea V.D. Wulp, 1877

Remarks. – Until recently this group of tipulomorphs, nearly related to the
Culicidea, was known only from representatives of the family Dixidae, which in
the present-day fauna have the character of a relict group and are known for the
first time from the Paleogene. The great similarity in the organization of the larvae
(in spite of the great differences of the winged insect) compelled many authors
to consider the contemporary dixids as only representatives of a special subfamily
or in the better case of a family close to the bloodsucking Culicidae. The inaccuracy
of this conclusion has been mentioned (p. 50). These tipulomorphs discovered
in the Karatau fauna are undoubtedly representatives of a group close to the dixids
by virtue of all their peculiarities.

Family Dixamimidae Rohdendorf, 1962
Rohdendorf, 1962, p. 314

Description. – The antennae of the male are 15-segmented, consisting of two
large segments and 13 segments of the flagellum; the latter gradually taper to the
apex, provided with a few short bristles. The proboscis is substantial with a pair
of three-segmented feelers. Eyes are large, kidney-shaped, and separated; there

are small ocelli. The thorax is of medium size. The legs are sturdy, running. Wings are wide, approximately equal to the length of the body. There is an intermedial cell and some intermedial transverse veins. RS branches out from R in the basal third of the wing. Transverse rm located at the middle of the wing. Wings with anal blade. The type of the family and the only genus is *Dixamima* Rohdendorf, 1962.

Comparison. — This Jurassic family differs sharply from present-day dixids in many features which by no means contradict the direct connections of the Dixidae with the Dixamimidae as descendants from ancestors. In its turn the Dixamimidae was definitely the direct derivative of the Liassic Eoptychopteridea together with the ancient, original Culicidea.

Genus Dixamima 1962
Rohdendorf, 1962, p. 314

Type of genus: *D. villosa,* 1962

Description. — Basal segments of the flagellum of the antenna of the male one-and-a-half times as long as wide; the distal segments are thinner, three to four times as long as wide. Second segment of the maxillary palp thickest, irregularly curved; third segment cylindrical, three times as long as wide. There are two ocelli. The anterior edge of the wing is straight. SC quite thin, long, reaching two-thirds of the length of the wing, straight, at the end bent and going into C which, in its turn, is very sturdy, thick at the anterior border of the wing, thinner at the edge of the apex and terminating at the end of the last branch of RS; R thick and almost quite straight; RS considerably thinner than R: main branch of RS is located at the middle between the beginning of RS and the end of SC; anterior branch of RS divides into two veins somewhat distal to the level of the end of SC, forming a characteristic, curved backwards fork of parallel veins; posterior branch also parallel with RS_2 and M_1; rm is located very close to the fork of RS, connecting the base of RS_3 and M; there are three intermedial transverse veins of which one is located immediately behind rm, being its continuation; another transverse vein (may be the base of M_3?) is located proximally from rm and finally the third vein lies at the middle between rm and the edge of the wing; the described transverse veins limit two intermedial cells, of which the distal is approximately twice as large as the proximal; two anterior medial veins form the fork of M_{1+2}, almost immediately coming out from the anterior outer angle of the intermedial cell, the posterior boundary of the intermedial cells is constituted by the vein M_3; the structure of the veins of the wing is not known: apparently there are straight veins CuA and An_1, of which the former unites with M_3 by means of an oblique MCu, having the form of the base of M_3. The abdomen is of six segments, which narrow slightly to the apex.

Composition. — It is described according to a single species, the type of genus.

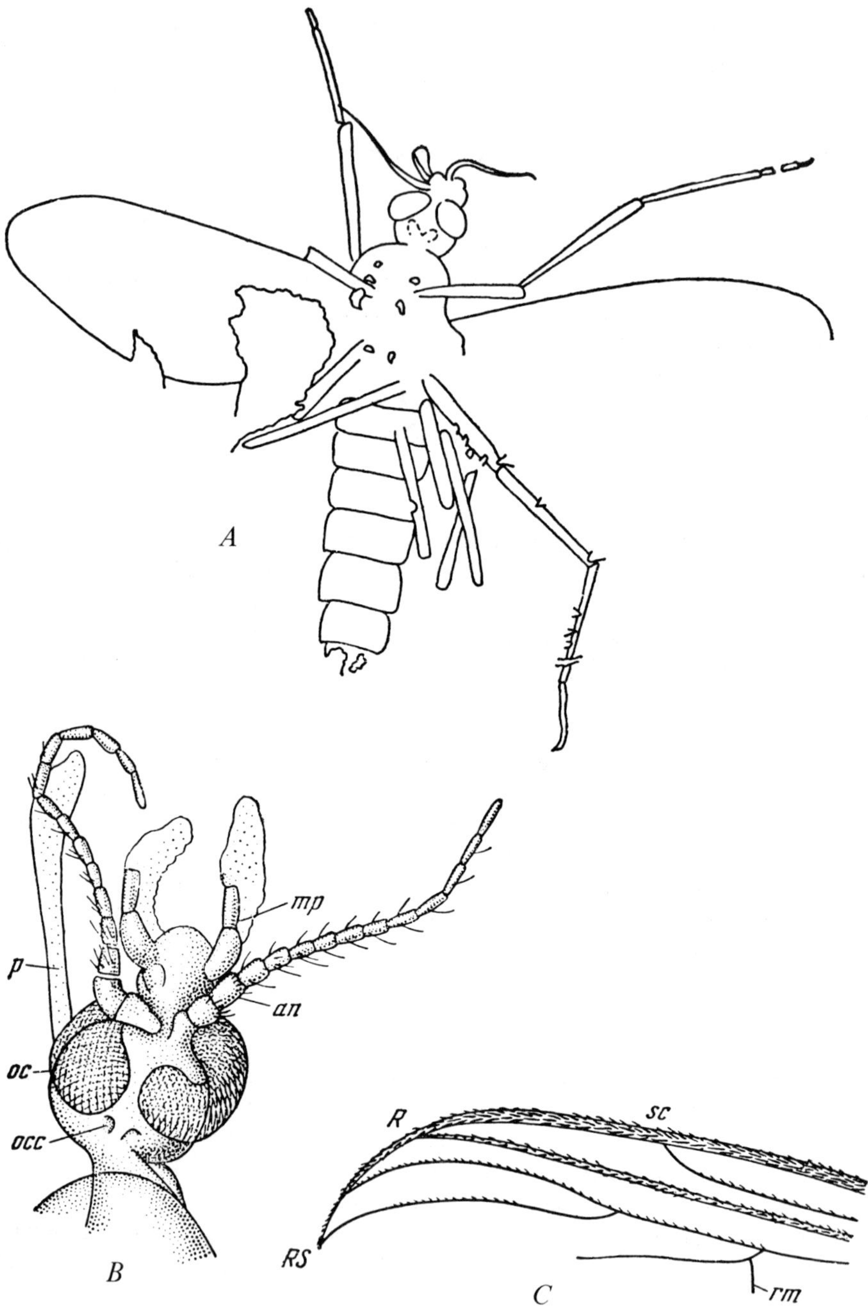

Fig. 69. Dixamimidae of the Middle Jurassic of Karatau. *A. Dixamima villosa* Rohd. General view. Paratype. Coll. PIN No. 334/167. *B.* The same, head of paratype. Coll. PIN No. 335/167. *C.* The same, part of left wing. Holotype. Coll. PIN No. 333/167. (*A.* according to Rohdendorf, 1962; *B.* original.) Abbreviations: *an* – antenna; *mp* – maxillary palpus; *oc* – compound eye; *occ* – simple eye; *p* – proboscis.

Dixamima villosa Rohdendorf, 1962 (fig. 69*A, B*)
Rohdendorf, 1962, p. 314, fig. 990

Holotype. – Remnant of male. Coll. PIN No. 333/167, Michailovka, Middle Jurassic.

Description. – The whole body is covered with quite thick, but not especially long hairs; the distance between the compound eyes is approximately equal to the diameter of the quite large simple eye. The legs are covered by thick bristles; femur approximately one-and-a-half times the thickness of the tarsi, of uniform width: almost all the divisions of the legs are parallel distally. Wings clear, without spots; C, SC, R and RS covered by macrotrichia, located on the thick veins more densely; the parts of C between SC and R on the one hand and R and RS, on the other are equal; the middle part of vein RS, is highly contiguous with the end of R_1 as a result of which the cell of R in this place is sharply constricted. The length of the body is 5.55, the length of the wing 4.00, length of thorax 1.60, length of the abdomen 3.45, length of the posterior femur is 1.30, of the posterior tarsus 1.60 mm.

Material. – Besides the holotype, in the same horizon of shale, approximately in the area of 25 m^2, five remnants of the bodies of males of different preservation were discovered. Coll. PIN No. 88/167, 334/167, 335/167, 336/167, and 337/167 Michailovka, Mount Kochkar-ata, district Ak-Tas, Ters-jeilan, Middle Jurassic of Karatau, B. Rohdendorf, 1937.

Superfamily Chironomidea Macquart, 1838

Remarks. – Quite numerous remnants of Karatau Diptera undoubtedly belonging to this superfamily consist of pupae or their exuviae; remnants of winged chironomids are relatively scarce and the majority are of very incomplete preservation. Altogether in the collections of the Paleontological Institute of the Academy of Sciences of the U.S.S.R. there are only 50 remnants of chironomids from the Jurassic of Karatau, of which approximately 35 belong to immature forms; below are described two forms of pupae and one of the winged insect. The state of preservation of the material does not allow us to determine with certainty the family of the described pupae. The description of the new family is given below according to the remnants of the winged insect, and is very incomplete.

Family Protendipedidae Rohdendorf, 1962
Rohdendorf, 1962, p. 317

Description. – Legs thin, tarsus of moderate length, anterior metatarsus not enlarged, considerably shorter than the tarsus. Antennae in the female consist of not less than seven segments. Head small, covered by a strongly protruding mid-thorax. Proboscis is very short. Wings equal to the length of the body, covered by a thick mantle of hairs (macrotrichia?); the form of them elongated with a narrow basal part apparently without a vane and scales. There is a phragma in the form of a backwardly directed hook; venation specialized (its structure is unknown). Ab-

domen in the female is elongate, conical, consisting of nine segments, of which the
three distal form the ovipositor.

Comparison and composition. – It is distinguished from the present-day Chironomidae by the thick cover of hairs (macrotrichia) on the wings, by their narrowed
base, thin legs, structure of the abdomen and legs. There is a single genus.

Genus Protendipes Rohdendorf, 1962
Rohdendorf, 1962, p. 317

Type of genus: *P. dasypterus* Rohdendorf, 1962.

Description. – The third to sixth segments of the antennae are cylindrical in
form, two-to two-and-a-half times as long as wide; head considerably narrower
than the thoracic section; rounded. Anterior femurs somewhat curved and weakly
thickened towards the end; anterior tarsi almost half as thick as the femur. Middle
femur the same as the anterior, tarsi noticeably shorter. Middle tarsi long, far
longer than the tibiae. Posterior femur extremely elongate, posterior basitarsus is
two-thirds thickness of femur; posterior tarsi as long as their tibiae. Venation of
the wings is not indistinct (as a result of the presence of hairs), but delicate (highly
contorted in the fossil). Three anterior segments of the abdomen (second to fourth)
equal in length and width, the next three (fifth to seventh) in the form of a trap-
ezium, narrower at its posterior end; the end of the abdomen consisting of three
segments is the ovipositor and carried on the eighth to ninth segments pairs of
peculiar appendages. It is described from a single species, the type of genus.

Protendipes dasypterus Rohdendorf, 1962 (fig. 70)
Rohdendorf, 1962, p. 317, fig. 1001

Holotype. – A female lying on the back (anterior legs are poorly preserved,
wings contorted and the venation is not clear). Coll. PIN No. 2452/457, Halkino,
Middle Jurassic of Karatau, A.V. Martynov, 1924.

Description. – Middle tarsus is more than twice as long as the middle tibia; its
first segment is one-third as long as the tibia and half as long as the second segment
which is approximately equal to the third, the fourth is somewhat shorter than
the first, the fifth is quite short. The length of the body is 3.25, wing 3.5, posterior
femur 1.35, posterior tarsus 1.25, width of posterior femur 0.063, posterior tarsus
0.038, width of abdomen is 0.65 mm.

Material. – The holotype.

Comparison. – The form of chironomid which is described below from its pupa
does not allow us to reach a decision concerning its family. The general appearance
suggests pupae of representatives of the Chironomidae: partly Orthocladiinae,
partly Podonominae.

Genus Eopodonomus Rohdendorf, g.n.

Type of genus: *E. nymphalis* Rohdendorf, sp. n.

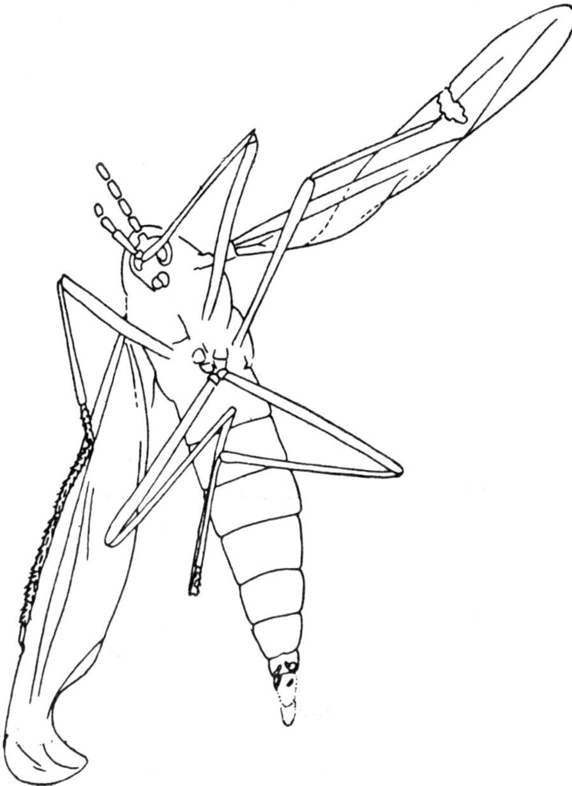

Fig. 70. Chironomidea of the Middle Jurassic of Karatau. *Protendipes dasypterus* Rohd. Holotype. Coll. PIN No. 2452/ 457. Length of body 3.25 mm. (According to Rohdendorf, 1962.)

Description. — Thoracic section somewhat shorter than half the length of the body; thorax convex, hanging over the head, prothoracic spiracle in the form of a long, slightly curved prolongation tapering to the end. The seven segments of the abdomen have angular projections at the sides; the first segment is the narrowest, strongly attached to the thorax, the second is wider, the third, fourth and fifth are the widest, and of equal length; sixth and seventh are not clear; the end of the abdomen has a pair of sharp cerci, backwardly directed and covered with bristles and a pair of projections of irregular form located at the sides of these.

Comparison. — The genus is established from fragments of the pupa of a single species, the type of the genus. The relationship of this form to the chironomids is not in doubt.

Eopodonomus nymphalis Rohdendorf, sp. n. (fig. 71*A*)

Holotype. — Remnant of male pupa lying on the left side of the thorax and the ventral side of the abdomen. Coll. PIN No. 2452/303K, Halkino, Middle Jurassic, A.V. Martynov, 1924.

Description. – Thorax indistinct, abdomen clear with a pattern in the form of transverse dark spots on the anterior edges of second, third, fourth and fifth tergites; cerci of characteristic form. Length of body 5.65, of thoracic section 2.4, of anterior spiracle 1.05, depth of thoracic section 1.75 mm.

Superfamily?

Family?

Genus *Pachyuronympha* Rohdendorf, g.n.

Type of genus: *P. karatauensis* Rohdendorf, sp. n.

Description. – Body curved owing to the strong development of the chief segments of the abdomen. Thoracic section small, less than one-third of the whole length of the pupa; the thorax is moderately convex with a well-marked transverse suture and shield projecting a little, coverings of legs and wings of moderate size. Segments of the abdomen without projections at the sides; segments one, two, three and four in profile with simple saddle-shaped depression, very large: the fifth segment is very narrow; the sixth segment from above has the form of a very wide, expanded in its lateral parts, biscuit-shaped tergite, considerably wider than the fifth and seventh, which in front is convex and with a broad hollow behind; end of abdomen without appendages.

Comparison. – The genus is established from fragments of the pupa of a single species, the type of the genus. The relatively moderate size of the thoracic section, together with the structure and size of the abdomen do not allow us to consider the form as a representative of the superfamily Chironomidea. It is possible to assign this genus to the Tipulidea, but this cannot be proved precisely.

Pachyuronympha karatauensis Rohdendorf, sp. n. (fig. 71*B*)

Holotype. – Remnant of pupa lying on the right side of the body, but the end of the abdomen is turned to the dorsal side. Coll. PIN No. 2452/303 J, Halkino, Middle Jurassic, A.V. Martynov, 1924.

Description. – Color light brown; the dark eyes stand out; there are dark markings on the posterior border of the back; the posterior edge of the anterior segment of the abdomen, the pattern on the second segment, the pleural borders of third and fourth segments are dark brown. Length of the body is 4.75 mm.

Material. – The holotype.

Superfamily Mesophantasmatidea Rohdendorf, 1961

Description. – Highly elongated wings with a narrow anal blade, devoid of a vane, wing scales and phragma, and with a sharply narrowed base; anterior border quite straight, strengthened by having C, SC, and R drawn together. Venation reduced but not displaced either to the anterior border or to the apex. One long anal vein drawn together with the cubital pair. Wing with large cells; C reaches

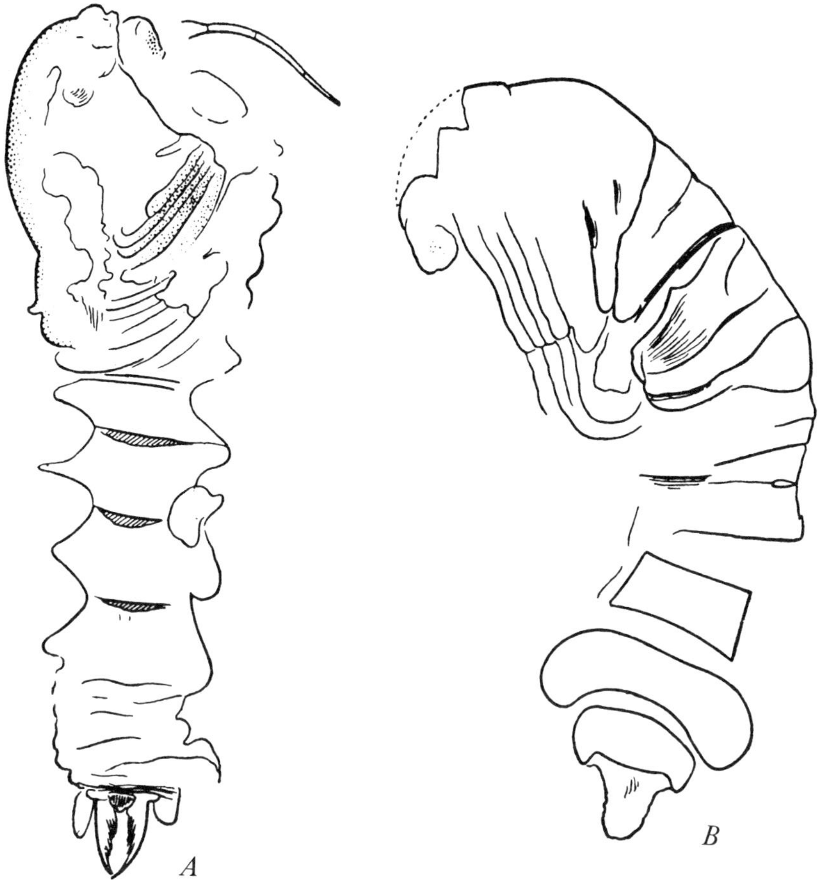

Fig. 71. Larvae of tipulomorphs of the Middle Jurassic of Karatau. *A. Eopodonomus nymphalis* Rohd.,
g.n., sp. n. Holotype. Coll. PIN No. 2452/303 K. Length of body 5.65 mm. *B. Pachyuronympha karatauensis*
Rohd., g.n., sp. n. Holotype. Coll. PIN No. 2452/303 J. Length of body 4.75 mm. (Original.)

only the apex of the wing, there are three branches of M and three branches of
RS.

Composition. – The superfamily includes a single monotypic family of the
Jurassic of Karatau.

Family Mesophantasmatidae Rohdendorf, 1962
Rohdendorf, 1962, p. 319

Description and composition. – R almost reaches the apex of the wing. RS
branches out from R at the middle of the wing and sends off in the first place a
simple anterior branch; the posterior branch of RS forms a fork at the very apex
of the wing but the branches of the fork diverge at a great angle. A sturdy straight

rm is located distal to the branching of the first anterior branch of RS. The three branches of M go into the posterior edge of the wing approximately at an equal distance one from another. CuA and CuP are very highly drawn together, poorly distinguished from one another. The end of CuA reaches the middle of the posterior edge of the wing; A goes into the posterior border immediately proximal to CuA. There is a single genus *Mesophantasma* Rohdendorf, 1962, the type of the family.

Comparison. — This family is characterized by somewhat peculiar features of the venation of the wing, which compel us to regard it as a representative of a very peculiar superfamily and to consider to what infraorder it can be assigned. Actually the elongate form of the wing, the delicate base of it, devoid of alula and axillary lobe, suggests the attachment of this form to the superfamily Tipulidea and especially to the most recent group, namely the Tipulidae. At the same time the greatly advanced reduction and mechanical improvement of the whole venation of the blade of the wing does not occur in any of the known forms of tipulids. The shortening of the vein RS_2, the reduction in the number of branches of M to three veins, the thickening of C, R and CuA in comparison with the thin RS and M suggest representatives of a younger infraorder of Diptera such as certain asilomorphs (for example the Systropodidae or the Cyrtosiidae of the bombyliids and the Vermileoninae of the tabanids). However the unusualness of the basal part of the wing prevents us from referring this form to the Asilomorpha. In fact the place of this remarkable insect in the system of the order remains an open question, not excluding the probability that *Mesophantasma* in reality was a representative of an altogether special infraorder which appeared at the base of some kind of ancient forms (dictyodipteromorphs or bibionomorphs) and which developed a parallel series of features similar to the tipulomorphs and asilomorphs.

Genus *Mesophantasma* Rohdendorf, 1962
Rohdendorf, 1962, p. 319

Type of genus: *M. tipuliforme* Rohdendorf, 1962.

Description. — SC reaches three-quarters of the length of the wing and bears an anterior branch possessing the appearance of a transverse vein and located somewhat proximal to the middle of the wing; R strongly drawn together with SC, on looking at the wing from above it covers this vein and only beyond the middle of the wing branches out from SC backwards; C thick and is complete to the apex of the wing between the two last branches of RS; RS begins as two veins branching from R at the middle of the wing and fusing soon into one unpaired trunk; the anterior branch of RS runs parallel with the end of R and is not less than three-and-a-half times the length of the common trunk of RS; posterior trunk of RS soon after the branching of the anterior branch sends on a straight and quite thick rm and gives a distal fork of sharply diverging (about 36 degrees) RS_2 and RS_3; the common trunk of the posterior branch of RS is twice as long as RS_2; the base of M far thinner than R and Cu and highly approximated to R; M_1 emerges at the posterior border of the wing, twice as close to RS_3 as

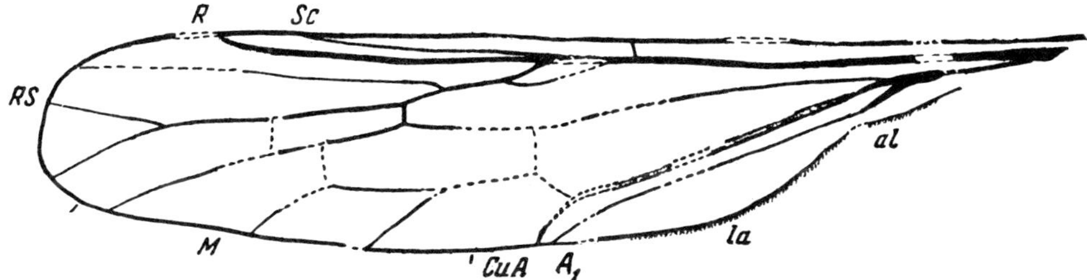

Fig. 72. *Mesophantasma tipuliforme* Rohd. (Mesophantasmatidae). Middle Jurassic of Karatau. Wing. Length 8.1 mm. (According to Rohdendorf, 1962.)

to M_2; the end of the last vein is located at the edge of the wing nearly exactly at the middle between M_1 and M_3; M_3 is located at the border of the wing much closer to M_2 than to CuA which is thicker than CuP and very closely drawn together with it, thinner before going into the edge of the wing. A_1 clear and thin, in its greater part parallel with Cu; this vein at the first description (Rohdendorf, 1962, p. 319) was incorrectly designated as CuP. The anal blade and vane are narrow, with delicate macrotrichia at the edge. The whole membrane of the wing with thick microtrichia. It is described according to a single species, the type of the genus.

Mesophantasma tipuliforme Rohdendorf, 1962 (fig. 72)
Rohdendorf, 1962, p. 319, fig. 1007

Holotype. – Remnant of left wing (parts of the base, its anterior border and large areas of the blade are destroyed. Coll. PIN No. 2452/560, Halkino, Middle Jurassic, A.V. Martynov, 1924.)

Description. – The distal parts of RS_3 and M_1 and also M_2 and M_3 are parallel; length of the wing is 4.8 times as great as its width. Veins dark brown, the membrane of the wing is greyish. Length of the wing is 8.1, its width 1.7 mm.

Material. – The holotype

Infraorder Bibionomorpha

Superfamily Fungivoridea Latreille, 1809
Family Archizelmiridae Rohdendorf, 1962
Rohdendorf, 1962, p. 326

Description and composition – R long, more than three-quarters the length of the wing; RS has two anterior branches: one in the form of a transverse vein between R and RS, the other oblique, going into the edge of the wing between R and the end of the chief branch of RS; the chief trunk of RS and M weak, reduced; rm very short. The head is large; the antennae are not longer than the thoracic section; the legs are thin, of the running type. The thoracic section is small. The abdomen is little enlarged. It is described from the single genus *Archi-*

zelmira Rohdendorf, 1962, the type of the family.

Comparison. — Features of the venation of this form indicate its nearness to the ancient pleciofungivorids (the presence of anterior branches of RS), on the one hand and to younger fungivoritids (elongated R) on the other. The peculiar reduction of the main sections of RS and M in part draws the described form near to the Cenozoic Ceroplatidae and related families which possess the same characteristic mechanical specialization of the venation. The features of the Archizelmiridae point to ancient connections of these insects with the original forms of the two named families which existed simultaneously with them; this alone indicates that direct relationships among them are unlikely.

Genus *Archizelmira* Rohdendorf, 1962

Rohdendorf, 1962, p. 326

Type of genus: *A. kazachstanica* Rohdendorf, 1962.

Description and composition. — The costal field is wide; SC delicate and short, shorter than half of R which is almost straight, sturdy, and at the end very slightly bent forward, before fusion with C; the place of branching of RS is poorly marked and lies apparently in the end of the basal half of the wing; the basal piece of RS from the place of branching from R to rm is very weak, poorly marked, its length is equal to the second section of this vein (from rm to the main anterior branch); the whole basal section of RS at the place of union with the transverse rm is as if drawn by it backwards and both of the basal sections of this vein are at an angle to each other, not being a mutual continuation; the third longest part of RS, between the two anterior branches, is located parallel with the end of R, but the field between these veins is narrow, equal to the width of the costal field at the level of the proximal branch of RS, which is located somewhat obliquely, straight; the distal anterior branch of RS is arch-like, bent outwards and coming out at the edge of the wing almost at the middle between the ends of R and the main (posterior) branch of RS; the medial trunk of M_{1+2} to the fusion with the transverse rm has the form of an arch-like vein uniting with the base of M_3, which, in its turn, also combines with CuA; in the large field of free membrane between the trunks R and CuA, is a trace of a weak longitudinal vein, the rudiment of the common trunk of M; the branch M_{1+2} is much thinner than RS; the common trunk of this vein presumably is equal to the length of M_1 or M_2; M_3 is almost entirely straight; CuA curved in an arch almost reaching the middle of the wing. The abdomen consists of seven segments. It is described according to a single species, the type of the genus.

Archizelmira kazachstanica, Rohdendorf, 1962 (fig. 73)

Rohdendorf, 1962, p. 326, fig. 1040

Holotype. — Remnant of female lying on the abdominal part of the body with extended wings and folded legs. Coll. SAGU No. 17, Karatau, Middle Jurassic.

Description. — Color dark: thorax black, abdomen and antennae brown, legs

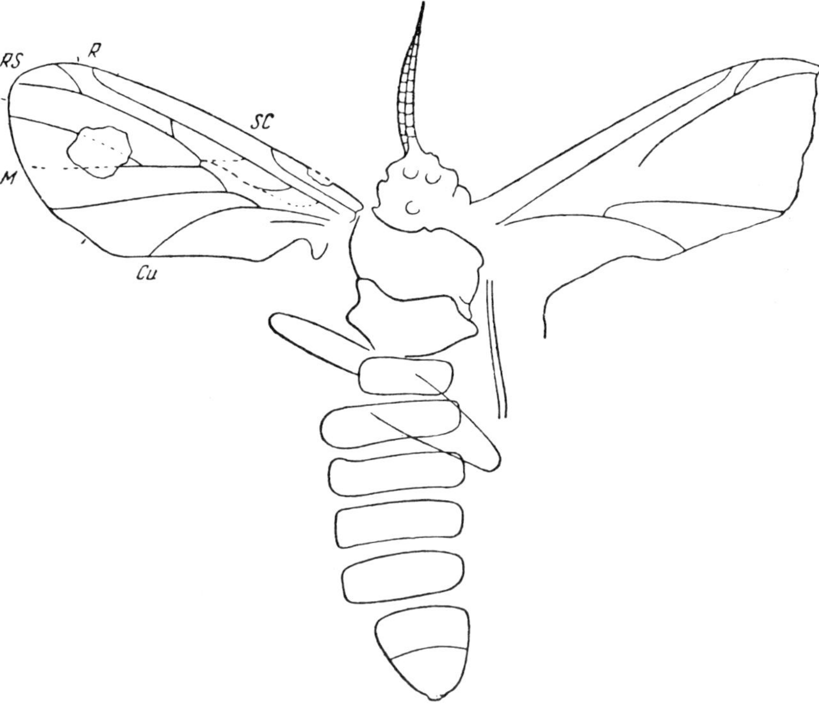

Fig. 73. *Archizelmira kazachstanica* Rohd., (Archizelmiridae). Middle Jurassic of Karatau. General view. (Original.)

light brown, wings clear. The transverse rm vein is almost three times shorter than the anterior proximal branch of RS. The length of the antennae is 1.0, wing 3.0, width of wing 1.3 mm. Measurements of the tergites of abdomen are: 3rd - 0.37, 4th - 0.32, 5th - 0.35, 6th - 0.35, 7th - 0.30, width of 4th tergite 1.0 mm. The length of the body of the fossil is 4.15 mm; in fact the length of the insect was less, since the segments of the abdomen as a result of decomposition during burial are greatly separated, and the real size of the insect did not exceed 3.5 mm.

Material. — The holotype.

Family Pleciofungivoridae Rohdendorf, 1946
Rohdendorf, 1946, p. 51

Remarks. — Below are described four species known earlier as genera for which I compiled a key and developed interrelationships (Rohdendorf, 1946).

Genus *Pleciofungivorella* Rohdendorf, 1946.
Rohdendorf, 1946, p. 54

Type of genus: *P. binerva* Rohdendorf, 1946.

Remarks. — The description of two new species of the genus from that same site in which was discovered the first species, the type of the genus, was quite unexpected. A preliminary determination of the new representatives of *Pleciofungivorella* as individuals of the first species (which was done at a glance!) proved to be incorrect. Differences in the venation of the wings, the size of the body, the thickness and proportions of sections of the legs all indicates the specific character of the distinctions of these fossils.

Key to the species

1 (2)   Length of body more than 2 mm; posterior femur at least twice as thick as the tarsus; the proximal anterior branch of RS unites with R almost at its very end, in the place of junction with C . . . .
. . . . *P. proxima* (p. 244)

2 (1)   Length of body 1.75 mm; the posterior femur not more than one-and-a-half times as thick as the tarsus 3

3 (4)   SC unites with C proximal to the level of the branching of RS from R; the proximal anterior branch of RS branches out from the main trunk of R at a slight angle, curved . . . .
. . . . *P. brevisubcosta* (p. 246)

4 (3)   SC longer and goes into C distal to the level of the branching of RS from R; the proximal anterior branch of RS branches out at a great angle, almost straight, not curved . . . .
. . . . *P. binerva* Rohdendorf, (1946, p. 55)

Pleciofungivorella proxima Rohdendorf, sp. n. (fig. 74*A, B*)

Holotype. — Remnant of male, lying on right side of body (head, part of the thorax, end of the abdomen and legs poorly preserved, posterior halves of wings crumpled). Coll. PIN No. 7/167, Michailovka, Middle Jurassic, B. Rohdendorf, 1937.

Description. — SC runs into C scarcely distal to the level of branching of RS from R; first part of RS almost twice as great as the length of rm, located obliquely to RS; second section of RS (from rm to the proximal anterior branch) more than three times as long as rm and nearly twice as long as the first part; proximal anterior branch of RS straight, situated obliquely and joining R almost before its very end; the distal anterior branch of RS slightly curved; the ptero-stigma is not clear; the third part of RS (between the anterior branches) is equal in length to the proximal branch; the last section of RS is noticeably shorter than the whole remaining part of the vein. The posterior tarsi is half as thick as the posterior femora; the length of the posterior femur is 0.65, of the posterior tarsus 0.80 mm. The thorax is moderately convex: its length is equal

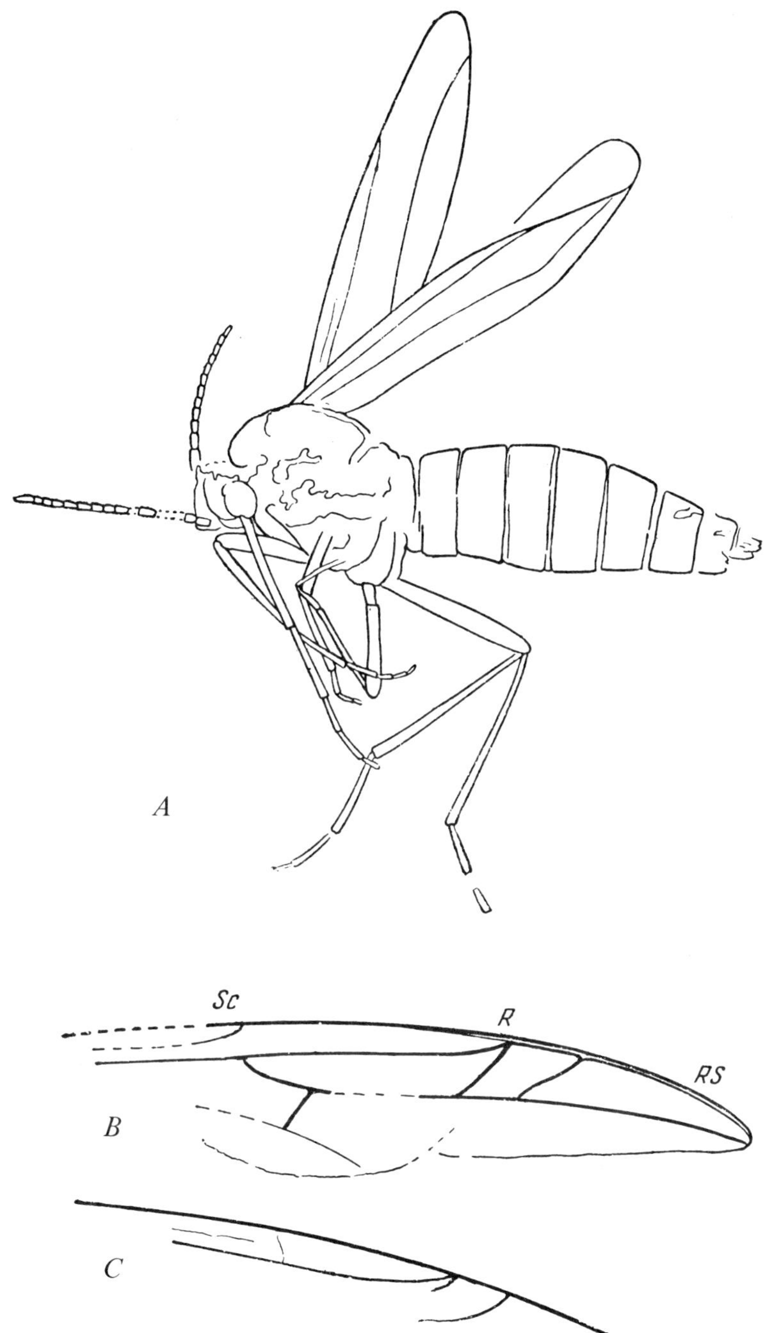

Fig. 74. *Pleciofungivorella proxima* Rohd., sp. n. (Pleciofungivoridae). Middle Jurassic of Karatau. Holotype. Coll. PIN No. 7/167. Length of remnant 2.4 mm. *A*. General view. *B*. Venation of right wing. *C* The same of the left. (Original.)

to 0.65, depth 0.70 mm; length of wing is 1.9 mm. Head with large eyes; the length of the antenna is 0.85 mm. Length of the abdomen is 1.65 mm; the abdomen is widest at the middle of the 4th segment. The total length of the fossil is 2.4 mm owing to inflation of the abdomen: the real length of the body of the insect probably is about 2.2 mm.

Material. — The holotype.

Pleciofungivorella brevisubcosta Rohdendorf, sp. n. (fig. 75)

Holotype. — Remnant of male lying on the abdominal part of the body (thorax, left wing and posterior part of right, legs and partly the abdomen crumpled and poorly preserved). Coll. PIN No. 8/167, Michailovka, Middle Jurassic, B. Rohdendorf, 1937.

Description. — SC goes into C considerably closer to the base of the wing than the level of the branching of RS from R: the first part of RS is three times as long as rm which is located almost perpendicular to RS; the second part of RS is only twice as long as rm and shorter than the first part; the proximal anterior branch of RS is very oblique to RS and presumably curved (the place of its connection with R is not clear); the part of RS between the anterior branches is more than one-and-a-half times as great as the second part; the distal anterior branch of RS curved in an arch; the last part of RS noticeably shorter than the whole remaining part of the vein; pterostigmal spot clearly separated. Posterior tarsi are at least two-thirds as thick as the femora and noticeably longer than them; posterior metatarsi very large, somewhat shorter than half of the tarsus and larger than all the remaining segments of the tarsi, of which the second is very short. The thorax is convex. The head has large compound eyes and big simple eyes; the antennae are longer than the thoracic section. The length of the body is 1.75 mm, wing length 1.63 mm.

Genus *Prohesperinus* Rohdendorf, 1946
Rohdendorf, 1946, p. 56

Type of genus: *P. abdominalis* Rohdendorf, 1946.

Remarks. —Up to now this genus was known only from a single species, the type of the genus. Below is described a second species from the same location.

Prohesperinus pedalis Rohdendorf, sp. n.
Prohesperinus sp.: Rohdendorf, 1951, p. 17, fig. 93

Holotype. — Remnant of female (?) lying on the right side of the body. Coll. PIN No. 2452/225, Halkino, Middle Jurassic, A.V. Martynov, 1924.

Description. — The first segment of the middle tarsus is the largest but noticeably shorter than half of the whole tarsus, i.e., than the sum of all the remaining segments. The second segment is smaller than half of the first and one-and-a-half times as large as the third, which is approximately as much again as large as the fourth; the fifth is the smallest. According to ab-

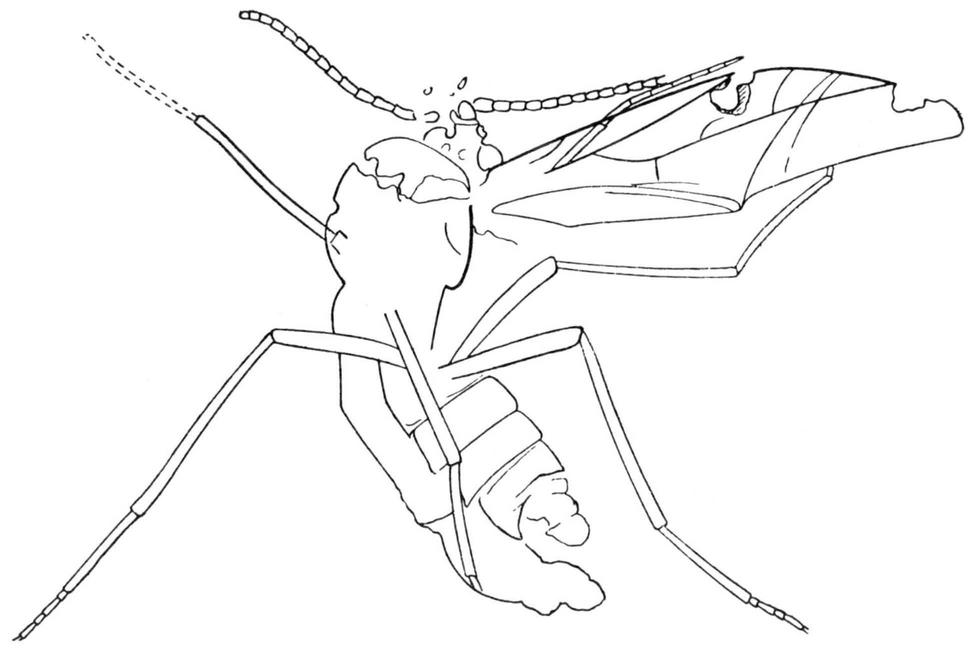

Fig. 75. *Pleciofungivorella brevisubcostata* Rohd., sp. n. (Pleciofungivoridae). Middle Jurassic of Karatau. Holotype. Coll. PIN No. 8/167. Length of remnant 1.75 mm. General view. (Original.)

Fig. 76. *Eohesperinus weberi* Rohd., sp. n. (Pleciofungivoridae). Middle Jurassic of Karatau. Holotype. Coll. SAGU, No. 69. Length of remnant 2.5 mm. (Original.)

solute size, the segments of the middle leg are arranged in the following order -
1, 2, 3, 4, 5 and according to relative sizes - 13: 6: 4: 3: 2. The first segment
of the posterior tarsus is almost equal to the sum of all the remaining segments;
the second segment is scarcely greater than one-third of the first and somewhat
greater than the third which is greater than twice as long as the fourth; the
last two segments of the leg are almost the same. According to absolute size,
the segments are distributed in the following order - 1, 2, 3, 4, 5 and according
to relative sizes — 16: 5.5: 4.5: 2: 2. The length of the anterior femur is 1.2,
of the anterior tibia 1.3, of the middle tibia 1.25, middle tarsus 1.33, of posterior tibia 1.7, of posterior tarsus 1.55 mm.

Comparison. — The preservation of fragments of the body and wings is
very poor; the structure of veins R and RS is very close to such a species
as *Prohesperinus abdominalis* Rohdendorf, from which the described species
differs in the thinner middle and posterior femora, having almost parallel
extreme outlines, narrow parallel tarsi and different proportions of the segments of the legs.

Material. — The holotype.

Genus *Eohesperinus* Rohdendorf, 1946

Rohdendorf, 1946, p. 60

Type of genus: *E. martynovi* Rohdendorf, 1946.

Remarks. — Until recently only a single species, the type of the genus, was
known. A second species is described below.

Eohesperinus weberi Rohdendorf, sp. n. (fig. 76)

Holotype. — Remnant of male (?) lying on the left side of the body; wings
straightened, legs drawn in (head and its appendages, thorax, legs, abdomen and
right wing poorly preserved). Coll. Geol. Inst. SAGU, No. 69, Karatau, Middle
Jurassic. The species was named in memory of the paleontologist V.N. Weber.

Description. — The venation differs from that of *E. martynovi* Rohdendorf
in the following features. SC terminates distal to the level of the anterior end of
rm; the ends of R and the anterior branch of RS are parallel, not coming together;
part of C between the ends of R and the anterior branches of RS somewhat more
than one-third the length of the anterior branch of RS; part of C between SC and
R is equal to the section of C between R and RS; the first part of RS (from R to
rm) almost twice as long as rm and noticeably less than one-third of the second
part of RS which is nearly one-and-a-half times as short as the last part of RS (between C and the anterior branch); branch M_{1+2} is more than two-and-a-half times
as long as the common part of this vein (from the fork to rm); the angle of divergence of the branches M_1 and M_2 is small (less than 45 degrees); the parts of the
trunk of M_{1+2} up to and after rm are of sharply different length; the proximal
part is short, equals approximately two-fifths of the distal; M_3 is located in the form
of a posterior branch of the common trunk of M. The length of the body is 2.5,
wing - 2.6 mm.

Comparison. – It is distinguished from *E. martynovi* by the structure of the medial veins and the position of the anterior branches of RS.

Family Pleciomimidae Rohdendorf, 1946
Rohdendorf, 1946, p. 61.

Remarks. – Below are described two genera: *Archilycoria* Rohdendorf, 1946 and *Megalycoriomima* Rohdendorf, 1962, which were insufficiently known up to now.

Genus *Archilycoria* Rohdendorf, 1946
Rohdendorf, 1962, p. 328

Type of genus: *A. magna* Rohdendorf, 1946

Description. – The wing is quite wide and truncated at the apex; it is twice as long as wide. C is sturdy, reaching the apex of the wing, and terminating at the middle between RS and M_1. SC is reduced, indistinct and does not unite with C (this vein is not represented in the illustration). R is straight, equal approximately to four fifths the length of the wing, sturdy. RS branches out from R at the level of the end of the basal fifth of the wing; rm weak and straight, approximately three times as short as the first part of RS; distal part of RS moderately curved. M is weak; fork of M_{1+2} longer than the second section of the common trunk of this vein; first part of M_{1+2} short, two-and-a-half times shorter than the second section; M_3 in the form of a branch of CuA, equal to M_2 in length but considerably stronger than it. CuA is uniformly bent in a curve; CuP well-noticeable along with the proximal half of CuA. The thoracic section highly convex, overhanging the head. Head with large compound eyes, apparently, protruding (tapered?) mouth organs and quite thin antennae, equal to the length of the head and thorax taken together (the number of segments is not clear). Legs sturdy, running, with well-marked spurs at the ends of the tarsi; femur moderately dilated, twice the thickness of the tarsi which are parallel at the outsides and only slightly expanded at the distal ends. The abdomen narrows slightly to the end.

Comparison and composition. – Related most nearly to the genus *Paritonida* Rohdendorf, a representative of a peculiar subfamily, differing in the longer R, the absence of SC and the shorter M_3. It is described according to a single species, the type of the genus.

Archilycoria magna Rohdendorf, 1946 (fig. 77*A*)
Rohdendorf, 1946, Table X, fig. 25
(nomen nudum); 1962, p. 328, fig. 1052

Holotype. – Remnant of female lying on the right side of the body with legs and wings stretched out (head, thorax, wings and legs in part poorly preserved). Coll. PIN No. 2452/359, Halkino, Middle Jurassic, A.V. Martynov, 1924.
Description. – Coloration of head, thoracic section and abdomen dark brown;

wings brown with black veins; legs light brownish-yellow. Posterior legs shorter, middle legs somewhat longer and anterior considerably larger than the tarsi. First segment of the middle and posterior legs is very long. Length of the anterior tarsi is 1.15, of the middle 1.5, posterior 1.7, anterior legs 1.35, middle legs 1.55 and posterior legs 1.6 mm. The length of the body is 4.89, the length of the wing 3.57, the length of the abdomen 2.07 and the width of the wing at the level of the end of M_3 is 1.41 mm.

Material. – The holotype.

Genus *Megalycoriomima* Rohdendorf, 1962
Rohdendorf, 1962, p. 326.

Type of genus: *M. magnipennis* Rohdendorf, 1962.

Description. – Wing large, truncated at the apex, somewhat more than twice as long as wide. C sturdy, reaching only to the end of RS. SC weak, poorly marked, not uniting with C. R almost completely straight, equal to approximately 0.7 of the wing, sturdy. RS branches out from R at the boundary of the first and second quarters of the wing, curved at the base and farther on almost parallel with R, uniformly and strongly bent backwards in its distal third, terminating nearly at the very apex of the wing. Transverse rm vague (presumably owing to poor preservation of the fossil). Also indistinct anterior medial branches of M_1 and M_2; there is a vein M_3 which is thin and uniformly bent back in an arch in its distal half, which branches from CuA somewhat more distal to the level of the branching of RS from R; the main part of vein M_3 is located very close to CuA, almost parallel with it. CuA sturdy in its greater part, is arch-like: the end of it is thinner and sharply bent backwards. CuP is not clear. The thoracic section is moderately convex, insignificantly hanging over the head. Head with large compound eyes; mouth organs not protruding, antennae apparently not longer than the thoracic section. The structure of the legs is unknown. The abdomen is large, slightly dilated in the middle, eight-segmented.

Comparison and composition. – The genus is quite peculiar and its relationship to others is not clear; apparently it is closest to the genera *Pleciomimella* and *Lycoriomima*, differing from them in a series of features (larger size, very thin and delicate vein rm, M_{1+2}, broad, larger wings). We have not excluded the possibility that the genus *Megalycoriomima* in reality appears as a representative of a special subfamily being sharply apart from the other genera of Pleciomiminae. It is described according to a single species, the type of the genus.

Megalycoriomima magnipennis Rohdendorf, 1962 (fig. 77*B*)
Rohdendorf, 1962, p. 326, fig. 1041

Holotype. – Remnant of male (?) lying on the left side of the body (legs, medial veins, antennae and in part the head not preserved). Coll. PIN No. 2452/421, Halkino, Middle Jurassic, A.V. Martynov, 1924.

Description. – Color of thoracic section and abdomen dark brown; the wings

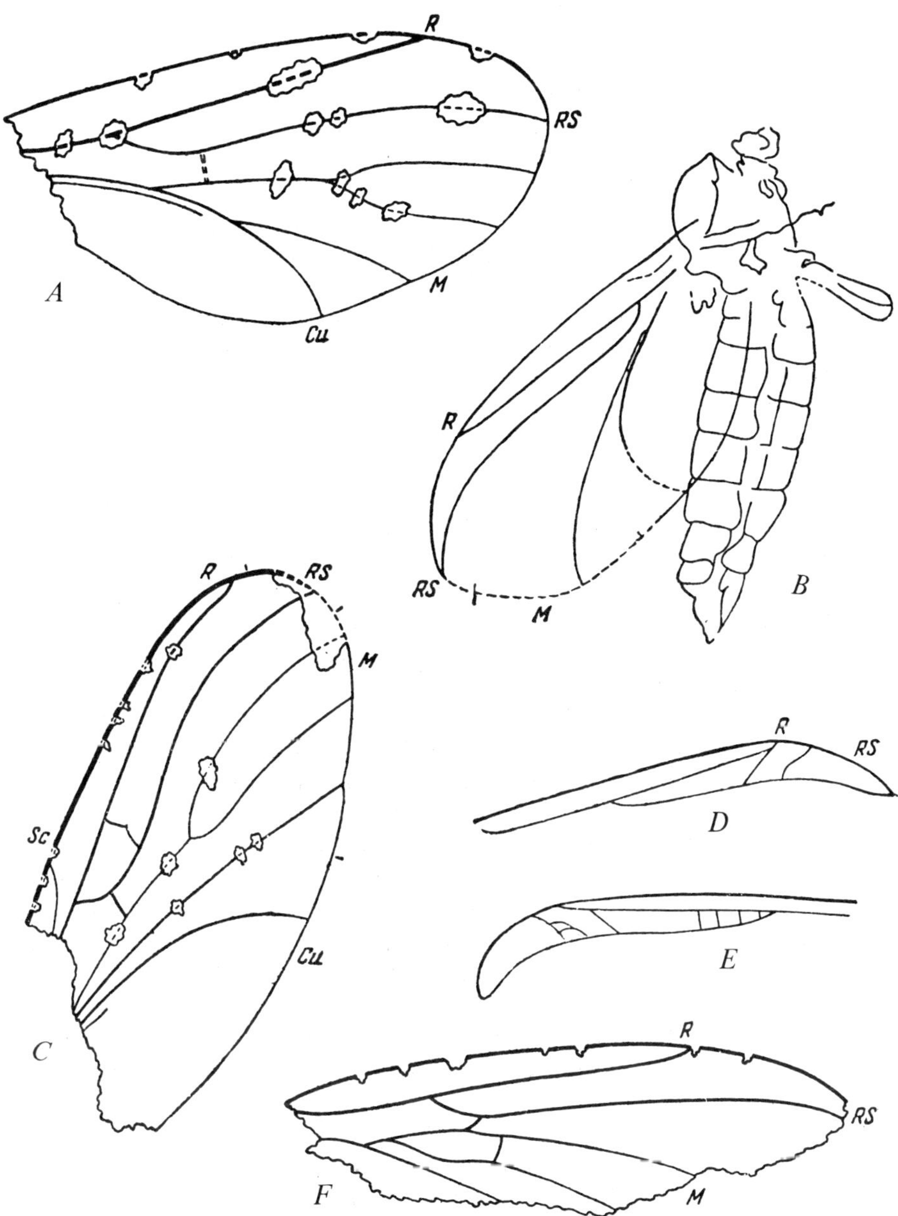

Fig. 77. Bibionomorphs of the Middle Jurassic of Karatau. *A. Archilycoria magna* Rohd., (Pleciomimidae). Wing. Holotype. Coll. PIN No. 2452/359. *B. Megalycoriomima magnipennis* Rohd., (Pleciomimidae). Holotype. Coll. PIN No. 2452/421. Length of remnant 4.1 mm. *C. Mesosciophilodes similis* Rohd., sp. n. (Fungivoritidae). Holotype.Coll. PIN No. 1/167. Length of remnant 4.3 mm. *D. Tipuloplecia breviventris* Rohd., (Tipulopleciidae). Right wing. Holotype. Coll. PIN No. 2452/601. *E.* The same, left wing. *F. Sinemedia angustipennis* Rohd., (Sinemediidae). Wing. Holotype. Coll. PIN No. 2452/665. (*A. B. D. E. F.* according to Rohdendorf, 1962; *C.* original.)

are clear, colorless. The length of the body is 4.1, head 0.35, thorax 1.14, abdomen 2.8, wing 3.3, width of wing 1.6, depth of thoracic section 1.0 mm.

Family Fungivoritidae Rohdendorf, 1957
Rohdendorf, 1957, p. 91
Allactoneuridae: Rohdendorf, 1938, p. 42; 1946, p. 75
(partially)

Genus *Mesosciophilodes* Rohdendorf, 1946
Rohdendorf, 1946, p. 77

Remarks. – Up to recent time there was known only one species of the genus, from a location near Halkino; below is described a new species from a site near Michailovka.

Mesosciophilodes similis Rohdendorf, sp. n. (fig. 77C)

Holotype. – Remnant of right wing (base of wing and part of the apex not preserved). Coll. PIN No. 1/167, Michailovka, Middle Jurassic, B. Rohdendorf, 1937.

Description. – SC uniformly arched, much thinner than C and R, going into C at the level of the connection of rm with RS; R in its greater part is straight and only at the end is bent backwards, following the bend of C, with which to some extent it runs parallel; RS branches out from R somewhat proximal to the level of the end of SC, going off from R at a very great angle (not less than 60 degrees) and further, bending irregularly, it runs to the apex of the wing, retaining on the whole its parallelism with R; the anterior branch of RS at the middle with a break and an offshoot, located proximally to the level of the fork of M_{1+2}; the second part of RS (from rm to the anterior branch) is only somewhat greater in length than the anterior branch and is one-and-a-half times greater than the chief part; transverse rm straight, located at an angle to RS and M, equal to approximately 0.6 of the chief part of RS; the medial branches are thinner than RS and rm; fork of M_{1+2} uniformly spreading out to the edge of the wing; the unpaired trunk of M_{1+2} (from rm to the fork) is 2.2 times as short as the branch M_2; the part of the common trunk lying proximal to rm (from M_3 to rm) is not shorter and presumably is longer than the distal part; M_3 sturdier, almost straight; CuA still sturdier (as is R), curved; CuP is noticeable only at the basal part of CuA; A is not clear. The length of the remnant of the wing is 4.3 mm (the real length is about 4.5 mm), the greatest width of the wing is 2.15 mm.

Comparison. – It differs from *M. angustipennis* Rohdendorf, 1946 by the greater size, more proximal location of the anterior branch of RS, shorter SC, more transverse location of rm and other features.

Family Tipulopleciidae Rohdendorf, 1962
Rohdendorf, 1962, p. 328

Description. – The wings were folded at rest one on the other at the back edge of the body, considerably longer than it. R equals three-quarters the length of wing, straight, without branches; RS has several anterior branches partly in the form of short transverse veins between R and RS, partly in the form of free branches uniting with C; the main trunk of RS sharply bent back; rm thin, poorly marked. The head is not especially large, freely projecting ahead; antennae short, only somewhat longer than the head and considerably shorter than the thoracic section. Thoracic part with a depressed back edge, massive; legs of the thin type, femora and tarsi differ little from one another according to thickness. The abdomen is short, not longer than head and thorax taken together.

Comparison and composition. – It is described from a single genus and the structure of the venation indicates the relationship of this peculiar form to the fungivorids. The elongation of the wings, the apparently thin legs, the significant reduction of the venation of the posterior half of the wing, the short abdomen - all indicate an extremely unusual historical development. Thin legs and large wings have very rarely appeared in the history of the infraorder Bibionomorpha (for example, the Bolitophilidea) whereas among the Tipulomorpha it turned out to be almost the rule. In the light of these facts, the features of the Tipulopleciidae are still more curious and indicate a similarity to tipulomorphs. The thin legs, and weak, long thrust wings reflect the limitation of the function of the winged phase of the insect - the decrease of running, of lifting flight, of rapid take-off, the improvement of passive grasping, the decrease in the importance of feeding. But at the same time these processes influence the improvement of the larval phase.

Genus *Tipuloplecia* Rohdendorf, 1962

Rohdendorf, 1946, p. 83 (nom. nudum); 1962, p. 328

Type of genus: *T. breviventris* Rohdendorf, 1962.

Description. – SC is not clear; the costal field is parallel to the margin, quite wide. R is parallel to C, going into it at a very acute angle. Between R and RS in the very fork are located two to four straight transverse veins, branches of RS; in the distal part of the radial field there are two oblique, curved or branching anterior branches of RS, one of which unites with R, another, the distal, with C; the distal section of the vein RS for some of its extent is located parallel to the end of R; the main part of RS is not less than three times as long as rm which is situated somewhat proximal to the middle of the wing. It is described according to a single species, the type of the genus.

Tipuloplecia breviventris Rohdendorf, 1946 (fig. 77, *D, E*)

Rohdendorf, 1946, p. 83, Table X, fig. 24

(nom. nudum); 1962, p. 328, fig. 1054

Holotype. – Remnant of male (?) lying on the ventral side of the body with folded wings and partially straightened legs (thoracic section, antennae and venation poorly preserved). Coll. PIN No. 2452/601, Halkino, Middle Jurassic, A.V. Martynov, 1924.

Description. – Color dark; head, thoracic section and abdomen dark brown, legs brown, wings lighter, but also noticeably colored. The length of the body is 2.85, head 0.3, thoracic section 1.0, wing 3.25 mm.

Family Sinemediidae Rohdendorf, 1962

Rohdendorf, 1962, p. 328.

Description. – The wings are shorter than the body, quite narrow. R is equal to two-thirds of the wing, without branches; RS curved, without branches; rm short, obliquely disposed; there is an intermedial transverse vein between the common trunk of M_{1+2} and the main part of M_3. The head is large, projecting; antennae approximately equal to the head and thorax. The thoracic section is small, with moderately convex back, not hanging over the head; legs of the running type, quite short with noticeably thickened femora and, presumably, having definite gripping qualities. The abdomen is long, eight-segmented (plus genital segments), tapered at the end and moderately dilated at the level of the fourth segment, length of abdomen exceeds twice the head and thorax taken together.

Comparison and composition. – It is described according to a single genus. All the organization of this insect indicates the attachment of it to the fungivorids and to the family Pleciomimidae in particular. However the features of the medial system of veins and the form of the elongated wings sharply distinguish this new group, and denote family rank. Appraising the features of the described family and comparing it with the Pleciomimidae, the elongation of the wings should be noted, while the thorax is relatively weakly dilated and, therefore, the development of dorso-ventral muscles is small. The lifting type of wings indicate powerful lowering and comparatively weak raising of the wings. Such a phenomenon undoubtedly had a place in many other pleciomimids (for example, species of the genera *Lycoriomima, Paritonida*), but in them it is accompanied by shortening of the wings and, consequently, by higher frequency of wing beat. The absence of reduction of the wing in *Sinemedia* indicates a great development of the forces of thrust; the lifting force is guaranteed by more powerful downstrokes of the wing. The short, running, presumably gripping legs together therefore with thrust wings suggest that these insects were active long-range fliers. It is quite possible that the sinemediids ecologically were connected with distinct nutritive substrates of their larvae. Of interest is the question concerning the imaginal feeding of this form. It is also very possible that the improvement of the locomotor features reflects the character of the feeding of the winged insect, as a predator, or at least as a nectar feeder. Insufficient preservation of the head of the fossil does not allow us to throw light upon this question.

Genus *Sinemedia* Rohdendorf, 1962

Rohdendorf, 1962, p. 328.

Type of genus: *S. angustipennis* Rohdendorf, 1962.

Description. – SC is indistinct; costal field wide in the main part and gradually narrows to the apex of the wing. C and R sturdy, joining. RS branches out from

R at the end of the first quarter of the wing; the first part of RS is three times
as long as rm; the latter is obliquely disposed, and RS at the junction with rm
forms a break so that the distal, greater part of RS turns out at an angle both to
rm and to its proximal part (first section). RS is parallel to R and gradually converges with C. The medial system of veins is far thinner and weaker than the
radial; the unpaired trunk of M_{1+2} is very long dividing into branches M_1 and
M_2 approximately at the level of the end of R; between the base of this unpaired trunk of M_{1+2} and M_3 there is a straight, thin but clear transverse vein,
longer than rm, which is located distal to the latter approximately at a distance
equal to its length. The basal part of M_3 is peculiarly bent, forming together with
the base of M_{1+2} a characteristic triangular cell; vein M_3 is sturdier than M_{1+2}.
CuA slightly curved, not especially sturdy, equal in thickness to vein M_3.

Composition. – It is described according to a single species, the type of the
genus.

Sinemedia angustipennis Rohdendorf, 1962 (fig. 77F)
Rohdendorf, 1962, p. 328, fig. 1055.

Holotype. – Remnant of female lying on the right side of the body with folded
legs and displaced wings, which lie one on the other (antennae, head, posterior
parts of wings and partially the abdomen are very poorly preserved). Coll. PIN
No. 2452/655, Halkino, Middle Jurassic, A.V. Martynov, 1924.

Description. – The color is very dark; antennae, head, thoracic section and
abdomen are black, wings are dark brown with black veins; legs are light brownish
yellow. The length of the body is 3.25, wing 2.25 mm; the abdomen is inflated
and consequently the real length of the body presumably did not exceed 3 mm.

Material. – The holotype.

Superfamily Bibionidea Newman, 1834

Family Protopleciidae Rohdendorf, 1946
Rohdendorf, 1946, p. 42

Remarks. – This family includes a total of three genera, one in the Liassic of
western Europe and two in the Jurassic of Karatau. Below is described a second
species of the genus *Mesoplecia* from a location close to Halkino; up to now there
was known only a single species from a site close to Michailovka.

Genus *Mesoplecia* Rohdendorf, 1938
Rohdendorf, 1938, p. 49.

Type of genus: *M. jurassica* Rohdendorf, 1938.

Key to the species

1 (2)   Anterior branch of RS slightly curved, parallel with the end of
          R; RS and M_1 convergent at the edge of the wing; distal part of common
          trunk of M_{1+2} (from rm to the fork) considerably shorter than rm;

length of wing 7.0 mm . . . .
. . . . *M. jurassica* Rohd., 1938

2 (1)   Anterior branch of RS sharply curved, at the end coming together with the end of R; RS and M_1 parallel; distal part of common
trunk of M_{1+2} equal to rm; length of wing 8.13 mm . . . .
. . . . *M. stigma* Rohd., 1962

Mesoplecia stigma Rohdendorf, sp. n. (fig. 78)
Rohdendorf, 1962, p. 329, fig. 1060, (nom nudum)

Holotype. – Remnant of female lying on the right side of the body with straightened wings (head and greater part of legs not preserved). Coll. PIN No. 2452/01,
Halkino, Middle Jurassic, A.V. Martynov, 1924.

Description. – C sturdy, scarcely going into the end of RS, slightly convex. SC
sturdy at the base, gradually tapers to the end, slightly arched, going into C and
connected with R by a short, straight transverse vein, located proximally to the
branching of RS; SC reaches almost to the middle of the wing. R straight, slightly
concave in front, is sharply bent just before going into C. RS branches out from
R at the boundary of the first and second thirds of the wing and bears one sturdy
anterior branch, going into C immediately distal to the end of R; first part of RS
is three-and-a-half times as long as rm, the second part equals 0.81 of the first,
the last section greater than both of the proximals taken together; anterior branch
of RS sharply bent at the base and at the end, in its middle part parallel with R
and sharply convergent with its end. There is a clear pigmented spot at the ends
of R and RS. The main part of M is very weak, almost lacking at its proximal end;
the branching of M_{1+2} from M_3 is located somewhat distal to the level of the
fork of R-RS; common trunk of M_{1+2} is divided almost in two by the vein rm
which is only slightly closer to the fork of M_{1+2}; the fork of M_{1+2} is very large;
many times greater than the distal part of the common trunk of these veins; all
the branches of M are uniformly diverging; their ends are located at an equal distance at the edge of the wing; from the very base of M_3 there branches out a
straight transverse vein mcu. CuA is not particularly sturdy. The thorax is moderately convex. The legs are not especially sturdy; the coxae, middle and posterior
femora and tarsi have numerous short bristles. The abdomen consists of seven
segments and the genital apparatus; the tergites are covered by projecting, quite
short bristles. The length of the body is probably about 10 mm. The length of
the wing is 8.13, of the thoracic section 2.32, of the abdomen 7.98 (distended),
depth of the thoracic portion (from the shield to the end of the middle coxa)
3.06 mm.

Material. – The holotype.

Superfamily Rhyphidea Newman, 1834
Family Protorhyphidae Handlirsch, 1906
Handlirsch, 1906, p. 487; Rohdendorf, 1962, p. 332

Remarks. – Up to now this family has been known only from the Lower Jurassic

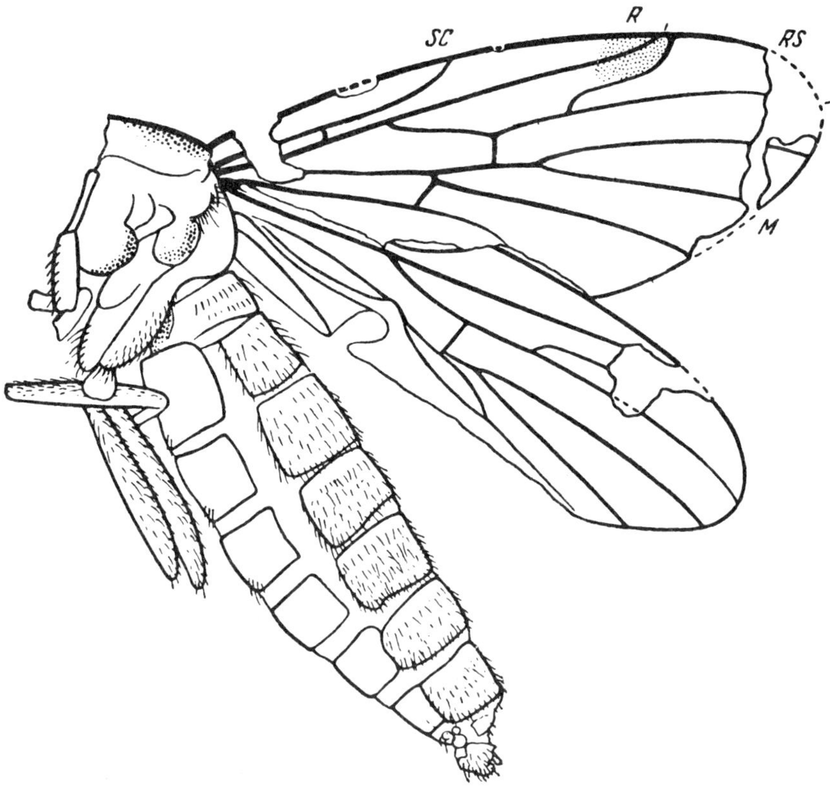

Fig. 78. *Mesoplecia stigma* Rohd., (Protopleciidae). Middle Jurassic of Karatau. General view. Holotype Coll. PIN No. 2452/01. Length of remnant about 8.0 mm. (Original.)

of western Europe and was determined by us in the composition of the fauna of the Upper Triassic of central Asia (see p. 203) and in the Middle Jurassic of Karatau. Two related genera are known — *Protorhyphus* Handlirsch, 1908, and *Archirhyphus* Handlirsch, 1939, the described Karatau species belongs to the latter genus.

Genus *Archirhyphus* Handlirsch, 1939.
Handlirsch, 1939, p. 102; Rohdendorf, 1962, p. 332.

Type of genus: *A. geinitzi* Handlirsch, 1939

Remarks. — The differences of the new species from those known from earlier Liassic consist in features of the venation and size: not having the material of the European species and assessing its characteristics entirley from the description and representation of the wing given by the author, I do not consider it practical to draw up a key for these species.

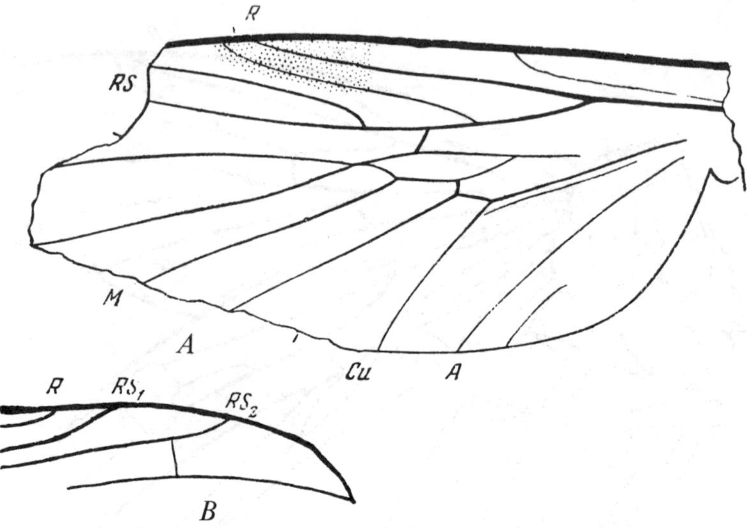

Fig. 79. *Archirhyphus asiaticus* Rohd., (Protorhyphidae). Middle Jurassic of Karatau. Holotype. Coll. PIN
No. 2452/334. *A.* Left wing. *B.* Apex of right wing. (Original.)

Archirhyphus asiaticus Rohdendorf, 1962 (fig. 79 *A, B)*
Rohdendorf, 1962, p. 331, fig. 1072 (nom. nudum).

Holotype. — Remnant of female lying on the left side of the body with extended
wings and folded legs (legs poorly preserved). Coll. PIN No. 2452/334, Halkino,
Middle Jurassic of Karatau, A.V. Martynov, 1924.

Description. — C is sturdy, SC weak, going into C at the level of the primary
fork of M; radial veins sturdy and dark; R straight; the main (posterior) trunk
of RS to the level of the end of R weak, but uniformly curved; both anterior
branches of RS sturdy; first part of RS more than twice as long as the second
which is almost equal to the third; the last part of RS considerably greater than
the three basal parts; the main part of the medial vein is very weak, reduced; the
intermedial cell is elongate, or irregular pentagonal form, the unpaired trunk of
M_{1+2} short, somewhat shorter than rm; between the branches of RS_2 and RS_3
there are indistinct transverse veins; the base of M has the form of a short transverse vein, while the transverse mcu is disposed obliquely forming as it were a
fork of CuA; all four branches of M are uniformly divergent, not particularly weak;
CuA sturdy, straight to the vein mcu and gradually bending in an arch between
this vein and the edge of the wing; CuP weak but highly noticeable. A_1 is sturdy
and straight, very weakly convergent with CuA; A_2 is scarcely noticeable. The
head is round, somewhat bent under the back edge of the thoracic section; the
antennae are not especially thin, 13-segmented; the last segment tapered, the
middle segment in profile almost square. The thoracic section is convex; legs are
running, quite thin. The abdomen is very large, considerably longer than the head
and thorax taken together, inflated in the middle, consisting of eight segments

and the genital apparatus. The length of the body is 3.85, thorax 1.25, wing 2.9, width of wing 1.05 mm.

Material. — The holotype.

Infraorder Asilomorpha

Superfamily Tabanidea Latreille, 1802
Family Eostratiomyiidae Rohdendorf, 1951
Rohdendorf, 1951, p. 85; 1962, p. 336

Description. — The body is big, more than 10 mm. The head is large, contiguous with the thoracic section which is bulky and moderately convex. The abdomen is short, little longer than the head and thorax taken together, consisting of six abdominal segments and the apical complex. The legs are short, belonging to the gripping, unarmed type. The wings have a straight anterior border, with pointed apex and convex posterior edge. C is sturdy at the anterior edge, thin at the posterior. There are several radiomedial transverse veins and besides that one transverse vein between R and RS, and between RS, and the common trunk of RS_{2+3}; all these transverse veins are located at the level of the end of SC. There are four branches of M which unite with the thin C. CuA and A_1 are separated.

Comparison and composition. — The form of the body, wing and venation indicate the attachment of this Jurassic dipteran to the superfamily Tabanidea. The unusualness of the venation (particularly the presence of sturdy transverse veins at the middle of the wing), the open medial and cubital cells, sturdy head, short legs and comparatively small aerodynamically quite perfect wings sharply distinguish this new group from all of the families known up to now. The described family is closest to the Cenozoic Coenomyiidae and partly to the Tabanidae and Acanthomeridae; it is possible that Eostratiomyiidae had related connections with the isolated family Nemestrinidae. It is described according to a single genus.

Genus *Eostratiomyia* Rohdendorf, 1962
Rohdendorf, 1951 (nom. nudum); 1962, p. 336.

Type of genus: *E. avia* Rohdendorf, 1962

Description. — C sturdy to the apex of the wing, tapering from the middle of the distance between RS and thin to the posterior border. SC thin, poorly noticeable, somewhat more than 0.4 of the length of the wing, curved at the end, drawn together with R and going into C. R sturdy and straight, about 0.6 of the wing length, somewhat thinner than C. RS branches out from R in the basal quarter of the wing; the main part very thin and delicate; RS_1 irregularly curved gradually converging with R and C; RS branching out from the common trunk of RS_{2+3} scarcely proximal to the level of the end of R, curved and going into C in front of the apex of the wing; RS_3 forms a straight continuation of the chief trunk of RS, slightly convex forward and terminating immediately behind the apex of the wing. Transverse vein between R and RS_1 stronger than other transverse veins,

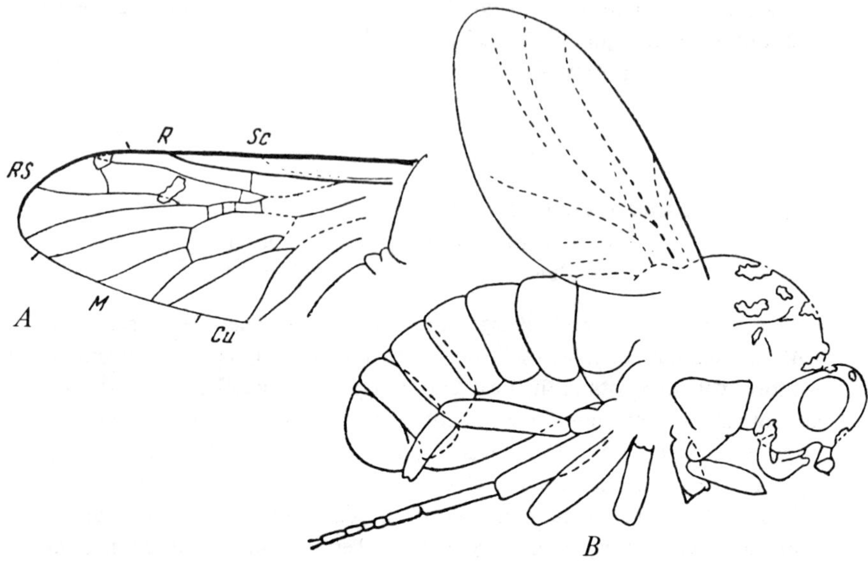

Fig. 80. Asilomorphs of the Middle Jurassic of Karatau. *A. Eostratiomyia avia* Rohd., (Eostratiomyiidae). Wing. Holotype.Coll. PIN No. 95/2511. Length 10 mm. *B. Eomyia veterrima* Rohd., (Eomyiidae). Holotype. Coll. PIN No. 2452/614.Length 2.25 mm. (*A.* according to Rohdendorf, 1951; *B.* according to Rohdendorf, 1962.)

straight and located almost exactly at the level of the end of SC; transverse vein between RS_1 and RS_{2+3} somewhat more distal than the preceding; there are four transverse rm veins; there is an irregularly curved thin transverse vein between the end of RS, and the middle of RS_2. There is a large discal cell which is convex in front and limited behind by a straight vein M_3; M_4 is separated from the discal cell, quite straight, and for the most part separated from the remaining branch of M. CuA is stronger than the medials, curved. The common medial trunk is almost straight, quite sturdy. A_1 is thinner than CuA, slightly bent.

Composition. — It is described according to a single species, the type of the genus.

Eostratiomyia avia Rohdendorf, 1962 (fig. 80*A*)

Rohdendorf, 1951, p. 79, fig. 33*B*

(nom. nudum); 1962, p. 336, fig. 1087

Holotype. — Remnant of male lying on the left side of the body with left wing up and legs drawn in (head, right wing and legs poorly preserved). Coll. PIN No. 95/2511 (positive) and 91/2511 (negative), Michailovka, Middle Jurassic, A. N. Turutanova-Ketova, 1932.

Description. — Transverse vein between RS_1 and RS_{2+3} separated from the transverse vein between R and RS_1 at a distance, less than the length of the vein;

three transverse rm veins located distal to the level of the transverse between
RS_1 and RS_{2+3}, approximately at an equal distance and a fourth rm lies considerably proximal to all the other transverse veins, very near to the fork of RS_1
and RS_{2+3}; common trunk of RS_{2+3} is 1.35 times as long as the common trunk
of RS (from R to the fork RS_1 - RS_{2+3}); the distal fork of RS_{2+3} long and narrow,
somewhat less than twice as long as its common trunk; RS_3 and M_1 almost parallel;
M_1 and M_2 diverging slightly; M_2 and M_3 sharply diverging; M_3 and M_4 convergent;
parts of the posterior edge of the wing between the named veins of unequal size:
the biggest part is between M_4 and CuA, then M_2 - M_3, M_1 - M_2, RS_3 - M_1 and
the smallest between M_3 and M_4. The length of the fossil is 11.75: head plus
thorax is 5.1, abdomen (inflated) 8.25, the real length of the abdomen is about
5.25, and of the insect about 10 mm. The length of the wing is 7.25, the width
of the wing is 2.6, the depth of the thoracic section is 3.5 mm; the length of the
anterior femur is 1.8, of the anterior tarsus 2.0, of the posterior femur 2.25, of
the posterior tarsus 2.25 mm.

Material. − The holotype.

Family Rhagionempididae Rohdendorf, 1938
[(nom. trans. Rohdendorf, 1962) (ex Rhagionempidinae
Rohdendorf, 1938).] Rohdendorf, 1938, p. 33; 1962, p. 336.

Remarks and composition. − This family was described for the first time only
as a subfamily of the Rhagionidae; a more intent investigation of the features of
the single representative of the group allowed us to understand its great unusualness a little more, and helped to throw some light on the family rank of the differences. The characteristic features of the family consist of the elongated form
of the head, of massive, large forward projecting, three-segmented antennae and
of some features of the venation: shortened C, reaching only to the apex of the
wing, the short fork of RS_{2+3}, four branches of M_1, going into the edge of the
wing. There is a single genus, *Rhagionempis* Rohdendorf.

Comparison. − The nearness of this family to the Rhagionidae and to other
representatives of the Tabanidea is shown in the character of the venation and
structure of the antennae; at the same time the form of the head, the costalization
of the venation (reductions of the costal vein around the posterior margin of the
wing) permit us to suggest connections of this form with the early ancestral groups
of other younger subfamilies of the Asilomorpha.

Family Rhagionidae Latreille, 1802

Remarks. − A single family of Jurassic asilomorphs is still living at the present
day. The three Jurassic genera of the family known to me belong to two sharply
isolated subfamilies, the differences of which were noted above in the key. The
subfamily Vermileoninae, represented in the contemporary fauna by two peculiar
genera (see p. 76) in the Jurassic fauna includes the genus *Archirhagio* which
differs very sharply from other Jurassic rhagionids, not less sharply than the
present-day genera *Vermileo* and *Lampromyia* from other contemporary genera

The unusualness of the wings, devoid of basal enlarged sections (wide anal region, vane and scales), the structure of the body, and the large sizes, compel us to consider a higher rank for the group Vermileoninae as being, in fact, an individual family and not merely a subfamily of the Rhagionidae. However the final decision may be taken only after a special study of material of the genera *Vermileo* and *Lampromyia* which I do not have at my disposal. Two other Jurassic genera of the family also resemble contemporary genera of the Rhagionidae. To determine more accurately the relations of the Jurassic and contemporary genera is difficult now in the absence of complete information concerning the structure of the venation of the present-day forms. The establishment by me in 1938 of the subfamily Protorhagioninae was premature: the reality of this grouping may be established only after a detailed review of the system of the family. Below I have described one new Jurassic genus of rhagionids. The differences between this and those known earlier are indicated in the key.

Key to the Jurassic genera

1 (2)   The part of C between R and RS_1 is less than half as long as the part of C between RS_1 and RS_2; the veins R and RS_1 gradually and quite strongly diverge; the fork of RS_1-RS_{2+3} without a transverse vein; rm located distal to the level of the end of SC; cell of Cu closed and with a clear pedicel, inasmuch as A_1 combines with Cu, and not with the edge of the wing .... *Rhagiophryne*

2 (1)   The part of C between R and RS_1 less than one-sixth as long as the adjoining distal part; the veins R and RS_1 are parallel or almost come together; the fork of RS_1-RS_{2+3} with a clear straight transverse vein; RM located considerably proximal to the level of the end of SC; cell of Cu is open: the vein A_1 does not unite with Cu .... *Protorhagio*

Genus *Rhagiophryne* Rohdendorf, 1962

Rohdendorf, 1951 (nom. nudum) 1962, p. 337, fig. 1091

Type of genus: *R. bianalis* Rohdendorf, 1962

Description. — The wing is 2.2 times as long as wide. C is sturdy down to the end of RS_3 or M_1, thin farther on. SC equal to two-fifths of the wing, sturdy, going into C. R is sturdy, curved in a slight arch, equal to half of the wing. RS branches out from R in the main quarter of the wing, thinner than R; first part of RS almost equal to the second which, in its turn, is somewhat greater than the third: the ratios of these parts equal 14: 13: 11. The transverse rm is sturdy and short, located at the middle between the levels of the ends of SC and R. The fork of RS_{1+2} is long and narrow, more than three times as long as the part of RS between rm and the fork; vein RS_2 at the base with angular break. The common trunk of M is weak and in places not clear. The first fork of M is located at the level of the first fork of RS_1, M_1 and M_2 almost parallel with one another

and RS$_3$. M$_3$ diverges sharply from M$_2$, noticeably, but slightly drawn together
with M$_4$ which is quite straight and unites with the long and sturdy mcu almost
at its base. CuA is sturdy and almost throughout uniformly curved in an arch;
CuP is noticeable only beside the chief half of CuA. A$_1$ almost straight and
uniting with CuA before its end. Highly noticeable is the clear, short vein located
behind A$_1$; it is parallel with it and presumably is A$_2$.

Composition. − It is described from a single species, the type of the genus.

Rhagiophryne bianalis Rohdendorf.
Rohdendorf, 1951, p. 79, fig. 33A (nom. nudum); 1962; p. 337.

Holotype. − Remnant of right wing, lying on the under surface (preservation
very good; only some sections of the border of the wing and the medial system
of veins and also the stem of the radius are destroyed). Coll. PIN No. 965/2,
Halkino, Middle Jurassic, A.V. Martynov.

Description. − It has a well-pigmented pterostigmal spot which envelops the
end of vein R; intermedial transverse vein between M$_2$ and M$_3$ is one-and-a-half
times as short as the main part of the vein M$_2$; the chief part of M$_4$ is almost
absent, since mcu joins with M$_4$ very near to the intermedial cell; pedicel of
cubital cell approximately equal to the transverse rm, very short. The length of
the wing is 4.5, the width of the wing 1.95 mm.

Material. − The holotype.

Superfamily Stratiomyiidea Latreille, 1802

Remarks. − This is the most ancient group of asilomorphs. First discovered
in the Liassic fauna, it includes in the Jurassic fauna of Karatau no less than four
individual families, all of which are extinct Mesozoic groups.

Family Archisargidae Rohdendorf, 1962
Rohdendorf, 1962, p. 334.

Description and composition. − Large sizes: wing 16 mm, narrow and long, its
length is four times as great as the width. The venation is costalized: anterior veins
considerably stronger than the remainder. The head, with large eyes is equal to
the width of the small thoracic section. The abdomen is elongate, apparently not
less than twice as long as the head and thorax; the segments of the abdomen are
1.8 - twice as wide as its length. The legs are quite thin but comparatively short.
There is a single genus, *Archisargus* Rohdendorf, 1938, from the Jurassic of
Karatau.

Comparison. − This insect, which served for a while as the type for the family
described now, was first referred by me on a purely formal basis (costalized wings)
to the family Stratiomyiidae. However, already the first attentive investigation
of this form clearly shows the absence of true connections with the named
Cenozoic family. The size and elongation of the wings raises doubts about the
presence of real costalization of the venation; it is possible that the absence or

slight development of veins in the greater part of the wing membrane is only an
apparent phenomenon and in reality this insect possessed weak but rich venation.
The unexpected large size of the insect sharply distinguishes it from all   other
Karatau Diptera. It is very difficult to determine the phylogenetic relations of
the Archisargidae; in fact the very attachment of this insect to a superfamily and
even infraorder remains uncertain. Furthermore, one must not forget that the
cubital area of the wing of *Archisargus* remains unknown. This is extremely important for the distinction of the two related groups - Diptera and Paratrichoptera
of course the attachment of the form being considered to the paratrichopterids
is very unlikely, but cannot be excluded completely.

Genus *Archisargus* Rohdendorf, 1938
Rohdendorf, 1938, p. 30.

Type of genus: *A. pulcher* Rohdendorf, 1938.

Remarks. — It is described according to a single species, the type of the genus.

Archisargus pulcher Rohdendorf, 1938
Rohdendorf, 1938, p. 31; 1962, p. 333, fig. 1072

Remarks. — Described according to a single remnant, the holotype.

Family Palaeostratiomyiidae Rohdendorf, 1962
Rohdendorf, 1951, p. 84 (nom. nudum); 1962, p. 334.

Description. — A small insect: wing about 2.5 mm, the length of it being
approximately 2.2 times as great as the width. The venation is very little costalized: the radial veins are only slightly stronger than the medial, C does not go
into the posterior border of the wing. The head is short and wide, slightly narrower
than the massive thoracic section. The wings are shorter than the length of the
body and longer than the abdomen; the latter is short and wide, shorter than the
head and thorax taken together, consisting of four segments, each of which is
not less than four times wider than its length. The legs are comparatively sturdy.

Comparison and composition. — In the first place this insect was accepted as
a representative of the family Stratiomyiidae, a component of a peculiar Jurassic
subfamily; in fact, however, the greater isolation of *Palaeostratiomyia* resulting
in abrupt integration and oligomerization of the body (reduction of the abdomen,
increase of the thorax and head) and the presence of little perfect venation, emphasize the unusualness and family rank of the differences of the Palaeostratiomyiidae from the Cenozoic Stratiomyiidae. However, they do not exclude the
possibility of directly related connections between these two groups as ancestor
and descendant. There is a single genus from the Jurassic of Karatau.

Genus *Palaeostratiomyia* Rohdendorf, 1938
Rohdendorf, 1938, p. 32.

Type of genus: *P. pygmaea* Rohdendorf, 1938.

Remarks. — It is described according to a single species, the type of the genus.

Palaeostratiomyia pygmaea Rohdendorf, 1938
Rohdendorf, 1938, p. 33.

Remarks. — Described according to a single remnant, the holotype of the species.

Family Eomyiidae Rohdendorf, 1962
Rohdendorf, 1962, p. 334.

Description. — A very small insect: wing about 1.5 mm, the length of it being 2.2 times as great as the width. Venation apparently moderately costalized: C is located only at the anterior edge, the remaining veins are delicate, indistinct. The head is large with compound eyes of moderate size, with very short antennae. Thorax substantial, moderately convex; wings considerably shorter than the whole body but longer than the head and thorax taken together; the legs are sturdy, not particularly short, and of the gripping type. The abdomen is little elongated, somewhat longer than the head and thorax taken together. There is a single genus from the Jurassic of Karatau.

Comparison and composition. — This most peculiar dipteran unfortunately cannot be studied with sufficient completeness owing to the poor preservation of the wings, the venation of which remains unknown. At the same time the general interest of this form is great: the structure of the head, thorax, and wings all suggest representatives of the youngest progressive infraorder, the Myiomorpha. The many-segmented structure of the abdomen, the massive thoracic section and, in particular, the remoteness of the anterior pair of legs from the head resulting from the large size of the prothorax compel us to refrain from the unconditional reference of the genus *Eomyia* to the myiomorphs and to place it in the infraorder Asilomorpha as a specialized representative of the superfamily Stratiomyiidea; the final solution must await discovery of the structure of the wings.

Genus *Eomyia* Rohdendorf, 1962
Rohdendorf, 1962, p. 334

Type of genus: *E. veterrima* Rohdendorf, 1962.

Description. — The depth of the head is less than its length; the compound eyes are equal approximately to the cheeks, their depth is greater than their length; the antennae are two-segmented (?), the first segment is shorter and narrower than the second (?), which has the form of a bulb and bears a short apical arista in the form of a bent bristle; there are three clear simple eyes; the

proboscis is small and short. The thoracic section in profile is more convex than
the head; there is a well-defined sternopleural sclerite (mesepisternite) which is
remote from the lower edge of the head; the femora are slightly thickened, only
insignificantly thicker than the tarsus; the anterior edge of the wing is straight
in the main half, farther on moderately convex, the venation is very poorly-
distinguishable, presumably very delicate; the posterior border of the wing is
highly convex, the anal blade is large. The abdomen with clearly separated segments of which the first and second are the longest, the sixth is the shortest;
the end of the abdomen ('seventh segment') is rounded at the tip.

Composition. – It is described on the basis of a single species, the type of
the genus.

Eomyia veterrima Rohdendorf, 1962 (fig. 80*B*)
Rohdendorf, 1962, p. 334, fig. 1081.

Holotype. – Remnant of male, lying on the left side of the body (legs and
venation of wings not preserved). Coll. PIN No. 2452/614, Halkino, Middle Jurassic, A.V. Martynov, 1924.

Description. – Color of the head, thorax and abdomen is black; the wings are
colorless; the femora are brown, the tarsi and feet are yellowish, light. The length
of the body is 2.25, of the wing 1.5 mm.

Material. – The holotype.

Superfamily Asilidea Latreille, 1802

Remarks. – This is widespread in the Cenozoic and is very rich in representatives
of the superfamily of asilomorphs. It appears for the first time in the Middle Jurassic fauna of Karatau in the form of representatives of the genus *Protomphrale*,
originally accepted by me as a member of the present-day family Scenopinidae,
which definitely is incorrect. This genus differs from the forms of scenopinids
known to me and belongs to a peculiar Jurassic family, although related and
close to the contemporary group. It is, however, essentially distinguished from
it by less costalization of the wings.

Family Protomphralidae Rohdendorf, 1962
Rohdendorf, 1962, p. 337.

Description. – The wing is short, little costalized; the chief trunk of RS is
removed from the anterior edge and is located almost at the middle of the membrane; there are four branches of RS. The head is apparently as in representatives
of the Scenopinidae. The thoracic section is very substantial, elongate; the legs
are of the gripping type, the wings are shorter than the body but noticeably longer
than the head and thorax taken together. The abdomen is short and thin, scarcely
longer than the head and thorax taken together. It is described according to a
single genus.

Genus *Protomphrale* Rohdendorf, 1938
Rohdendorf, 1938, p. 39; 1962, p. 337.

Type of genus: *P. martynovi* Rohdendorf, 1938.

Remarks. — There is a single species, the type of the genus.

Protomphrale martynovi Rohdendorf, 1938
Rohdendorf, 1938, p. 140; 1962, p. 338, fig. 1094.

Remarks. — It is described according to a single remnant, the holotype of the species.

General character of the Middle Jurassic fauna

Examining the composition of the dipterous fauna of the Middle Jurassic which is known to us from two locations in Karatau (64 species), its unusualness and differences from the Liassic on the one hand and from the Upper Jurassic on the other, may readily be noted.

The general composition of the Middle Jurassic fauna is composed of the representatives of four infraorders, 12 superfamilies, 26 families and 55 genera. Comparing the Upper Jurassic Diptera with more ancient forms it is necessary to note the chief appearance of representatives of a series of groups which reach maximum development later, and are widespread in the present-day fauna. Such, first of all, are the superfamilies Dixidea (genus *Dixamima* of a peculiar family), Scatopsidea (genus *Protoscatopse* of a peculiar family), Asilidea (genus *Protomphrale* of a special family) and the Tabanidea (six genera of four families). Particularly characteristic of the Middle Jurassic fauna is the sudden appearance of diversified tabanids by which it differs sharply from all the more ancient fauna. Comparing the family composition of the fauna it is necessary to note its unusualness: the majority of the families are found only in the Middle Jurassic fauna and only a few (Architipulidae, Pleciofungivoridae and Protorhyphidae) are spread also in the Triassic and Liassic or only in the Liassic (Pleciomimidae) or Liassic and Upper Jurassic (Protopopleciidae). They are thus positive leading groups of fauna for the whole Jurassic period. Another characteristic of the Middle Jurassic fauna consists of the presence of representatives of groups unknown among any other fauna, either more ancient or younger. Such are two superfamilies of tipulomorphs (the Tanyderophryneidea and Mesophantasmatidea) and four families of this infraorder, three families of fungivorids, three of bibionids and one of the scatopsids: i.e., seven of 12 families of Bibionomorpha are known only from the Jurassic fauna. Still more striking is the family diversity of asilomorphs of nine families of which seven are known only from the Middle Jurassic.

Summing up, one can note the overall similarity of the Jurassic fauna, in infraorders and the vast majority of superfamilies, with the Cenozoic fauna; on the other hand the family composition is sharply distinguished from the Cenozoic,

somewhat closer to the Liassic and Triassic and in its basis peculiar to the indicated part of geological time.

Examining the composition of the individual infraorders it is necessary first
of all to note that only nine species of the Tipulomorpha have been described.
In addition, however, this infraorder includes the greatest number of superfamilies,
either ancient, known from the Triassic (Tipulidea and Chironomidea), or new
(Dixidea). Finally, in this same infraorder, there are peculiar superfamilies known
only from the Middle Jurassic, referred to above. All this confirms the wide development of tipulomorphs in Middle Jurassic times, and the variety and great
distinctions of this fauna from that of the Cenozoic.

Bibionomorphs of the Middle Jurassic are represented by 45 species, 12 families,
and four superfamilies, forming the most vast and diverse group of Diptera. The
superfamily composition of the infraorder differs little from the Liassic (the new
superfamily Scatopsidea makes its appearance), approaching closer to the Cenozoic fauna. It is very interesting that, in spite of the greater number of forms,
no peculiar extinct Jurassic superfamilies of bibionomorphs have been discovered
in the Middle Jurassic fauna; the unusualness appears only in the family composition. These Diptera are comparatively well studied and therefore the absence of
some superfamilies, widespread in the Cenozoic, is definitely much more reliable
than it was with the tipulomorphs. Such an absence is that of the superfamily
Cecidomyiidea, an obvious derivative of fungivorids, which appeared apparently
in the later Middle Jurassic. The absence of the superfamily Bolitophilidea is
considerably less real and presumably results from incomplete data. The connection
between some of the Upper Jurassic families of the Fungivoridea and the corresponding Liassic and even Triassic groups is very clear; such, for example, are numerous
genera of the family Pleciofungivoridae, quite abundantly represented in the Triassic
and Middle Jurassic, and simultaneously the source of many other families of
fungivorids and, presumably, of all the superfamilies of Bibionidea.

Comparing the character of the bibionomorphs and the tipulomorphs of the
Middle Jurassic, one can obviously establish the younger, which noticeably
approaches the Cenozoic composition of the first infraorder, clearly differing
from the more ancient tipulomorphs.

To consider the 10 species of the Middle Jurassic Asilomorpha, belonging to
the same number of genera, eight families and three superfamilies, is still premature. It is important to note the presence of only 'new' superfamilies, which
are represented also in the contemporary fauna, together with the unusualness
of the family composition already recorded above. Two families, Rhagionidae
and Nemestrinidae, in common with the Cenozoic fauna, the first of which is
the most richly developed in the Middle Jurassic, now possess clearly relict features.

The ecological features of the Diptera of the Jurassic of Karatau have been
examined by me previously (Rohdendorf, 1938, 1946, 1947a), and in order not
to repeat myself, I shall first of all note only the chief conclusion concerning the
character of the whole buried faunistic complex which, apparently, represents
an 'experimental group' of the real Jurassic fauna existing in Karatau. The accidental
character of the composition of the Karatau Diptera is determined by the properties

of the features of burial - by the drift of flying insects with the wing on the surface of very shallow reservoirs (or of semi-liquid drying clay silt). There seems to
have been no transfer of insects by streams of water. All this influenced the exceptional diversity of the fauna preserved (there was an almost complete absence
of series of one form only) and the exceptional quality of the preservation which
provides paleontological documents of unusual value in paleographic and phylogenetic studies. For instance, entire insects with indestructible extremities were
usually only dilated as a result of decomposition on the substrate. As a consequence of the features of burial there was an overwhelming predominance in the
faunistic complex of terrestrial forms of Diptera not linked with reservoirs.

A detailed ecologic-functional selection of individual groups of Middle Jurassic
fauna emerges from this investigation; the features of most superfamilies have
been already analyzed by me during the study of contemporary Diptera; some
fossil superfamilies of tipulomorphs have been already illuminated in the descriptive
part. Noticeably more fruitful should have been the selection of functional features
within individual superfamilies, the evaluation of the features of the separate
families and genera. However such a detailed analysis is beyond the frame of the
scheme of the present work.

Upper Jurassic Diptera of Western Europe

The Jurassic Diptera of western Europe were discovered 100 years ago when,
for the first time, Brodie (1845) and later Westwood described some species of
bibionomorphs and tipulomorphs from Upper Jurassic deposits in England. Such
antiquity of the beginning of the study of Jurassic western European Diptera
should have been able to establish the scientific maturity of this division of
palaeoentomology. However, knowledge of the fossil Jurassic fauna of the Diptera of western Europe remains at quite a low level: in fact, for 70 to 80 years,
no entomologists developed any interest in Mesozoic Diptera. None attempted
to revise the described types of Brodie, Westwood, Gibel, and Handlirsch, let
alone to search for new fossils. All substantial descriptions and representations
of Diptera of the Upper Jurassic available in western European literature are
inaccurate and cannot be accepted as scientific material of full value.

As a consequence the systematics of western European material proved to be
completely vague in the first monograph of the Jurassic insects of Europe. Handlirsch (1906, 1908) attempted to clear up the systematic features of the Diptera
of the Upper Jurassic of England and Germany, but was not able to obtain great
results. In fact, he only described new forms in a recording way. As a result this
is a list of the species and genera of Diptera distributed by Handlirsch among
four or five families of the contemporary fauna of Diptera.

I have attempted to determine and evaluate the figures and accounts of the
dipterous forms of the Upper Jurassic deposits of western Europe made by
different authors. In a series of cases more or less positive results were obtained,
but in others the connections of the described form remained vague.

Altogether Handlirsch described 20 species of insects which belonged to the order Diptera; presumably only 18 species actually are dipterous. Below is presented a list of the described forms.

List of the Upper Triassic Diptera of Western Europe

Infraorder Tipulomorpha
  Superfamily Tipulidea
    Family Architipulidae Handlirsch, 1906
      Genus *Corethrium* Westwood, 1854
      C. pertinax Westwood, 1854

  Superfamily Chironomidea
  Family?
      Genus *Hasmona* Giebel, 1856
      H. leo Giebel, 1856

      Genus *Dara* Giebel, 1856
      D. fossilis (Brodie, 1845)

      Genus *Bria* Giebel, 1856
      B. prisca (Brodie, 1845)

      Genus *Asula* Giebel, 1856
      A. dubia (Brodie, 1845)

      Genus *Chironomopsis* Handlirsch, 1906
      C. arrogans (Giebel, 1856)
      C. extinctus (Brodie, 1845)

      Genus *Pseudosimulium* Handlirsch, 1906
      P. humidum (Brodie, 1845)

Infraorder Bibionomorpha
  Superfamily Fungivoridea
    Family Fungivoritidae Rohdendorf, 1962
      Genus *Thimna* Giebel, 1856
      T. defossa (Brodie, 1845)

      Genus *Pseudadonia* Handlirsch, 1906
      P. fittoni (Brodie, 1845)

      Genus *Cecidomium* Westwood, 1854
      C. grandaevum Westwood, 1854

Genus *Bibionites* Handlirsch, 1906
B. priscua (Giebel, 1856)

Genus *Sciophilopsis* Handlirsch, 1906
S. brodiei Handlirsch, 1906

Superfamily Bibionidea
Family Protopleciidae Rohdendorf, 1946
Genus *Simulidium* Westwood, 1854
S. priscum Westwood, 1854

Family?
Genus *Thiras* Giebel, 1856
T. westwoodi Giebel, 1856

Infraorder Asilomorpha
Superfamily Tabanidea
Family Nemestrinidae Macquart, 1834
Genus *Prohirmoneura* Handlirsch, 1906
P. jurassica Handlirsch, 1906

Superfamily?
Family?
Genus *Empidia* Weyenbergh, 1869
E. wulpi Weyenbergh, 1869

Family?
Genus *Remalia* Giebel, 1856
R. sphinx Giebel, 1856

In addition to these 18 species Handlirsch, in a list of the Upper Jurassic Diptera, still places as representatives of the family Psychodidae two species of the genus *Psychodites* Handlirsch, 1906 [*P. kenngotti* (Giebel) and *P. egertoni* (Brodie) of the Upper Jurassic of England]. From the description of these fossils I cannot agree with the determination of their order and I think that these insects in reality are Homoptera; the final decision on this question will be possible only after a study of the original material.

The poor state of knowledge of most of the Upper Jurassic Diptera of England and Germany makes it impossible to describe them in detail, much less to arrange them into a key where we show their relationships. The arrangement presented above is no more than my own appraisal on the available evidence, and there is no point in discussing it further.

Altogether, therefore, in this fauna 17 genera are so far known, eight of which belong to four already-known families, six to one known superfamily and three genera to three families the identification of which is difficult; one of them almost certainly is unknown up till now.

It is very important to compare this Jurassic fauna of western Europe with
the well-studied Karatau fauna. As already mentioned above (p. 219), my primary
assumption concerning the synchronism of these faunistic complexes proved to
be inaccurate. Comparing the features of the western European fauna with the
Karatau it is possible to find easily authentic demonstrations of their different
age corresponding to Upper Jurassic and Middle Jurassic.

The main characteristic of the Upper Jurassic Diptera of western Europe turns
out to be the absence of ancient tipulomorphs, which are richly represented in
the Lower and Middle Jurassic faunas together with an abundance of representatives of the superfamily Chironomidea, to which belong almost half of the species
and genera (seven species, six genera). This circumstance first of all, of course,
indicates the peculiarities of the formation of the locations, but simultaneously
shows also the youthfulness of this fauna as a whole.

Also interesting is the multiplicity of fungivorids, namely fungivoritids to
which belong five species of five genera; this group is known to have been the
source of the young Cenozoic families - Fungivoridae and Sciaridae. A more
precise comparison of the relations of the five European genera of the family
with the Karatau forms is very important, but impracticable without revision of
the original collections. The relatively small variety of Fungivoridea as a whole
is curious in comparison with the great variety of this superfamily in the more
ancient Middle and Lower Jurassic faunas first of all in Karatau. Also interesting
is the presence of two representatives of the bibionids belonging to two families
of which one was almost certainly new, and which is up to now unknown from
Karatau. On the whole the composition of the fauna of bibionomorphs of western
Europe differs noticeably from that of Karatau which is more ancient.

Of the few representatives of asilomorphs only one species can be determined
more or less exactly. It belongs to a family unknown in the fauna of Karatau
and which was widespread in the Cenozoic fauna, namely the Nemestrinidae,
the most peculiar of contemporary tabanids, the ancient nature of which has
long been obvious. The finding of Nemestrinidae in the fauna of the Upper
Jurassic matches the general character of this fauna, which we still know insufficiently but which more closely resembles the Cenozoic than the Middle Jurassic
Karatau fauna.

The Upper Jurassic Diptera of Eastern Asia

Until recently every location of insects of the Upper Mesozoic fauna discovered
in Zabaikal, China and Mongolia were attributed to Cretaceous time. These are:
Turginsk, described in the past century from many locations in Zabaikal (Eichwald, 1865; Rohdendorf, 1957), and Lajan described by Grabau (1923); Ping
(1928), and Cockerell (1927) from northern China and Mongolia. These deter-
minations of age had little basis until recently. At the same time the composition of these eastern Asiatic faunas, very different from the known western
Mesozoic faunistic complexes, were little known. The Diptera of Lajan are two

species of the genus *Chironomoptera* (superfamily Chironomidea; a more exact determination of the status of this genus is impossible!) described by Ping (1928).

The difficulties in determining the age of the Turginsk-Lajan faunistic complex will apparently be overcome eventually. The discovery of the diverse fauna of Turginsk age in Zabaikal, including remnants of insects of different orders, permits us to secure reliable data for the characteristics and comparisons of this fauna. It is necessary to mention the results of the study of the main, most common Turginsk insect - the mayfly *Ephemeropsis*. As indicated in the investigation of O.A. Chernova (1961) this genus is closest to the Upper Jurassic *Hexagenites*, which allows the author to affirm the Upper Jurassic age (marl) of this fauna.

The Diptera in the composition of this fauna, not counting the reports of Ping from Lajan, were not known. Only recently large new collections from the Zabaikal locations were obtained, among which was discovered diverse material on the remainder of the Diptera. The investigation of these new collections from Zabaikal is a problem for the near future.

Cretaceous Diptera

The insects of the Cretaceous have until now remained the least studied. This is because the Cretaceous species of Diptera are counted only in units and were discovered, moreover, only in the boundary formation of the system - in the lowest parts of the Lower Cretaceous of Mongolia and China, and in the Laramie formations of North America, in strata transitional to the Paleocene. In fact the real Cretaceous insect fauna remains one of the most disappointing gaps in the paleontological chronicle; the formation of Tertiary and present-day faunas is one of the most important problems of palaeoentomology. Its resolution will be influenced by the study of Cretaceous insects which, although they have not been exposed up till now, almost certainly exist. I am convinced of this because numerous finds of Cretaceous plants have been made.

The known 'Cretaceous' fauna of eastern Asia, which included remnants of Diptera, mostely turned out to be more ancient, Upper Jurassic (see above, p. 272). In regard to dipterous insects, all the vast series of Cretaceous deposits remain completely unknown.

Of an altogether different character are the Diptera in the Laramie formations, exposed in amber in the vicinity of Cedar Lake in Canada (Carpenter, 1937); these insects already have a purely Cenozoic appearance and belong to different genera of the families Chironomidae and Ceratopogonidae, the majority of which are also contemporary. These scarce data cannot throw light on the features of the fauna over so long a section of geological time.

Tertiary Diptera

The Tertiary Diptera were described very long ago - the first literature data

known to me are dated 1742 and deal with insects of Baltic amber. [12] Altogether
to the present, upwards of two thousand species of Tertiary Diptera have been
described from quite a large number of very irregularly studied locations. Remnants of the Tertiary fauna of Diptera are scarce and among them there stands
out sharply the fauna of Baltic amber known from a series of locations of the
Kaliningrad district and Lithuanian SSR.

Not everything that has been done on the Tertiary Diptera can be considered
and evaluated in this work. The evolution of the different groups of Diptera, the
infraorders, superfamilies and families - all these big questions can be solved only
by an investigation of Mesozoic material. Such a sudden limitation of the value
of Tertiary fossils is the inevitable conclusions arrived at by studying the oldest
Tertiary fauna, namely the Paleogene (Baltic amber). The Tertiary fauna of
Diptera consists almost entirely of representatives of present-day families; the
differences between this fauna and that still existing are basically the absence
of some contemporary families, (for the most part rare also in the present-day
fauna) and its different ancestral species composition. The number of families
of which the representatives are known only from Tertiary deposits is very
small - in fact, there are only two poorly known representatives of the fungivorids,
the isolation of which possibly arose as a result of error in the descriptions of
these fossils (families Necromyzidae and Mycetophilitidae).

Inaccuracy and superficiality of the descriptions of Tertiary Diptera still more
depreciates the value of the published material of the last hundred years. Only
very recently have isolated investigators begun the redescription of the old collections of some faunas; such, for example, are the works of Statz (1940) for the
Upper Oligocene locality of Rott in western Germany and James (1937, 1939)
for the Miocene locality of Florissant in North America. But these few investigations of course do little to clear up the existing confusion in the investigation of
Tertiary Diptera.

The great similarity of the Neogenic fauna of the Diptera (for example the
Miocene of Florissant) to the contemporary fauna, together with the similarity
of more ancient Paleogene faunas (for example, the fauna of Baltic amber),
indicates the importance of the study of Tertiary Diptera for the illumination
of the phylogeny of the separate families and genera of the order. For clarification
of the historical development of the present-day families and genera of Diptera
the Tertiary records are positively necessary and their accumulation is the most
important problem of paleontology; at the same time, for the illumination of the
phylogeny of the higher taxa of the order, the data for Tertiary faunas are of
little use.

The Tertiary fauna of Diptera cannot be analysed as fully as the older faunas
because so little is known about it. It is much more useful to note differences
between the tertiary and contemporary faunas, analyzing concisely the composition of the individual superfamilies and families. It is simplest to enumerate the
separate families of the order according to the list noted in Part 1 (p. 9-14).

The individual infraorders Nymphomyiomorpha, Deuterophlebiomorpha and
Blephariceromorpha in the present-day fauna have not yet been discovered in

12. In 1742 remnants of tipulids from Baltic amber were incompletely described and pictured by Sendelius
under the name '*Tipula*' (Handlirsch, 1907, p. 1004).

Tertiary faunas, showing that, as at the present day, these families were uncommon in Tertiary times. The present-day ecological features of these insects, which are inhabitants of high altitude terrain were, presumably evolved in Tertiary time and the preservation of fossils of high mountain organisms is very improbable (Efremov, 1940, 1950).

The contemporary superfamilies of Tipulomorpha not yet discovered in Tertiary faunas are rare relict groups, poor in species - the Pachyneuridea, Rhaetomyiidea and Orphnephilidea; all the rest of the superfamilies are represented in different faunas of the Paleogene or Neogene.

The superfamily Tipulidea is found in almost all faunas. The most numerous in species are Limoniidae (more than 130 species of approximately 30 genera) and Tipulidae (about 80 species); the remaining families are represented by a few species - such are the Trichoceridae (four species) and Ptychopteridae (two species of two genera). The family Tanyderidae has not yet been discovered in Tertiary faunas.

The superfamily Psychodidea is represented by three families only in the fauna of Baltic amber. Very interesting is the presence of a comparatively large number of species of the subfamilies Trichomyiinae (12 species of two genera) and Sycoracinae (four species of two genera), at present rare and narrowly distributed relict forms. The chief family, Psychodidae, includes about 20 species and three genera. The Phlebotomidae, as also the Nemopalpidae, are each known by a single poorly described species.

The superfamily Culicidea is known in Tertiary faunas from remnants of both families - the Culicidae (seven species of three genera) and Chaoboridae (four species of two genera). Remnants of Tertiary blood-sucking mosquitoes have undergone a study recently and there have been established the presence of five species of the genus *Culex* (from Upper Eocene and Lower Oligocene deposits of western Europe and North America), of one species of *Aedes* (from the Lower Oligocene of White Island) and of *Taeniorrhynchus* (from the same deposits). One of the fossil species (*Aedes protolepis* Cockerell from the Oligocene) was represented in the form of a large series of both sexes and was redescribed in detail by Martini and Edwards, and the attachment of it to the contemporary subgenus *Armigeres* (Martini, 1929) was indicated.

The superfamily Dixidea, poor in forms in the contemporary relict group, is represented by species of the genus *Dixa* from Baltic amber (two species) of the Oligocene of White Island (one species) and of the Lower Oligocene deposits of France (one species).

Like the tipulids, the superfamily Chironomidea was not worked up completely until recently. Representatives of the family Chironomidae (about 100 species), Ceratopogonidae (about 35 species) and Simuliidae (approximately 10 species) were discovered. A majority of the descriptions are inaccurate and do not allow us to judge the relationships of the individual families.

All the contemporary superfamilies of the infraorder Bibionomorpha are represented in the Tertiary fauna. The peculiar monotypic superfamily Bolitophilidea was found in the fauna of Baltic amber in the form of one remnant

which has not been identified more precisely as a representative of the genus
Bolitophila.

The superfamily Fungivoridea is represented in the fauna of Baltic amber
very richly, by almost all present-day families; up to now only the Ditomyiidae
(relict Diptera poor in species) is unknown. Very important is the presence among
Tertiary fossils of representatives of the rare families in contemporary fauna
Allactoneuridae (to me known still as undescribed species from Baltic amber),
Manotidae, Lygistorrhinidae (genus *Palaeognoriste* Meunier from Baltic amber,
Diadocidiidae (six species of two genera), Macroceridae (eight species of one
genus), and Mycetobiidae (eight species of two genera); this testifies to the great
development of these Diptera in Paleogene time. Most diverse are representatives
of the Fungivoridae (more than 200 species of 42 genera), Sciaridae (75 species
of nine genera) and Ceroplatidae (23 species of three genera). Quite special families
apparently produce representatives of the genus *Necromyza* Scudder (Miocene
of western Europe) and the genus *Mycetophilites* Burmeister (Upper Oligocene
of western Europe); these forms are very insufficiently described and to judge
accurately concerning their features is impossible.

The superfamily Cecidomyiidea is represented in the Tertiary fauna (chiefly
of Baltic amber) by species of all three contemporary families - Cecidomyiidae
(38 species of 18 genera), Lestremiidae (14 species of five genera) and Heteropezidae (four species and three genera). These data are very preliminary and in
the future will be expanded considerably.

The superfamily Bibionidea in Tertiary faunas is represented very abundantly
while in the fauna of Baltic amber these insects are relatively scarce. Most rich
in species are the Penthetriidae which are found in almost all known Tertiary
locations (over 100 species of two genera). In second place stand the Bibionidae
(about 90 species of the genera *Bibio, Bibiodes, Philia*). The last family, Hesperinidae, is represented by a single species of the extinct genus *Mycetophaetus*
Scudder of the Miocene deposits of North America.

The superfamily Scatopsidea is represented by only five species of the Scatopsidae from Baltic amber and the Lower Oligocene of France: two other relict
families up to now have not been discovered in a fossil state.

The superfamily Rhyphidea, one of the most ancient and important groups
of Diptera, is represented comparatively poorly in Tertiary faunas, by approximately 10 species of two genera of the family Rhyphidae.

The big and important infraorder Asilomorpha is represented in the Tertiary
fauna by all the contemporary superfamilies and by the vast majority of families;
only some of the relict families Acanthomeridae, Apioceridae, Cyrtosiidae and
Solvidae, poor in species, remain undiscovered.

The superfamily Tabanidea is represented by approximately 45 species of
almost all families, the most abdundant being forms of the families Rhagionidae
(about 25 species of five genera), Tabanidae (about 10 species of four genera),
Nemestrinidae (six species). The family Coenomyiidae is represented by one
species. This composition of the Tertiary fauna of tabanoids is very noteworthy;
in spite of the abundance of existing horseflies and the relative scantiness of

other families, in the Tertiary fauna these latter are well represented, even some
which are rare nowadays, such as the Coenomyiidae.

The superfamily Stratiomyiidea also developed peculiarly in Tertiary time.
Together with an abundance of species of Stratiomyiidae (about 25 species of
11 genera) there are representatives of the rare, at present relict, groups Xylophagidae (seven species of five genera) and Rachiceridae (one species). The Acroceridae are represented by one poorly-described species. The abundance of
Xylophagids is very interesting.

The superfamily Asilidea, richly developed at present, is represented in Tertiary
faunas by almost all families and the most numerous Asilidae (about 30 species
of 15 genera) and Therevidae (collections still insufficiently studied, but presumably includes not less than 10 species). The other families (Mydaidae and Scenopinidae) are known from Tertiary faunas by a few poorly investigated remnants.

The superfamily Bombyliidea in Tertiary deposits has been discovered in a
comparatively moderate number of species of three families. The greatest number
of species (16) belong to the Bombyliidae (approximately six genera) and a few
species to the Usiidae and Systropodidae; the presence of these latter families,
which are poor now, is very noteworthy.

The superfamily Empididea, rich in forms now, is also very abundantly represented in the Tertiary fauna, chiefly in Baltic amber (that definitely is connected
with the characteristic ecology of these insects, inhabitants of the forest and
moist habitats, of the stems and stalks of plants). All three families are present
and the richest in species are Dolichopodidae (about 70 species) and Empididae
(approximately 60 species); the Hilarimorphidae were found as a single form.

The peculiar infraorders Musidoromorpha and Phoromorpha are represented
by a few species of the families Lonchopteridae and Phoridae, still very insufficiently studied and in their majority discovered in the fauna of Baltic amber. Neither
the parasitic groups of phoromorphs nor the termitoxeniomorphs have been
found up to now in the fossil condition.

The most diverse contemporary infraorder Myiomorpha is represented in the
Tertiary faunas by comparatively numerous remnants of the majority of superfamilies of the present-day fauna; until now only fossil remnants of the relict
superfamily Somatiidea and the intestinal gadflies, the Gastrophilidea are not
known. However, in fact, the existing paleontological data on the history of the
myiomorphs is much poorer. Very many big superfamilies of this infraorder,
although found in the fossil state, are in so small a number and of such poor
preservation that altogether they do not yield accurate identification even to
family.

The superfamily Platypezidea is represented by remnants of the Platypezidae
in Baltic amber and the Oligocene of North America and western Europe: fossil
remnants of the Sciadoceridae are unknown.

The superfamily Syrphidea is represented chiefly in the Neogene of America
and Europe. Such are about 65 species of approximately 25 genera of Syrphidae
and about five species of two genera of Pipunculidae.

The superfamily Conopidea is known only from two species of two genera

of Conopidae: the family Stylogastridae has not been discovered up to now in
the fossil state.

The superfamily Trypetidea, one of the largest in the whole infraorder in
Tertiary deposits up to now, is represented in all only by single remnants of
three species of Trypetidae and Otitidae; the families Pyrgotidae, Tachiniscidae,
and Platystomatidae are unknown in the fossil condition.

The superfamily Psilidea is represented by individual remnants of representatives
of the families Psilidae, Micropezidae and Diopseidae; the contemporary Cypselosomatidae, Megamerinidae and Nothybidae, which are poor in species, are unknown in the fossil state.

The superfamily Heleomyzidea is the most completely represented in the
fossil condition. There are known fossil remnants of the diverse families Heleomyzidae, Anthomyzidae, Sciomyzidae and Sepsididae, now numerous in species,
as well as the relict ones poor in species, Coelopidae (namely Dryomyzinae). Only
one, the now relict family Rhopalomeridae, is unknown in the fossil condition.

The superfamily Sapromyzidea, at present diverse and rich in species, is little
represented in the fossil state. Such as exist are isolated species of Sapromyzidae
and Lonchaeidae. The contemporary Piophilidae and the tropical Celyphidae are
absent in collections of Tertiary insects.

The superfamily Borboridea, at present a group quite rich in species, is represented by species of the families Agromyzidae and Borboridae. Fossil remnants
of the quite diverse contemporary Milichiidae are unknown up to now, and the
same is true for the now relict, specialized Cryptochaetidae.

The superfamily Drosophilidea, a vast present-day group, as well as the superfamily Chloropidea are represented by a few species of all families - Ephydridae
Drosophilidae and Chloropidae.

At the present time it is possible to draw up very few generalizations on the
basis of the existing material. First of all there are the important findings in the
Tertiary deposits of remnants of present species of relict families; such findings
confirm the important phylogenetic significance of these groups. Such, are the
Platypezidea, the more diverse Heleomyzidea (Dryomyzinae, Heleomyzidae,
Anthomyzidae and others) and the Psilidea (Micropezidae and Diopseidae); the
last superfamily, furthermore, bears witness to the tropical character of the
observed faunistic complexes.

The investigation of a group of the superfamily Calyptrata, the Tertiary remnants of which up to now are very little known, cannot give substantial indications.
Although in the contemporary fauna the superfamilies of calyptrates include a
very large number of species (almost as many as the acalyptrates!), the number
of their fossil remnants is very low. It is sufficient to say that up to now there
have been described only about 20 species of not exactly distinct representatives
of the Anthomyiidea, Muscidea, Sarcophagidea, Oestridea and Tachinidea.

The superfamily Glossinidea is known through one species only, discovered in
the Miocene fauna of the Florissant (North America). In itself this fact offers
exceptional interest in view of the presence of these Diptera in the present-day
fauna only in Africa. The blood-sucking glossinids and the connection of them

with different hoofed mammals permits us to examine their geological history from various angles (p. 121, 302).

One species of the superfamily Hippoboscidea, scarce in the contemporary fauna, has been found in the Upper Oligocene deposits.

Representatives of the last three infraorders of parasitic Diptera - Streblomorpha, Nycteribiomorpha and Braulomorpha have not been found in the fossil condition up to now.

The short review made of the present-day data on the composition of the Tertiary faunas of Diptera show firstly the great apparent similarity to the contemporary fauna, and secondly the need for intensive investigation and revision of all that has been done up to now. Tertiary species of both the Paleogene (for example, Baltic amber) and the Neogene have been frequently described, but the value of these descriptions **is** very small as a result of their superficialities and inaccuracy.

PART III

PHYLOGENESIS OF THE DIPTERA

The investigation of the contemporary fauna of Diptera and the known fossil remnants, chiefly of Mesozoic forms, from locations of central Asia and Europe allowed us to clear up the phylogenetic relations of the chief groups of the order.

The unusualness of the evolutionary history of the Diptera, like that of all other holometabolous insects, lies in the fact that it completely contradicts the known principle 'ontogenesis repeats phylogenesis' or, in another version, 'ontogenesis repreats the ontogenesis of ancestors', in relation to the postembryonal development of these insects. That is to say that in the perfecting of individual development there is the production of a larval stage which lives in different conditions than the winged insect, lives in the most different sheltered media and is capable of feeding on quite different substances than the adult dipteran. All the history of the Diptera shows that improvement and adaptation of the larval stage from the moment of escape of the insect from the egg sheaths were among the most important and deepseated changes in their historical development. Changes in the larval stage of Diptera are often far more significant than those of the winged insect; related forms often differ more as larvae, than as adults. This influences a quite particular importance of the features of the stages of development of Diptera during an investigation of their phylogenetic relationships and system. Exceptional attention to the features of the larvae as the most important evidence of phylogenetic closeness is widely prevalent in the foreign literature (Hendel, 1937; Hennig, 1948, 1950) and, based on this conclusion, reorganization of the systematic schemes of the order (for example, the artificial association of groups clearly distant one from another on the basis of a similarity in the organization of the larvae) bear witness to the formal application of the 'biogenetic principle'. The failure to understand the chief features of the historical development of these insects makes this conclusion incorrect.

This process — the improvement of the stages of development in the history of the most diverse groups of Diptera — often took place by similar paths, as a result of which there occurred many monotypical features that appear parallel. The general similarity of this process was brought about by conflicts between the need for protection and for movement in dense or liquid substrates; between the need for respriation in conditions poor in oxygen and the structure of the metameric respiratory system; between the need to live on liquid or too-dense nourishing materials and the presence of the mouth apparatus of the gnawing type, and undoubtedly between some other contradictory tendencies. The conflicts resulted in similar changes in the form of the body (loss of legs), in the respiratory organs (restriction of them to the ends of the body), in feeding (reduction of the mouth organs and the development of extra-intestinal digestion), and in sheaths (their thickening, armament and the development of a puparium). Only a methodical, accurate examination of the processes of the historical development permit us exactly to evaluate the observed changes of the larval stage of different Diptera and to throw light upon the true phylogenetic connections of the groups.

Summarizing everything said, it is natural to infer the very great importance of the study of the stages of development of the Diptera: the organization of larvae and pupae testifies to the character of the historical development of the given group, and illuminates one of the most important determining features of the history of these insects. Stages of the postembryonal development of Diptera provide the first evidence of the direction of the evolution of these insects, indicating the paths of the solution of problems in development, including the problems of the recent, last stages in their history.

Existing schemes of the phylogeny of the Diptera

The most detailed review of the phylogenetic relations of the families of the order was made by Hendel (1937) using all available data from other authors (Handlirsch, 1906-08; Speiser, 1908; Lindner, 1925-28; Edwards, 1926; Tillyard, 1933, and others). This outline of phylogeny suffered one very great deficiency which substantially lessened its importance. The author not only investigated all the features of the organization purely formally, but still declared and adhered to the dogma of 'straight-line development (orthogenesis) of some systems of organs', 'regressive tendencies of development, . . . developing in different directions', the presence of 'distinct reduction of orthogenesis' (Hendel, 1937, p. 1872), that larvae possess 'their own phylogenesis' (!) and that 'the natural system must be based only on the morphology of adult forms' (the same, p. 1873). If to this is added the expression of Hendel concerning the appearance 'of approximately analogous mutations in different places of the genealogical tree' (p. 1872) which according to the author explains the development of the phenomena of convergence, then for us the importance of all these considerations, which only distort reality, become completely clear. I consider it necessary to state in detail the views of Hendel because his work has assumed a peculiar importance in the systematics of contemporary Diptera.

Hendel, an Austrian entomologist, investigated the system of different groups of Diptera in the first four decades of our century and provided a whole series of detailed monographs on many families of the order. One of the last of his works was an outline of the Diptera for Kükenthal's zoology; it can be said without exaggeration that his concrete systematic investigations are among the best on the basis of exactness and the latitude of the scope of material. However, Hendel was a scientist to whom a regular materialistic philosophy was foreign. It is obvious that if the works of Hendel, as a systematic diagnostitian, possess great value, then it is also necessary to say of the conclusions of Hendel the evolutionist, that they are not substantiated and are far from the materialistic

conception of the processes of the historical development of organisms. Everything said reinforces my view of the conclusions of Hendel: only new facts established by him are of value. All his generalizations require a doubly critical
analysis and his attempts at an explanation of the phenomena must simply be
discarded.

Hendel's examination of the phylogeny of Diptera is conducted by comparing
the individual groups one with another, and by a consideration of various morphological features. No 'genealogical tree' or phylogenetic scheme is constructed,
nor are questions of the divergence of the different families and superfamilies
really touched upon.

Hendel considers the superfamilies Tipulidea, Rhyphidea, Fungivoridea,
Bibionidea and 'Liriopeoidea' (part of the Tipulidea and all the Psychodidea),
to be the most ancient groups, which is close to the truth although not very
exact. Considering the composition of the superfamily Fungivoridea, Hendel
notes the closeness to it of gall midges, Cecidomyiidea being its side branch,
and the artificiality of the group 'Zygoneura' of Enderlein. Hendel considers
the group Bibionidea as being closely related to the Fungivoridea (which is evident), but enlarges it to include the Scatopsidea and Pachyneuridea, although
he notes their significant differences; such wide limits of the Bibionidea are
unjustified.

Quite mistakenly some tipuloids (Ptychopteridae and Tanyderidae) are united
with the Psychodidea (which Hendel assumes is a single family) in the common
formal group 'Liriopeoidea'; the argument cited is the presence of many ancient
features, but that by no means demonstrates a real closeness of these groups.
The artificiality of this association is apparently partly clear to the author himself, who notes a similarity of the Psychodidae with the Culicidae.

There are described in the large trunk of the 'Culicoidea' three of our superfamilies — Culicidea, Dixidea and Chironomidea; the closeness of these Diptera
does not provoke doubts, but the union of these groups, which are far removed
from each other, into one superfamily is certainly not substantiated.

Hendel correctly keeps the peculiar relict group Orphnephilidea separate in
the form of an independent superfamily in the same way as the still more peculiar
Blephariceridae. No conclusions about the phylogenesis of these Diptera are
made, however, unless one considers his assumptions concerning the closeness
of the orphnephilids to the chironomid complex as such. By this means he achieves
a synopsis of all 'nematocerous' Diptera; the big superfamilies Tipulidea and
Rhyphidea remained unnoticed, apart from quite short conclusions concerning
their closeness to the rest of the groups.

In reviewing the relations of the infraorder Asilomorpha, Hendel starts with
the Rhagionidae, Xylophagidae and Tabanidae (Tabanidea) noting their closeness. A second close group, the Stratiomyiidae, Solvidae (= 'Stratiomyioidea'),
is characterized, combining with the first through a connection between the
Xylophagidae and Solvidae. Finally the third group — Nemestrinidae plus Acroceridae (= 'Nemestrinoidea') — also converges with the first two, taking into
consideration the most 'highly specialized forms', the families, 'located at the
top of the genealogical tree' ('Spitzenfamilien'); all this complex Hendel desig-

nates the 'Homeodactyla' and contrasts the remaining asilomorphs as the 'Heterodactyla'. In this scheme only the association of all the named families is accurate; the treatment of the nemestrinids and acrocerids as 'upper groups', related one to another, and the assertion of immediate connections between the Xylophagidae and Rhagionidae ('Erinnidae' of Hendel - a completely artificial, unreal grouping!) are altogether erroneous.

The group 'Heterodactyla' includes, according to Hendel, three superfamilies: the 'Therevoidea' with the Therevidae, Apioceridae and Scenopinidae, the 'Asiloidea' with the Asilidae, Mydaidae, Bombyliidae (= Bombyliidea of our scheme) and finally the 'Empidoidea' with the Hilarimorphidae, Empididae (s.l.) and Dolichopodidae. The 'Therevoidea' is considered as the primitive superfamily, the 'Empidoidea' as the 'upper' group; the first superfamily converges with the Rhagionidae, and he concludes that both groupings of 'brachycerous' Diptera split off from the so-called common trunk still earlier than the 'hematocera' Rhyphidae, Fungivoridae and Bibionidae. The approximation of the empidids with the dolichopodids as components of a peculiar, clearly separated superfamily is quite correct. It is otherwise concerning the two other 'superfamilies' of Hendel's scheme; the separation of the Therevidae and Apioceridae from the Asilidae and Mydaidae is not justified, neither is the approximation of the complex of bee flies (Bombyliidea) to the gadflies and the mydaids. This review clearly shows the shortcomings of the purely morphological approach. Such a method did not allow us to detect the isolation of the bee flies, a not less characteristic group than the Empididea, which are clearly characterized by features of development and ecology. Also completely unjustified is the assignment of families among the 'primitive' (Therevidae, Asilidae) and the 'secondarily changed' (for example, the Apioceridae) which is based virtually only on the features of the venation of these forms, sometimes rich and non-mechanized, at other times strongly costalized and reduced. It is quite obvious that the venation of the blade of the wing is only in small measure a determining feature in the history of the Diptera.

Hendel considers the derivation of the Phoridae and correctly comments on their connections with the Clythiidae (Platypezidae) through the 'intermediate' group, the Sciadoceridae: he provisionally places all three named families into one superfamily the 'Phoroidea'. No decision is taken on the peculiar Lonchopteridae. It is necessary, however, to say that Hendel considers the 'suborder' Cyclorrhapha to be real and includes in it both the 'Phoroidea' and 'Musidoroidea'. The origin of the whole 'suborder Cyclorrhapha' is assumed to be from the 'forerunners the Therevidae', the basis for such a conclusion being the absence of empodia and hairly spurs on the tarsus; that, no doubt, is incorrect.

The consideration of the relationships of the main superfamilies of the Myiomorpha is extremely superficial; in fact Hendel discusses only the features of the ptilinum and the medial and anal systems of veins. It is affirmed that the 'most primitive' ('ursprünglichsten') cyclorrhaphous Diptera are the Syrphidae because of the presence of a long cubital cell and of the absence of a ptilinum. Next are the Conopidae, which are supposed to have served as ancestors of the

Acalyptratae and Calyptratae. This whole scheme is incorrect: the real original myiomorphs, the Platypezidae, the ancestors of which gave rise both to the Syrphidea and to other superfamilies of myiomorphs, were completely disregarded in the discussion. On the other hand the features of the Conopidae were clearly 'overrated' and this peculiar, highly specialized group proved to be also supposedly the original group for the vast complex of all the rest of the Myiomorpha. Here Hendel's scheme is completely incorrect.

Hendel's outline finishes with an examination of the known complex of parasitic Diptera, the so-called 'Pupipara'. He properly affirms the obvious polyphyly and convergent nature of this group and assumes connections of the Nycteribiidae with the cave Acalyptrata (Borboridae). Connections of the Hippoboscidae with the intestinal gadflies, the Gastrophilidae, are poorly substantiated and the relations of the Braulidae with other cyclorrhapha are completely vague. The Streblidae remained unknown to the author.

Another pertinent work is that of the English entomologist Edwards (1926) which includes a review of the phylogenetic features of the Diptera and is devoted only to the 'Nematocera'. It touches on the questions of the origin and connections of the oldest groups of the order and has a definite value as a result of the detailed consideration of features, the comparative rigor of the account and the absence of any attempt at theorizing. The objectivity of Edwards permits us to utilize his concrete conclusions relatively completely. Like Hendel the author analyzes in detail the individual features of the Diptera and comes to the following conclusions. At the beginning of Jurassic time or still earlier the Diptera were divided into three chief trunks, [13] the complex: Fungivoridea + Bibionidea + Scatopsidea + Cecidomyiidea, the complex: Ptychopteridae + Tanyderidae + Psychodidea + Culicidae + Dixidae + Chironomidae and, finally, the complex: Tipulidea (Trichoceridae + Tipulidae and related forms). The second complex shows a series of links between the families Psychodidae, Tanyderidae and Ptychopteridae and the tipuloid complex. In the group of this complex consisting of the Culicidae + Dixidae + Chironomidae + Ceratopogonidae possibly it is necessary to include also the Simuliidae and Orphnephilidae. The Rhyphidea constitute an isolated branch which is related to the original forms of the second complex. The specialized 'brachycera' (Asilomorpha) and Cylorrhapha separated later from the primitive members of the second complex which were close to the Rhyphidea. The features of the Blephariceridae, Deuterophlebiidae, Simuliidae and Orphnephilidae remained obscure for Edwards and he does not come to any definite solution concerning the phylogeny of these forms. These general conclusions show the importance of the investigation, which comes quite close to the correct solution of the problem of the phylogenesis of the ancient Diptera. The isolation of the Fungivoridea + Bibionidea + Scatopsidea + Cecidomyiidea, i.e., of our infraorder Bibionomorpha (without the Rhyphidea), is quite correct. The division into two individual trunks of the infraorder Tipulomorpha is not necessary. A closer approximation of the Rhyphidea to the tipulomorphs than to the bibionomorphs also is not correct. It is remarkable that Edwards had doubts about connections of the Simuliidae,

13. I use the names of the groups accepted now and not the original designations of Edwards, who used many family names now rejected as synonyms.

which are undoubtedly a member of the superfamily Chironomidea. As far as
the dating of the phylogenetic processes is concerned, the indication of the
separation of the chief ancient groups of the order in 'the beginning of the
Jurassic period' or still earlier is very interesting. It is necessary to keep in mind
that at the time of Edwards' work the phylogeny of the nematocerous Mesozoic
Diptera was very little known — present-day data show that the boundary between Triassic and Jurassic was exactly such a moment during which the formation of the chief groups of the order took place. For its time the scheme under
discussion quite satisfactorily solved the problem of the phylogeny of the Diptera.

I shall not analyze other phylogenetic studies on the Diptera which are concerned either with individual organs (works of Lamer, Crampton and other
authors), or with only separate groups of the order. A phylogenetic scheme for
the Diptera was produced by Handlirsch (1908, p. 1270) but this now has only
historical interest. It was set up on the basis of Brauer's system of the order and
on very inadequate data on the Mesozoic fauna; new facts--the finding in the
Jurassic of Karatau of real Rhagionidae, further of Protomphralidae, Palaeostratiomyiidae and Protocyrtidae completely violates the Handlirsch scheme.

A special position in the phylogenetic investigations of the dipterous insects
is occupied by the works of the German entomologist W. Hennig (1953, 1955,
1958). This investigator worked a great deal on the concrete systematics of
many families of myiomorphs, and paid special attention to the theory of
phylogenetic systematics. It is not possible here to examine at length all the
conclusions and generalizations of Hennig, the chief significance of which, in
my opinion, consists of determining the qualitative distinctions of the features
which changed in phylogenesis. It is necessary to appraise highly the conclusions
of Hennig because they noticeably surpass all other phylogenetic generalizations
in dipterology. The chief deficiency of Hennig's conclusions consists of the very
little attention to the ecological-functional analysis of the observed traits of the
structure of insects; unfortunately this markedly lessens the value of his syntheses.
However, on the whole, the strictness of the analysis of phylogenetic relationships
is a great success for Hennig. Many of the conclusions and terminological suggestions of this author are completely correct and may be accepted and used.

Hennig's term 'sister groups' ('Schwestergruppen') is very important. By this
term he designates two taxa which originated as a result of the evolutionary divergence of the original ancestral group; the presence of just a pair of 'sister
groups' indicates the moment of divergence, being accomplished as a result of
the solution of a concrete conflict in the process of historical development of a
given phylogenetic branch. The presence of just one pair of descendants after the
separation of the new taxon from its ancestral group regularly appears in the
phylogenetic analysis of any group of animals. In the face of this any assumptions
concerning the simultaneous derivation of many group-descendants, the so-called
'radiation' form which in fact does not exist, fall away. This conclusion of Hennig
is supplemented by me and is definitely very important in every concrete analysis
of phylogenesis.[14]

14. Actually, if we assume the nature of the process of evolution to be a continuous solution of
problems interchanging one with another, the solution of the problem is the basis of the derivation of
the new group, the basis of the moment of divergence. Thus inasmuch as the concrete conflict may have
only two diametrically opposite solutions, just two results are necessarily derived, that is, dichotomous
divergence.

Another of Hennig's suggestions is an improvement of phylogenetic terminology. Plesiomorphy and apomorphy are terms produced by him, which correspond to concepts which had long been known, but had no precise names. 'Plesiomorphy' and 'plesiomorphic' traits denote ancient, original or, as they are often called, 'primitive', 'archaic' features which are preserved in organisms from their ancient ancestors. Frequent ambiguity and error in naming these phenomena completely justifies the introduction by Hennig of new and convenient terms. The other term, the antithesis to the first, namely 'apomorphy' and the adjective 'apomorphic', denotes new, recently occurring, secondary traits. Hennig writes about the adaptive character of apomorphs using occasionally the special term ('Apookie'); it is obvious that it is scarcely necessary to mention the adaptive significance of new traits of organization, having in view the clearly universal, general character of adaptations. Of course even the plesiomorphs possess, or rather possessed, adaptive importance during their derivation. The importance is exactly in the primary and secondary nature of the features; by this the plesiomorphs are distinguished from the apomorphs.

The actual phylogenetic generalizations of Hennig were based on a series of monographic investigations. Such, first of all, is the composite work of three volumes on the larval forms of Diptera (Hennig, 1948-52), the importance of which is in its complete summary and attempt at the phylogenetic-systematic analysis of the material. In this monograph the proposed systematic scheme of the whole order in many of its parts scarcely requires correction and is partly accepted by me. However, the inevitable one-sidedness of considering only larvae led Hennig to a series of incorrect conclusions. The second large monograph of Hennig is devoted to an analysis of the venation of the wings of Diptera and its significance in systematics (Hennig, 1954). Being, on the whole, a precise investigation, this analysis of the features of the wings of Diptera has one very substantial deficiency--in it he scarcely touched on the aerodynamics of wings and body so that the functional importance of the wing and its parts remained completely unilluminated. For example, he did not examine completely the structure of the base of the wing (its basial, which carries the greatest load during the stroke); he does not take into consideration the variation of the absolute sizes of the body in the process of reduction of the venation but this is exceptionally important to any consideration of the structure of wings.

Finally, in his last big monograph, devoted to the phylogenetic relationships of the majority of the groups of myiomorphs, Hennig (1958) examines the system of the most difficult group of Diptera, namely, the vast complex of acalyptrates, including about 10 superfamilies and more than 30 families. The value of this investigation is very great. In it for the first time, in fact, the relation of these various small Diptera are examined seriously and successively with sufficient completeness. The author earnestly endeavoured to compare the traits of the organization of these Diptera; he paid special attention to the structure of the chetae of the head, the postabdomen of females and males and the structure of the basal part of the wing. However,

in spite of the absolute authenticity of all the facts under discussion, most
of them being new and described for the first time, in many respects the
conclusions of his investigation are incomplete. Hennig does not sufficiently
examine the venation of the blade of the wing; he does not take into account
the processes of reduction. At the same time, this is the direction in which
are accomplished the phylogenetic changes of these small Diptera in many
superfamilies and families. Another of his deficiencies is his inattention to the
ecological data, in particular to the trophic specializations of the different
groups of acalyptrates, inasmuch as the trophic specialization undoubtedly
gives a clue to an understanding of the systematic relationships of these insects.

Concluding the examination of Hennig's investigations it is nevertheless
necessary again to repeat the high value of these studies: everything done
by this author is quite an important step forward in the investigation of
the order Diptera and places the latter among many other orders of insects
in one of the first places.

It is necessary to note at the conclusion of my review of the phylogenetic
studies that the understanding of the phylogeny of these insects proceeded
in two directions little related one to another. On the one hand there were
carried on comparative anatomical studies (Hendel, Edwards, Crampton and
especially Hennig), which allowed them, in the whole, to make quite thoroughly
clear the relations of the chief groups of the order. On the other hand there was
the construction of phylogenetic schemes based on the paleontological documents
which were employed during the drawing up of a genealogical tree (Handlirsch);
this second, though it was the most reliable and direct path could not, however,
be productive because of the scarcity of materials on fossil Diptera.

The discovery of many and varied Diptera in the Mesozoic deposits of Kazachstan
and central Asia allowed the actual history of the order to be clarified with far
greater completeness and authenticity. Besides the employment of new facts, the
use of a different method of investigation came into being, namely an examination
of the evolutionary history of the Diptera together with a study of the connection
between an organism and the conditions of its existence, the clarification of the
nature of problems and their solutions in the history of the different groups of
Diptera and the illumination of real governing features. All this allowed us to
understand and clarify the chief processes of the historical development of these
insects.

<h3 align="center">Phylogenetic relations of the principal
groups of the order Diptera</h3>

During my investigation of the features of the historical development of the infraorder Bibionomorpha a scheme of the phylogenetic relations of the families of
this group was already indicated (Rohdendorf, 1946). Its basis was the study of
the Middle Jurassic fauna of Karatau and a thorough utilization of comparative
anatomical data on contemporary Diptera. In passing, during clarification of the

relationships of the different bibionomorphs I touched on a scheme also for other groups of Diptera, the relations of which were represented in the form of dichotomous branching lines of the chief trunk of the genealogical tree (l.c., p. 84, fig. 95). Attention was directed to the Bibionomorpha, (called in this work 'Nematocera Oligoneura'), and the relations of other groups of Diptera were not considered; furthermore, the absence of any materials on more ancient faunas of Diptera inevitably forced us to confine ourselves to an assumption concerning the age of those or other groups, and to indicate, largely without proof, the times of divergence and appearance of the different families and superfamilies. Such suppositions concerning the splitting off from the Bibionidea of the fungivorids 'at the boundary of the Permian and the Triassic' (l.c., p. 88), concerning the separation of the trunk of all the bibionomorphs from the rest of the Diptera 'not later than the middle of the Permian period' (l.c., p. 87)–are arbitrary and, as we shall see, are presumably inaccurate. I am convinced of this by the investigation of more ancient Upper Triassic Diptera which, until recent times, remained unknown and which possessed archaic features to a high degree, sharply distinguishing them from later forms. The presence of a series of peculiar forms in the composition of the Diptera of the Upper Triassic fauna indicates the possibility of the appearance of many new groups of the order during Triassic time and removes the necessity to suppose the Permian or, even more, the Carboniferous origin of the chief groups of the order as was indicated by me first in 1946.

The ancestral group of all the Diptera (fig. 81) was apparently the Upper Triassic infraorder Dictyodipteromorpha, the representatives of which were discovered for the first time in the Upper Triassic deposits of central Asia. These Diptera were undoubtedly quite diverse and belonged to not less than three individual superfamilies; of these superfamilies, one shows direct connections with forms of later groups which were widely prevalent. Such are the Dictyodipteridae, which are bound up with the ancient Bibionomorpha (family Pleciodictyidae), and the Tipulomorpha (family Tipulodictyidae). These original divergences of two main branches of the order of Diptera depended on the development of an aquatic way of life for the larval forms or at least improvement of a terrestrial one (see above, p. 6). The time of the separation of these infraorders from the dictyodipteromorphs can be referred with great probability to the middle of the Triassic period, as witness the composition of the fauna of the Upper Triassic. In this Upper Triassic fauna, besides the ancestral groups of the new infraorders (Bibionomorpha, and Tipulomorpha), there were also other, younger, highly changed forms.

Before proceeding to a survey of the further paths of the phylogenetic development of the Diptera it is necessary to note that apart from the three infraorders (the original Dictyodipteromorpha and their derivatives, the Tipulomorpha and Bibionomorpha) mentioned in the composition of Upper Triassic fauna, a representative of an entirely peculiar group of Diptera was discovered, whose connections with other groups is not clear. In any case it must be quite remote, and merits a systematic rank equal to an aberrant infraorder. Such a group are the Diplopolyneuromorpha (p. 139), the structure of whose wings indicates great mechanical perfection and, in that way, the length of their history. The separa-

tion of the diplopolyneuromorphs from the dictyodipteromorphs is obviously
more ancient and falls presumably in the Lower Triassic. As already noted, it
is impossible to say anything about the directions of the evolution, historical
development or the nature of the problems which influenced the derivation of
these peculiar insects.

The further history of the Diptera consists, in fact, only of the fortunes of
two of the infraorders--the aquatic tipulomorphs and the terrestrial bibionomorphs, particularly in the separation from the last of the most important
infraorder of Diptera, namely the Asilomorpha. This process terminates apparently at the end of Triassic time when, from the forms related to the
rhyphids (of the already separated superfamily Bibionomorpha), the first
asilomorphs separated out--representatives of the Tabanidea and Stratiomyiidea.
It is not yet possible to indicate definitely the original group of tabanids, the
closest to the ancestral forms. The appearance of tabanids and stratiomyiids
in the existing paleontological records took place very rapidly. Such are the Lower
Jurassic Protobrachycerontidae and numerous Middle Jurassic asilomorphs which
differ sharply from the bibionomorphs and may only in part be connected with
the rhyphids, namely with the Protorhyphidae, discovered in the Upper Triassic
deposits of central Asia. The protorhyphids are still known only from remnants
of wings, the venation of which exhibits apomorphic features--far-reaching
mechanical improvement (small size of the discal cell; location of branches of RS).
Some asilomorphs (for example the Eostratiomyiidae or the Nemestrinidae)
possess venation, which indicates considerably less mechanical improvement and
more plesiomorphy (the multiplicity of rm and other features) that does not
allow us to consider them as direct descendants of the protorhyphids.

There remains the peculiar, up to now poorly known group, the Eoptychop-
teridea--of Lower Jurassic, Liassic tipulomorphs--which have a very large discal
cell, together with a peculiarly constructed radial system of veins. These are
known up to now only from the Liassic of western Europe and their position
among Diptera is obscure. On account of the structure of their wings they show
peculiar traits of similarity with the bibionomorphs, occupying a kind of 'intermediate' place between the latter and tipulomorphs. It is quite probable that
after further investigation such a similarity to the bibionomorphs will turn out
to be superficial. The eoptychopterids were probably the ancestors of the Dixidea
and possibly of the Culicidea; these conclusions follow from investigations of the
plesiomorphic venation of the wing in which no displacement of transverse veins
toward the apex of the wing is seen which is so characteristic for the Tipulidea
and which is absent from the dixids and culicids. To say anything precise about
the derivation of these important groups of tipulomorphs is impossible at present
owing to the absence of data concerning the structure of the body. One can only
assume that the separation of the dixids and culicids originated not later than
Lower Jurassic time.

Before examining the phylogenesis of the infraorder Bibionomorpha, the
chief group of Diptera which gave rise to the younger infraorders widespread
in the Cenozoic, we shall discuss the history of the trunk of the Tipulomorpha

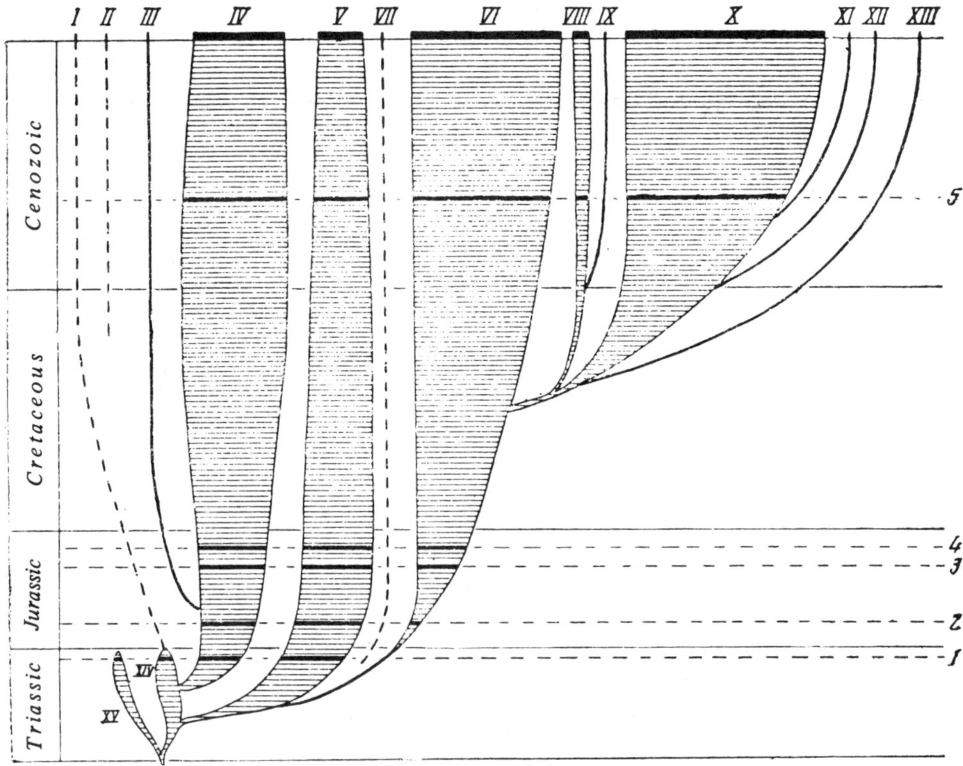

Fig. 81. Diagram of the phylogenetic relations of the infraorders of Diptera represented on the background of the geochronological scale; the width of the bands indicates the number of species. The ordinate is Triassic, Jurassic, Cretaceous and Cenozoic from bottom to top. For designations, see page xiii. (According to Rohdendorf, 1962, with alterations.)

which is monogenetic and considerably less diverse (fig. 82). Arising in Triassic times from some kind of closer unknown dictyodipteromorphs, these insects first become known to us in the Upper Triassic fauna from five separate superfamilies, of which two or three (Tipulidea and Chironomidea, perhaps also the Rhaetomyiidea?), still survive, and two or three others which are known only from the Triassic. Concerning these latter groups, it is first of all necessary to refer to the Tipulodictyidea which have the most simply constructed, little-mechanized wings. They stand closest to the original forms of the Tipulidea on the one hand and are descendants of some kind of dictyodipteromorphs on the other. The Eopolyneuridea, another Triassic superfamily of tipulomorphs, is comparatively diverse and, like the tipulodictyids, shows numerous plesiomorph features inherited from their ancestral forms (an abundance of transverse veins and branches of the medial system, little mechanical adaptation). It is considerably more complicated to clear up the relations of the eopolyneurids with succeeding younger Diptera.

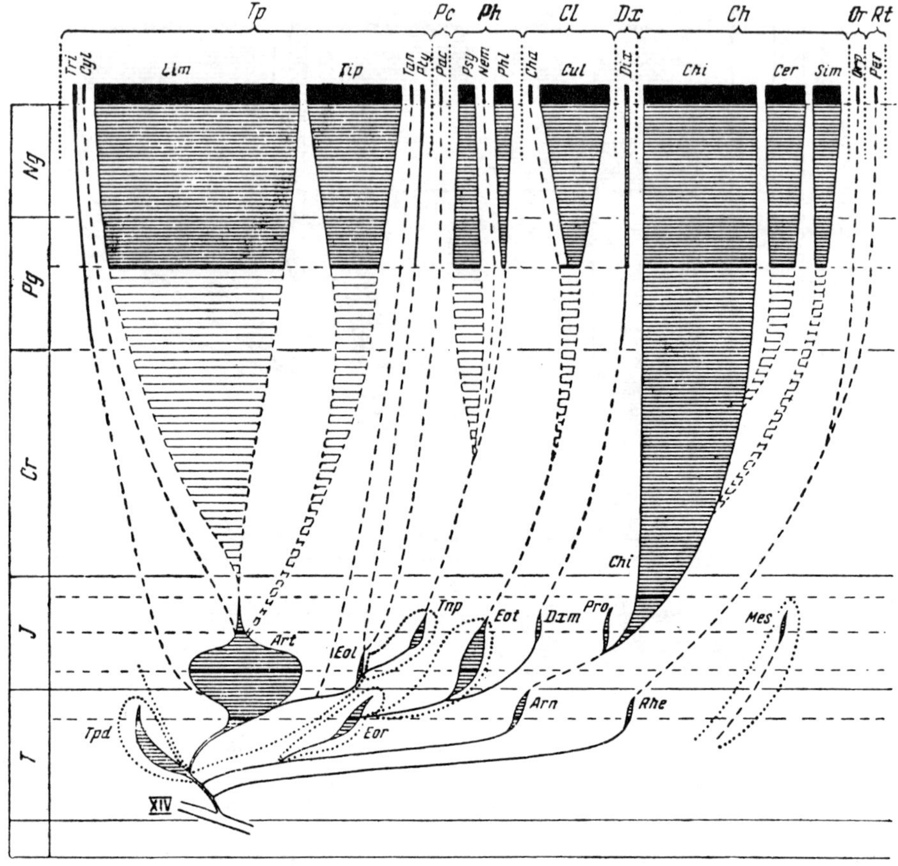

Fig. 82. Diagram of the phylogenetic relations of the superfamilies (dotted lines) and families (continuous and discontinuous lines) of the infraorder Tipulomorpha, represented on the background of the geochronological scale. For designations, see page xiii. (Original.)

Most of these Triassic tipulomorphs apparently separated from the remaining Diptera, not being the ancestors of any Mesozoic or Cenozoic groups. There is little basis for assuming connections between the eopolyneurids and the sharply separated superfamilies of tipulomorphs or even infraorders of the present-day fauna (as, for example, the Deuterophebiomorpha, Blephariceromorpha or, finally, the Musidoromorpha) although, because there are some similar traits in the venation, it is possible. But then there is some similarity in the venation of representatives of the Musidoromimidae and Cenozoic Musidoromorpha, but this can by no means be the basis for linking these groups. The problem will only be solved after a study of Mesozoic faunas of Diptera that are still insufficiently known.

The Rhaetomyiidea, the last Triassic superfamily of tipulomorphs, is very peculiar. The chief trait of the rhaetomyiids consists of a great development of the venation of the blade of the wing, immediately suggesting some of the asilomorphs. The derivation of this superfamily is not clear and presumably it is necessary to look for it among the ancient Tipulodictyidea; however this supposition is only provisional because of the high degree of apomorphy of the wing. The further history of the rhaetomyiids is interesting: there is some similarity with the superfamily Orphnephilidea, the representatives of which (species of a few genera of the single family Orphnephilidae) are rare, relict contemporary Diptera and possess most peculiar apomorphic venation which allows us first of all to consider them as descendants of the rhaetomyiids. But particularly interesting is the recent discovery of the relict Australian Perissommatidae (Colless, 1962), presumably a distant descendant of the Triassic Rhaetomyiidea (see p. 60).

The history of the Tipulidea and Chironomidea, living until the present-day epoch and for the first time discovered in the Upper Triassic fauna, is of comparatively little interest because of the insufficiency of material.

The most ancient Triassic Tipulidea are represented by the family Architipulidae which is the immediate ancestor of the contemporary complex of Tricoceridae + Limoniidae + Cylindrotomidae + Tipulidae and presumably Tanyderidae. A more careful analysis of the interrelations and history of this trunk of tipulomorphs awaits more knowledge both of the quite diverse contemporary forms, and of the fossils, especially the Mesozoic, a great number of which are described quite superficially and then only on the basis of remnants of single wings. The Jurassic fauna of tipulids is quite diversified and includes not less than three families, including here the Architipulidae and the little-studied family Eolimnobiidae, known from a single lower Liassic find. The eolimnobiids almost certainly were ancestors of the contemporary Ptychopteridae, because of the great similarity in venation, which is also plesiomorphic and does not disclose displacements of the transverse rm towards the apex of the wing. Finally, the last Jurassic family of tipulids, the Tanyderophryneidae, in a peculiar way, is apparently the ancestral group for the superfamily Psychodidea, although this is still insufficiently discussed. The small size of these insects combined with the regressive changes that have taken place in present-day psychodids do not allow us to assert with confidence that they are connected with Jurassic tanyderophryneids which are similar to them only in the structure of the venation of the mechanically primitive wings. The isolation of psychodids from other tipulomorphs is very great and gives no possibility of determining the paths of their phylogenesis.

It remains to examine the history of the contemporary, sharply separated relict superfamily Pachyneuridea, paleontologically altogether unknown. The features of the wings of these Diptera relate them to some bibionomorphs. In connection with this the pachyneurids until recently were considered almost representatives of the family Bibionidae or at least of the superfamily Bibionidea. Only investigation of the plesiomorph basial of pachyneurids showed their cardinal differences from the whole complex of Bibionidea + Fungivoridea and their closeness to the tipulids (Rohdendorf, 1946, p. 90). Apparently these

peculiar tipulomorphs are descendants of the Eolimnobiidae or of forms related to them; the known similarity in the structure of the radial branches
and the central position of the transverse vein rm points to that. This conclusion is obviously provisional and it may be confirmed only by a study of
new fossil material and knowledge of the individual development of representatives of the pachyneurids, up to now completely unknown.

Concluding the survey of the phylogenetic relations of the tipulomorphs
we still have to examine the history of the superfamily Chironomidea, one
of the most ancient among the whole infraorder. The venation of these Diptera, reflecting apomorphic tendencies, namely the development of lifting
flight and costalization by means of shifts of the veins and decrease in the
venation of the posterior half of the wing, indirectly illuminate their history.
The fact is that the delicate and long wings of the chironomids are very poorly
preserved during fossilization and their remnants in fact consist only of some
bodies, legs and pupae; these paleontological documents allow us only in
very small measure to throw light upon the phylogenesis of the known families. In the Upper Triassic fauna were discovered two peculiar forms of chironomids which differ greatly from the known Cenozoic families, being their
distant ancestral forms; the similarity of the Triassic, substantially plesiomorphic Architendipedidae, to Cenozoic families of chironomids is comparatively small but the closeness of them to the original dictyodipteromorphs
and partly to the eopolyneurids is quite evident. The latter proves a very ancient
Triassic division of the whole superfamily from the first forms, probably very close
to the initial tipulomorphs. The relatively numerous Jurassic and even Cretaceous
finds of chironomids allow us to determine only their attachment to the superfamily; the relations of the Cenozoic family may be therefore described only on
the basis of a study of their systematics, without reinforcement and confirmation
of these conclusions from the paleontological documents. Without analyzing in
detail the relationships of all the groups of the superfamily, it is necessary to note
as the most plesiomorphic family of mosquito-midges, the Chironomidae, to which
apparently belong some Upper Jurassic forms (from Mongolia and China) besides
numerous Paleogene species (of Baltic amber). Moreover, the families in the Paleogene amber fauna are represented by the Ceratopogonidae and Simuliidae which,
however, in the Cretaceous are still not discovered, if one leaves the Upper Cretaceous Laramie ceratopogonids (Carpenter, 1937) out of consideration. The
relative closeness of all families of chironomids to one another is well known: the
Chironomidae are somewhat closer to the Ceratopogonidae and less close to the
Simuliidae.

It is necessary still to mention the most peculiar Middle Jurassic Mesophantasmatidae, which are an apomorphic and sharply-separated superfamily in the system
of this infraorder. Known from the structure of the large, highly elongated wings,
devoid of anal veins and of the majority of the branches of M and RS, this dipteran differs sharply from all other tipulomorphs. The assignment of it to this
infraorder can be disputed without obtaining new material; any assumptions concerning the phylogenetic relations of the Mesophantasmatidae are premature.

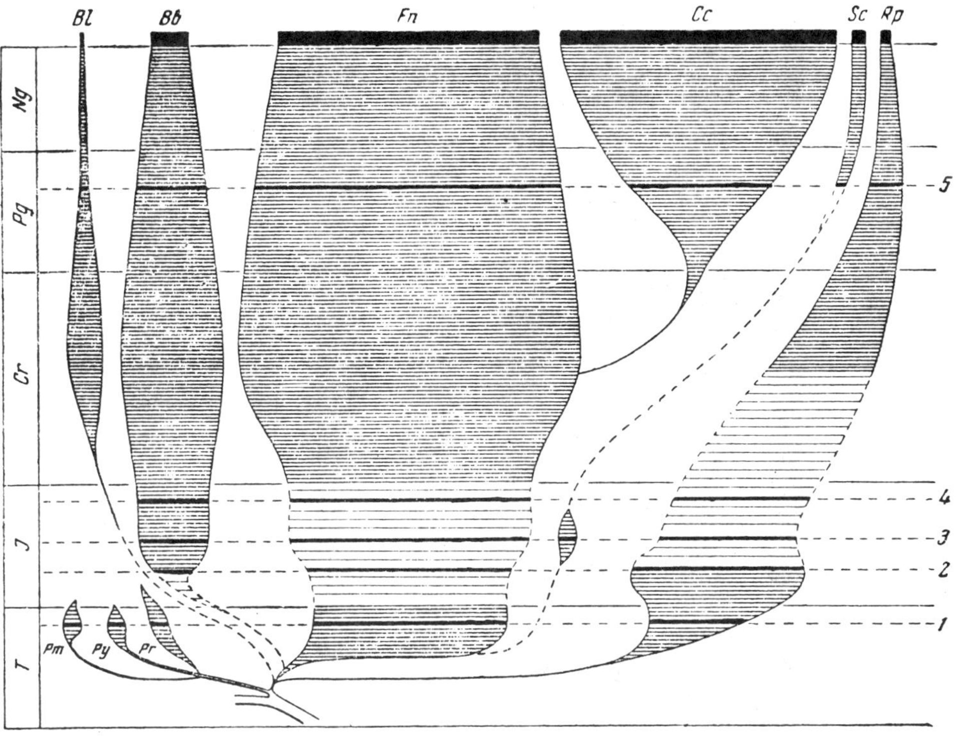

Fig. 83. Diagram of the phylogenetic relations of the superfamilies of the infraorder Bibionomorpha, represented on the background of the geochronological scale. (Original.) For designations, see p. xiii.

Passing now to a consideration of the phylogenesis of the infraorder Bibio-nomorpha (fig. 83), it is first of all necessary to refer to my earlier investigation (Rohdendorf, 1946), the chief conclusions of which still stand. The main con-clusion of this phylogenetic analysis was the establishment of the relations of the chief superfamilies and families of bibionomorphs of the Jurassic and Cenozoic faunas. The discovery of a new Upper Triassic fauna supplements the phylogenetic scheme of the infraorder by new branches of a different importance. Some of the Triassic superfamilies bear the character of ancient groups (Pleciodictyidea and Protoligoneuridea), the closest to the original forms of the whole infraorder directly connecting it with the infraorder of dictyodipteromorphs, to which one can add the Pleciodictyidea. Other Triassic superfamilies (the peculiar Phragmoligoneuridea) on the contrary are already highly specialized groups. Especially interesting is the discovery in the Triassic fauna of primitive representatives of the superfamily Rhy-phidea, family Oligophryneidae, directly connecting the Rhyphidea with the Fungi-voridea; up to now this connection remained vague and conjectural. The finding of Oligophryneidae is very important and refines the phylogenesis of this superfamily.

As already mentioned above the most ancient representative of the asilomorphs,

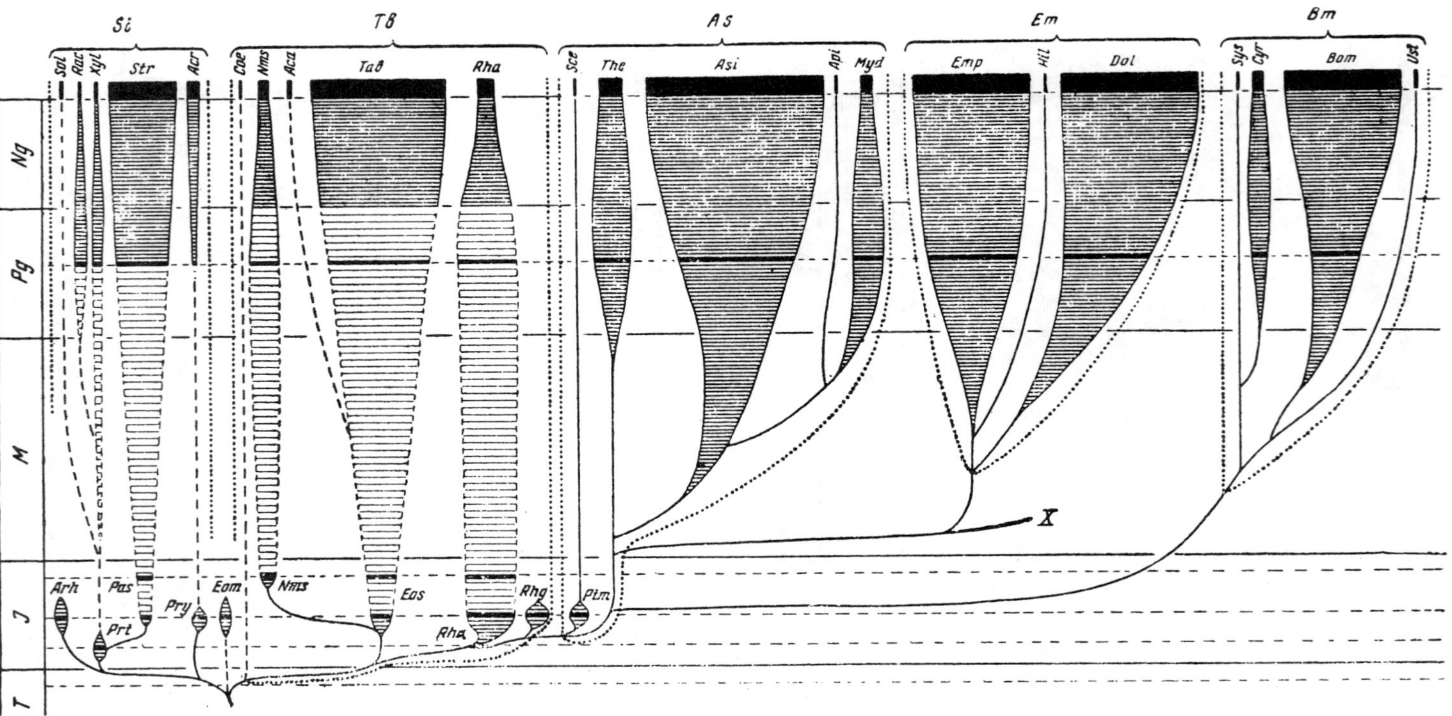

Fig. 84. Diagram of the phylogenetic relations of the superfamilies (dotted lines) and families (continuous and discontinuous lines) of the infraorder Asilomorpha, represented on a background of the geochronological scale. (Original.) For designations, see p. xiii.

Protobrachyceron liasinum Handlirsch, was discovered in deposits of the Lower Liassic of western Europe. This form, known only from the impression of a wing, is undoubtedly very close to the Cenozoic Solvidae, Rachiceridae and Xylophagidae, especially to the two first, being presumably a representative of their original ancestral group. At the same time the finding of this dipteran shows the ancient nature of the superfamily Stratiomyiidea, which makes its appearance in the lowest parts of the Liassic (fig. 84). Considerably richer is the fauna of the Middle and Upper Jurassic, in the composition of which were discovered representatives of different families not only of stratiomyiids but also of tabanids and some asilids (see above, p. 219-224). The ancient nature of all of these groups is therefore indicated, first of all of the Stratiomyiidea and Tabanidea, two of the earliest original superfamilies of the whole infraorder Asilomorpha. The distinction between the Jurassic representatives of these two superfamilies is sometimes quite puzzling as a consequence of their recent divergence from ancestral forms, the first asilomorphs: the separation of these superfamilies took place probably at the end of Triassic time.

Analyzing the phylogenesis of the Stratiomyiidea it is necessary to note that in Upper Triassic time there separated out from the common trunk of the superfamily ancestors of the peculiar Acroceridae, narrowly specialized parasites of spiders: already in the Karatau fauna we find the highly changed Protocyrtidae, which is close to some present-day genera of the family. The chief divergences in the history of the whole superfamily which gave rise to the main groups, originated at the beginning of Jurassic time, in early Liassic when three families arose: the Palaeostratiomyiidae (direct ancestors of the Cenozoic Stratiomyiidae), the Archisargidae (specialized, extinct group in the Jurassic) and the Protobrachycerontidae, already referred to above and which are close to the ancestors of the Solvidae, Rachiceridae and Xylophagidae. The further history of the Stratiomyiidea is little known. In Tertiary time all contemporary families are represented, and the distribution of the individual families differs little from that of the present day. Species of Xylophagidae and Rachiceridae, which are now rare, relict families, were comparatively abundant.

The first representatives of the Tabanidea appear in the fauna of the Middle Jurassic of Karatau. Such, first of all, are some genera of the Rhagionidae partly pertaining to subfamilies still extant in the present-day fauna (Vermileoninae), partly to extinct subfamilies, presumably Mesozoic (Protorhagioninae). Other Karatau tabanoids belong to peculiar extinct families: the Eostratiomyiidae, a group giving rise to the beginning of the Tabanidae and Acanthomeridae and which are, moreover, close to the original forms of the Nemestrinidae; the peculiar family Rhagionempididae, quite separate from the remaining tabanoids and apparently near to the ancestral forms of the younger superfamilies of Asilomorpha and, simultaneously, of all the remaining infraorders of Diptera — Phoromorpha, Musidoromorpha, Myiomorpha and the parasitic groups which are their derivatives. The features of different families of tabanoids permit us to divide them into two groups, the divergence of which was over presumably in Liassic time. The first group includes the trunks of the families Nemestrinidae, Eostratiomyiidae + Acanthomeridae + Tabanidae and Coenomyiidae; these tabanoids are the most diverse and in the contemporary fauna constitute the obvious

majority (as a result of the rich development of families of horseflies, Tabanidae). The
second stem of tabanoids includes in fact only one family, the Rhagionidae, a very
ancient group, related to the peculiar Rhagionempididae, up to now known only
from the Jurassic. The history of both ancient superfamilies of asilomorphs-tabanoids
and stratiomyioids - is partly similar: in these two groups the main processes of divergence originated in Jurassic time and later on the groups changed comparatively
little. Both superfamilies include in the present-day fauna a series of relict groups
which are becoming extinct, representatives of the majority of which in Tertiary
time (amber fauna) were more diverse. Such are the stratiomyioid Xylophagidae,
Rachiceridae, Solvidae, and the tabanoid Coenomyiidae.

The line of descent closest to the ancestors of the Rhagionempididae produced
the first asiloids, namely the Protomphralidae discovered in the Jurassic fauna of
Karatau; the present-day Scenopinidae are the changed descendants of this Jurassic family which is still known only from a single poorly preserved fossil. The discovery of *Protomphrale martynovi* Rohdendorf proved to be very surprising and
is an excellent illustration of the role of paleontological documentation in studies
of animal evolution. Until recently little-studied Scenopinidae seemed to be a
secondary, narrowly specialized apomorphic relict group and its relation with
other families of asilomorphs remained quite unclear. The finding in the Jurassic
of a dipteran which shows signs of being close to the scenopinids at a time when
asiloids were absent, indicates the peculiar position of this family, undoubtedly
linked with the original forms of all the recent groups of Diptera, first of all
the younger superfamilies of the infraorder. The necessity for further investigation of the organization of the contemporary scenopinids, and of the closest
descendants of the original forms of the whole trunk of the Asilidea, and simultaneously of all the remaining new groups of Diptera is obvious.

Examining further the phylogenetic tree of the asilomorphs after the branching
of Protomphralidae + Scenopinidae, we see that the diagram (fig. 84) shows a
series of divergences originating in the second half of the Jurassic and the beginning
of the Cretaceous. This part of the diagram is based principally on systematics because the paleontological evidence is very meagre.

The further history of the Diptera indicates that at the end of Jurassic time the
first representatives of the Bombyliidea diverged and at the very boundary of the
Jurassic and Cretaceous the stem of the Diptera divided into the first asilids and
empidids.

Examining the phylogenetic relationships of the families of the Asilidea it is
natural to consider the Therevidae as the most plesiomorphic descendants of the
first asilideans, from which separated the ancient Asilidae, in their turn giving
rise to the branches Apioceridae and Mydaidae. It will be possible to throw more
light on the true phylogenesis of the Asilidea only after a detailed review of the
contemporary system of the superfamily, which comprises for the most part forms
widespread in tropical regions.

Little, too, can be said about the phylogenesis of the superfamily Bombyliidea,
which until recently was considered to consist only of a single family of bee flies;
in reality, however, there are no less than four separate groupings in the composi-

tion of this superfamily, which are undoubtedly related but separated one from another and presumably independent families. The final clarification of the systematics of the Bombyliidea can come only from comparing the organization of the various groups of the family; the solution of this problem is more feasible than that of the asilids, because bombyliids are abundant in the desert zones of Asia and so are accessible to investigators. Apparently these parasitic asilomorphs diverged during Cretaceous time and by the beginning of the Tertiary period all the chief groups of the superfamily were present.

Examining the further divisions of the phylogenetic tree of the order Diptera, it is necessary to note the derivation of two important branches, the families Empididae and Dolichopodidae which form the superfamily Empididea; the divergence of branches of this superfamily originated undoubtedly not later than the middle of the Cretaceous period and by the Paleogene both families are sharply different and diverse groups. The phylogeny of the Empididae is a complex problem because of the great variety of this family, consisting, as it does, of a series of sharply-separated subfamilies, for example, the Corynetinae, Atalantinae, Empidinae, Hybotinae and others. Dolichopodidae is approximately as abundant in species, differing only in its greater monolithic nature, the absence of isolation or lesser isolation of the subfamilies. Comparing the history of both these chief families of empidideans it is justifiable to assume that the Empididae are older and more plesiomorphic, which explains their greater diversity in the fauna of Baltic amber in comparison with the Dolichopodidae.

From the original forms of the Empididae, probably at the beginning of the Cretaceous period, there separated out the branch Myiomorpha, an infraorder of dipterans that is the youngest, the most diverse and the richest in species, being the chief component of the present-day fauna (fig. 85). Apparently the Cretaceous period was the time of formation of the main groups of the infraorder, the principal superfamily of which was already added at the beginning of the Paleogene. The closest to the original forms of the myiomorphs is the very plesiomorphic superfamily Platypezidea, represented now only by the small relict family of mushroom midges, the Platypezidae, and the relict Sciadoceridae of the southern hemisphere. These families are remnants of the once more widespread and diverse primary superfamily of the infraorder which gave rise to other superfamilies and even infraorders. As the source of the derivation of these original myiomorphs, the ancient Platypezidea, were some Empididea which perfected development by means of the production of a puparium, i.e., changing the character of the moult of the mature larva (the moment of derivation of the superfamily is shown in fig. 85/1.

The most ancient myiomorphs presumably arose from a still unknown extinct family of the Platypezidea which adapted to larval life in an aquatic medium or at least a liquid substrate, while the size of the body of the winged insect increased, giving rise to the first representatives of the superfamily Syrphidea. Evidence for this is the approximation of the posterior spiracles and development of metapneustic larvae simultaneously with a rich venation of the wings in the adult. The syrphids in the Cenozoic fauna are represented by two families of very unequal size

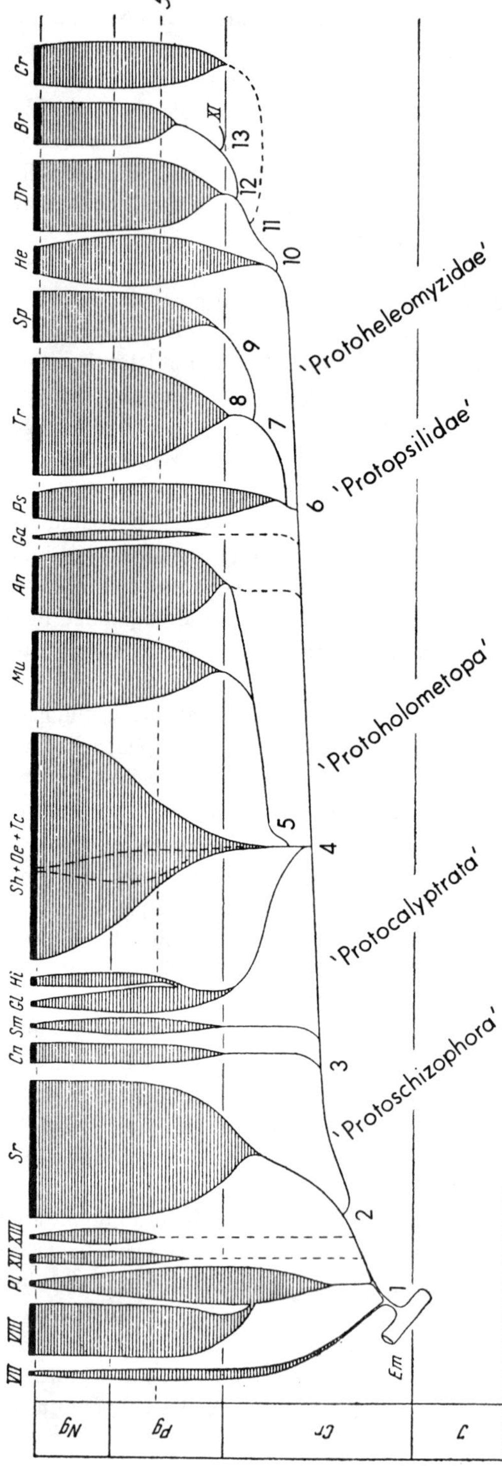

Fig. 85. Diagram of the phylogenetic relations of the superfamilies and some families of the infraorder Myiomorpha, represented on the background of the geochronological scale. (Original.) For designations, see p. xiii. The numbers on the phylogenetic scheme denote the moments of divergence and their nature (see text, p. 299-307).

and significance. The hover flies, or Syrphidae, are a diverse group of large Diptera, the larvae of which are either phytophagous or predatory, while the Pipunculidae are narrowly specialized, comparatively small and scarce Diptera, the
larvae of which are parasitic in the body of cicadas (Homoptera, Auchenorrhyncha).
Both these families are close to one another and presumably became distinct at
the very end of the Cretaceous; the original group almost certainly were hover
flies.

Simultaneously with the first Syrphidea branching off from the ancient Platypezidea there split off the original ptilinum-bearing myiomorphs ('Protoschizophora'), now absent from the fauna and presumably comprising a peculiar superfamily, the near descendants of which are the present-day relict Somatiidea and
Conopidea. The separation of these 'Protoschizophora' as a sister group of the
Syrphidea, originated by further improvement of individual development and absence of increase of body size. The new feature consisted of the development of
a frontal bladder in the adult insect, which allowed it during emergence from the
pupa (and hence from the puparium) to go successfully through layers of soil or
an accumulation of plant remnants, in the depth of which the larva could pupate
with greater safety. The absence of an increase in body size permitted it to improve
the venation of the wings quickly and better its aerodynamic qualities. The acquisition of the effective protective adaptation of a frontal bladder proved to be a
highly beneficial new character in the historical development of the myiomorphs.
The aerodynamic improvement of the venation (the development of a free elastic
membrane of the posterior border as a result of reduction of the anal cell and
decrease of the basal cells) undoubtedly determined all the further turbulent
blossoming of this infraorder (fig. 85/2)

The closest present-day descendants of the 'Protoschizophora', the superfamilies Somatiidea and Conopidea (fig. 85/3) are characterized by relict groups with
reduced venation, large basal and anal cells (especially the Somatiidea) and poorly-
developed ptilinum. The Somatiidea are still very insufficiently investigated;
their ecology remains unknown to me (see p. 108). The Conopidea are comparatively numerous and possess many apomorphic traits such as parasitism; they
also bear traits of the specialization and structure of the abdomen. In the minute
Stylogastridae, furthermore, there developed well-marked apomorphic characters
in the aerodynamic specialization of the wings.

Later on the group 'Protocalyptrata' split off from the 'Protoschizophora'
brought about by enlargement of the body, together with improvement in the
powers of flight (fig. 85/4). This is shown by the development of additional
wing blades (the wing and thorax scales) and by reduction of the venation.
Furthermore, the development of deep grooves on the mid-dorsum indicates
a great development in the flight muscles of the calyptrates. There is no doubt
that these processes indicate the development of great activity on the part of
the winged insect in connection with increased contact with different animals,
first and foremost with vertebrates and insects. The capacity for rapid movement
not only in flight, but also in running, proved to be an important asset in this
line of evolution. The early connections of the ancient protocalyptrates with

other animals presumably originated with provision for the larvae of high caloric, protein food; the original feeding was apparently necrophagia. More complex were the relations with the vertebrates, which consisted not only in necrophagia but also in the utilization of different excretions of living animals. So the ancient protocalyptrates developed the habit of blood-sucking, which led to the appearance of myiomorphs in which 'pupiparity' began to develop, i.e., retention of the larvae in the body of the mother, where they are fed in the 'uterus', and the birth of the completely formed larvae or even pupae (puparia). Such are the superfamilies Hippoboscidea and Glossinidea, undoubtedly related groups, which are distinguished one from another by greater (Hippoboscidea) or lesser (Glossinidea) degree of reduction in the capacity for active flight, and the close connection with the vertebrates on which they feed. At present, to detect plesiomorphic traits in these myiomorphs is quite difficult: such, for example, is in the glossinids the presence of a large number of tergites of the abdomen (six) and in both superfamilies little development of the sclerotization. Apomorphic traits are entirely diverse, especially in the hippoboscideans, which have greatly advanced processes of parasitic adaptations for living on the body of the host (mammal or bird). Analysis of the conditions which influence the derivation of these blood-sucking pupiparous myiomorphs at present can only be indicated. Apparently the development of pupiparity could have been accomplished only under conditions of the excessive, rich nourishment of the female fly simultaneously with the delayed birth of larvae, together with the impossibility of their feeding on the nourishing substrate. Such a situation might arise in bloodsuckers, the larvae of which developed on the temporary substrates (for example, small corpses in desert-like, highly isolated conditions) the search for which by male flies was protracted. For more complete and precise determination of the paths of evolution of these pupiparous myiomorphs it is necessary to examine different large-larvae-bearing Diptera: as, for instance, some calliphorids (Mesembrinellinae) or muscids.

After the separation from the protocalyptrates of the most ancient bloodsucking myiomorphs the further history of this stem resulted in the separation of the superfamilies Anthomyiidea and Muscidea on the one hand, and the superfamilies Oestridea, Sarcophagidea and Tachinidea on the other (fig. 85/5). Insufficient knowledge of the systematics of these myiomorphs prevents us from drawing any reliable conclusions about the lines of evolution and the chief conditions which determined their development. Apparently the development of plant-eating by the larvae, imaginal predation and decrease in size played an important role in the formation of the anthomyiideans and presumably of the muscideans. On the other hand the development of different forms of parasitism and necrophagia governed the historical development and formation of the oestrideans, sarcophagideans and the tachinideans. These general conclusions, of course, are very inaccurate: in reality the history of all the most diverse calyptrates is far more complex and will form the theme of widely based investigations in the future.

Returning to a consideration of the history of the protoschizophorids it is

necessary to assume that after the branch of protocalyptrates had split off,
another part of the trunk proved to be a 'sister group' of protocalyptrates, namely
the ancient acalyptrates, which one can arbitrarily designate as 'protoholometo-
pans' (Protoholometopa). The chief reason for their formation was a decrease in
size, associated with a reduction of venation and the development of saprophagia
and phytophagia. This hypothetical group of myiomorphs apparently did not
leave direct descendants in the contemporary fauna. There is a possibility of a
connection with these protoholometopan superfamilies of present-day calyptrates, namely the Anthomyiidea, which are quite isolated from all other calyptrates, sharply differing in their larval phytophagia. Another derivative of the
protoholometopids is probably the isolated superfamily of intestinal gadflies,
the Gastrophilidea, which is undoubtedly connected with the acalyptrates. As a
result of the development of parasitism in the larvae, characterized by many
apomorphic traits, we cannot precisely determine its connection with other
myiomorphs.

The further history of the protoholometopids consists of a divergence at
the base of adaptations to phytophagia and associated different oviposition
habits, and little development of the sclerotization of the body on the one hand,
and on the other the development of saprophagia (feeding on rotting materials)
with stronger sclerotization. As a result there appeared the first psilids or 'Proto-
psilidae' and the first heleomyzids or 'Protoheleomyzidae' (fig. 85/6). It is
necessary to note that the chief stem of the protoholometopids changed least
in the protoheleomyzids; the protopsilids underwent the greatest changes. For
the illumination of the phylogenetic relations of the acalyptrates the most important work is that of W. Hennig (1958) in which the relations of the different
groups of this vast complex of myiomorphs are examined critically. I have made
use of Hennig's chief conclusions and have partly supplemented them; my greatest
criticisms are of the small attention he pays to the structure of the wing, and of
his view of the size of the family taxon and the absence of ecological-functional
appraisal. He established the relations of some families, subfamilies and genera and
often substantially differed from other authors. He leaves us in no doubt of his
authenticity.

Although studies of the group Acalyptrata have been considerably more extensive
than those of the Calyptrata we are far from solving the real relations between all
superfamilies. The proposed phylogenetic schemes therefore are essentially preliminary and later on they will probably be substantially supplemented and changed.

From the contemporary representatives of the Psilidea it is impossible to show
a direct, least-changed descendant of these original protopsilids; most plesiomorph
traits can be observed in the rare tropical Nothybidae, Megamerinidae and Diopseidae,
in which the features of extreme specialization are not less numerous as is natural
for relict groups. In any case, the superfamily Psilidea, as a whole, must be assumed
as a direct descendant of the 'protopsilids'. A 'sister group' for the Psilidea proved
to be a complex of two, now big superfamilies, Trypetidea and Sapromyzidea
(fig. 85/7). The evolution of the last superfamilies is still not clear; they are probably
combined with different ecological groups: the trypetids, inhabitants of open ter-

rains, and the sapromyzids which are basically forest insects. Such a general characteristic is based on many traits of these Diptera; for example, the clear coloration
of trypetids, their phytophagia and frequent monophagia, and, by contrast,
the frequently dark or monochromatic coloration of the sapromyzids together
with their predominant schizophagia or even predation. There are numerous exceptions to the characteristics outlined in the scheme of these superfamilies (for
instance, the parasitic tendencies of the trypetids and Pyrgotidae, and the phytophagia of the sapromyzoid Lonchaeidae). The exceptional diversity of the sapromyzids and trypetids in the tropical zone, and our ignorance of the ecological
features of these Diptera indicates that an attempt to make an exact generalization
in this area would be premature (fig. 85/8, 9).

The protoheleomyzids were undoubtedly the direct ancestors of the contemporary
diverse superfamily of Heleomyzidea; in its composition are numerous relict
groups which are poor in species, and the probable descendants of the ancient protoheleomyzids (for example, the families Rhopalomeridae and Coelopidae). There
are also many known relict subfamilies in other families which have a number of
species: such are the Heleomyzidae in which out of six subfamilies five are clearly
relicts. Many heleomyzids possess plesiomorph features; such, for example, is the
incomplete development of the ptilinum in the Sciomyzidae, the simple structure
of the forehead in the Dryomyzinae and Heleomyzinae, and, finally, the absence
on their wings of any kind of reduction or displacements of veins. This all shows
the ancient nature of this superfamily, undoubtedly one of the most plesiomorphic,
little-changing groups of acalyptrates. Many groups of this superfamily, our subfamilies, are considered as separate families in the schemes of Hendel and Hennig.

The formation of the present-day superfamily of heleomyzids was brought into
being as a result of divergence, the separation of the first heleomyzids and the group
of 'protodrosophilids'. The basis of this process was the derivation of an abrupt reduction of body size and, dependent on this, a change in the venation of the wings
and their form. This decrease in the size of the protodrosophilids was presumably
caused by a scantiness of nourishing substrates for the larvae and a high temperature
aided by rapid development.

This process probably did not result in a total change of the ecological conditions;
both the original protoheleomyzids and their descendants the Heleomyzidea and
the protodrosophilids undoubtedly dwelt in forested, shady terrains, overgrown
with plants. The scantness of nourishing substrates presumably resulted in their
settling their larvae on fungi, perhaps on small corpses of invertebrates or any
other 'microstations'. This phylogenetic process definitely proved to be one of the
very important moments in the history of the myiomorphs; as a result of it there
developed the three big, youngest superfamilies of acalyptrates: the Drosophilidea,
the Borboridea and the Chloropidea, and also a special infraorder of parasitic
Diptera, the Braulomorpha (fig. 85/10, 12, 13).

The characteristic superfamily Chloropidea is probably a descendant of the
protodrosophilids. The phylogenetic relations of the chloropideans have, up to
now, remained insufficiently explained; I follow the assumption of Hennig regarding
their splitting off from the original drosophilideans. He conclusively declined all

other possible sources of derivation for these peculiar acalyptrates. Apparently the traits of early development of the larvae's oligophagy on grassy plants and also predation and parasitism, the reduction of the sclerotization of the body, together with the development of moderate displacements and reductions in the venation of the wings, all testify to the path of development of the chloropideans, the origin of which, however, is still not clear (fig. 85/11).

A further transformation of the protodrosophilids took place with the separation from them of the superfamily Borboridea and their conversion into the main stem of the contemporary superfamily Drosophilidea. The basis of this process was a decrease in body size and the development of connections with soil and decomposing substances as substrates for the development of larvae (schizophagia). At the same time the very characteristic process of a reduction in the venation of the wing (on which is formed the broad posterior outer membrane devoid of venation) took place. The sclerotization of all sections of the body is well developed with no tendency toward reduction. Into the composition of the contemporary borborideans enter four families, of which the Borboridae and Milichiidae are definitely close to one another and which are well characterized by the general traits of the superfamily.

I refer to the large and characteristic young borboridean family, Agromyzidae, which has sharply-expressed phytophagy in the larvae (as a rule monophagy), a weaker development of the sclerotization, particularly of the head, and an entirely different type of reduction in the venation of the wings. All these traits essentially distinguish the agromyzids from other borborideans and permit us to doubt their relationship. It is necessary to assume a very recent derivation of the agromyzids, dependent on deep-seated modifications of their organization (perhaps, as a result of phytophagy).

The descendants of the chief stem of the protodrosophilids, the superfamily Drosophilidea, underwent the process of divergence soon after its derivation, which split it into two chief families, the Drosophilidae and the Ephydridae. The basis of this process during the formation of ephydrids was the adaptation to reservoirs as the habitat of the larvae and the development of predation by the adult insects. Drosophilids retained the terrestrial way of life of the larvae, which became schizophagic, or partly phytophagic-fungiphagic. Some species are predatory in colonies of coccids. Schizophagia of the drosophilid larvae apparently originated in gorging on yeast cells in fermenting plant materials abundant in carbon.

The derivation of the peculiar parasitic braulids, or honeybee lice is not clear, but they undoubtedly form a separate infraorder of Diptera, the Braulomorpha. Apparently these Diptera, which are apomorphic to a high degree, are closest to the original forms of the superfamily of borborideans: a decrease in size, a habitat on the body of the bee, inquilinism (feeding on the nutriments of the host), all led to an extreme morphological change in the braulids, and this hampers the determination of their phylogenetic connections. The depth of the distinctions of the braulids, however, leaves no doubts about their extreme separateness, and compels us to attribute them to the rank of infraorder.

Still more complex is the solution of the question concerning the phylogenesis

of the parasitic Streblomorpha and Nycteribiomorpha. The extreme adaptations
to the parasitic way of life of these insects, together with the development of pupiparity, almost completely deprived these Diptera of any kind of plesiomorphic
traits which could throw light upon their phylogenetic relations. Apparently it is
necessary to assume that both these groups derived from the original myiomorphs,
from representatives of the contemporary superfamily Platypezidea which are now
extinct. It is necessary to regard the Nycteribiomorpha as more ancient in comparison with other parasitic Diptera such as the Streblomorpha on account of their
much higher specialization. Both infraorders are highly distinguished one from
another; the similarity between them consists of a similar biology (common hosts
- bats) and some parts of the structure of the larvae (two pairs of posterior
spiracles); it is possible that, systematically, both infraorders should be united
and should be considered as superfamilies. That, however, would be possible only
with new data on the organization of these insects.

There remains to be considered the phylogenetic connections of two infraorders,
Musidoromorpha and Phoromorpha. These are also connected with the original
myiomorphs. As already indicated by Hennig the Musidoromorpha may be con-
trasted with all the remaining cyclorrhaphous Diptera (Myiomorpha) in the nature
of a 'sister group'; speaking more precisely, the first, original forms of the chief
myiomorphs underwent divergence as a result of which there split off the first
myiomorphs. These were the platypezideans, which produced the present myiomorph
larva devoid of head skeleton, and the Musidoromorpha which retained still part
of the head skeleton of the larva, together with other features already of a purely
apomorphic character: large pharynx, absence of costalization of the venation and
well-developed, peculiar sclerotization. At present the musidoromorphs have a
clearly relict character (a single family with one genus and 20 species), although
they are by no means rare insects, being widely distributed in all regions of the
earth. The connections of the infraorder Phoromorpha are more distinct; it is
directly connected with the superfamily Platypezidea, being a specialized group
characterized by a whole series of apomorphic traits. Very characteristic for the
phoromorphs are traits of improvement in the organs of locomotion: abrupt
strengthening of the wings, their strong costalization and the development of
very powerful running (jumping) legs that indicate the complete adaptability
of these Diptera to dwelling in conditions of the upper layers of soil, in plant
residues and other analogous sheltered habitats. It is necessary to say that there
are still little-known features of the feeding and development of the phoromorphs;
undoubtedly a knowledge of these traits would explain much. The connection of
Phoromorpha with the platypezideans is indisputably demonstrated by the structure of the relict Sciadoceridae, most peculiar platypezideans, something like an
'intermediate link' between the myiomorphs and the phoromorphs.

The survey made of the phylogenetic relations of the different groups of Diptera
was established on the study of existing data concerning the organization and
history of these insects. As is shown, the study of the individual infraorders and
superfamilies is very different. That is why the authenticity of the conclusions
is not of the same value. The surveys of the infraorders Bibionomorpha and Asilo-

morpha are comparatively complete. Less reliable are the data on the history of the Tipulomorpha and Myiomorpha. The information on the Phoromorpha and the majority of separate infraorders of parasitic Diptera which are poor in species is very meagre. Some infraorders are so unusual that it is not possible even to relate them to any other Diptera: such, for example, are the Deuterophbiomorpha, Blephariceromorpha and Nymphomyiomorpha.

PART IV

METHODS OF STUDY OF THE HISTORICAL DEVELOPMENT OF ORGANISMS

The study of the historical development of dipterous insects was conducted on the basis of the fullest possible analysis of all sides of this phenomenon. The process of the development of organisms was examined not only in connection with changes in the environment and as being variable in time, but particularly as the solution of specific problems. This turned out to be very fruitful and allowed us to arrive at an understanding and explanation of numerous facts of historical development, including the possibility of answering such questions as why the given organisms appeared and what was the cause of their appearance. This part of the work is devoted to a summary of these factors.

First of all let us examine the significance of the evidence from branches of biological knowledge giving concrete examples of evolutionary problems in development, their nature, the ways in which they were solved in the past and how they compare with more recent situations. The second section is concerned with conclusions and contains an examination of how all the varied problems of historical development may be tackled.

The value of different kinds of evolutionary evidence in historical development

As already indicated in Part I of this book (p. 8), an understanding of the progress made in studies of the natural laws in the historical development of organisms leads on to an understanding of how conclusions and valid inferences can be drawn and hence to a wider knowledge of how the variation of organisms is possible. The completeness of the functional-ecological understanding of the morphological variations in phylogenesis and a methodical accurate appraisal of all these phenomena was of particular importance (Rohdendorf, 1950a, 1950b, 1959b).

Among the many biological disciplines which provide data concerning the evolution of living beings, a special place is occupied by paleontology, since this is the only one which furnishes direct historical evidence on the changes of organisms with time. All other sources give only indirect evidence of the evolution of the organic world although, as we know, these are also very broad and diverse. Such, first of all, is the knowledge obtained during investigation of the existing diversity of organisms of present-day fauna and flora, knowledge which is furnished by systematics. Very important and diverse data on the evolution of organisms are provided by different individual biological disciplines - morphology, physiology, embryology and ecology.

How can all these evidences of historical development be evaluated? The answer to this question has to be as exact as possible, having in view the regular scantiness of any kind of investigation into the phylogeny and historical development of organisms which, as a rule, are based on the data of any one discipline. We will examine therefore the significance of the different evidence in the historical development of organisms.

Evidence from paleontology.

Paleontological documents, fossil remnants of once living organisms of past geological epochs, are substantial demonstrations of changes of the organic world in time with indisputable authenticity concerning the historical development of organisms.

Can we, by using paleontological material, decide whether it is ever possible to reach a complete solution of the relations of the organism with its habitat, and the nature of the process of its development? Paleontological remnants are often extremely imperfect. From most complicated, highly organized animals there are preserved as fossils only a few simple organs or even mere impressions of the general outline of the body. It is also known that burial of remnants, at least of terrestrial organisms, often ends up with their discovery under conditions very distinct from the normal places of their habitat. All this makes it very difficult to understand the relations of an extinct organism with the surroundings in which it probably existed when it was alive. Only in some cases is the preservation of fossil remnants so complete that it is possible to determine the functional features of the organs of the animal. Sometimes the remnants of extinct organisms are buried in the same places in which they lived and in that way there may be some possibility of throwing light upon their interconnections with other phenomena and objects. On the whole, however, all the fragmentation, incompleteness and superficial nature of these data is clear. Thus, the insufficiency of paleontological evidence for the illumination of evolution in all its fullness is obvious.

Do the fossil remnants of organisms of the geological past which are obtained by paleontological investigation permit one to throw light upon the very important parts of every process of development? Is the gradual growth of the quantitative changes and the rapid, intermittent change of their qualitative variations which occur in the process of the solution of conflicts peculiar for all phenomena? Speaking briefly, do paleontological data allow one to throw light upon the obligatory sides of every process of development? For clarification of the process of historical development of the real entire organism it is necessary to have knowledge of all its chief features of activity and these may not be obtained on the basis of the study of incomplete fossil remnants (i.e., taphonomically) of selected organs by chance. The explanation of conflicts in the historical development of the vast majority of extinct organisms as a result of this stumbles upon very great difficulties and by contemporary methods of investigation is simply impracticable in the majority of cases.

The chief value of these data consists of their historical nature, illustrations of actual modifications which originated in time: these facts are completely reliable and indisputable. Paleontological documents which indicate interconnections of phenomena and processes of development are unreliable. In these matters paleontology is clearly insufficient. Very rich fossil remains of many extinct groups of animals which illuminate changes in the composition of fauna during very protracted parts of geological time, can be fully classified in spite of their imperfection. In some cases the fossils found permit us to judge the ontogenesis of organisms and hence their phylogenesis; such, for example, are some higher molluscs. In these cases similar classifications grow into phylogenetic systems of organisms; these can be argued historically and often allow us to consider the paths of historical development in a valuable way. In the absence of any straightforward evidence of phylogenetic and ontogenetic modifications, however, the classifications of paleontological documents themselves represent very valuable schemes, which paint a diversity of the organic world and permit us to compare and distinguish individual faunistic complexes that always possess great and many-sided significance. In fact, a study of the system of organisms based on the investigation of only a few organs, if of sufficient depth, always reveals in some measure the actual relationships: the incompleteness of remnants naturally limits the depth of the illumination of the system. The study of fossil remnants of any degree of completeness is important and is always necessary: any detailed basic classifications of fossil remnants are important stages in the study of the evolution of the organic world-- incompleteness of remnants should by no means retard their study and classification. However, all conclusions about the ecology of the extinct organism require comparison, analogy with organisms living now. When such a comparison is impossible, the structures of a fossil remnant are almost always unintelligible.

The study of fossil material without the complete use of data on present-day organisms is unreliable and leads only to phylogenetic schemes of limited accuracy and depends entirely on the completeness of the preserved remnants. Very many such schemes of 'phylogenies' are, in reality, only comparative morphological series of the evolution of separate organs. Our conclusions about these series as reflections of supposedly real phylogenies are often exaggerated.

Therefore the solution of the problems of paleontology in the evolution of living organisms is possible only in a very close working connection with the whole complex of biological knowledge, primarily with zoology. The precise and clear understanding of this most important principal is the chief guarantee of success.

Evidence from the systematics of organisms living now.
Before we consider the indirect evidence in the historical development of organisms which we obtain during a study of contemporary animals and plants, it is necessary to appraise the existing diversity of organisms, the whole variety of the contemporary fauna and flora, composed of a vast number of species, genera, families and other taxa. This huge field of biological knowledge is generally completely ignored during any kind of phylogenetic investiga-

tion. If the diversity of organisms is mentioned, then it is only in passing
during an account of the features of those or other groups of organisms.
That branch of biology called systematics, which directly studies the
diversity of the organic world, is too often considered a 'third rate' science,
a pragmatic collection of rules and habits for the distinction of organisms. One
can draw this conclusion from contemporary biological literature. The neglect
of the scientific value of the data of systematics and the disregard of the facts
concerning the diversity of organisms have clear roots in the erroneous state-
ments made about vague variability, about the independent derivation of new
traits from the conditions of life, and finally about the mutation process as
the only way in which anything new appears in evolution. From these false
assumptions springs the denial of any natural law in the diversity of organisms,
which is rather regarded as being of a purely random character, and the supposition that systematics is not a subject for scientific investigation.

All these conclusions are completely wrong. The existing diversity of the
organic world, the study of which is the chief problem of systematics, is
the main expression of the whole of evolution which takes place in harmony
with changes in the conditions of the life of the organisms. Such an appraisal
of the diversity of organisms is only possible from the position of scientific
biology; inattention to, or disregard of the facts of systematics is equivalent
to a denial of the interconnection between the phenomena of nature and
their modifications in time.

The study of the variety of organisms is still at a very low level. In the vast
majority of cases the chief goal of a systematic investigation is a determination
of the distinctions between organisms and not of their real interrelations. A determination of the differences of contemporary organisms is almost always constructed purely pragmatically on the basis of the external features of the organism
which are the clearest, most easily observed; the classifier-systematist is little
interested in the significance of these diagnostic features in activity. Ever since
Darwin's time, this superficial diagnosing of organisms has been the cause of
doubt and neglect in this important branch of general biology.

The true importance of the diversity of the organic world consists of the
way in which it exactly reflects complex historical changes of organisms. This
determines also the nature of the systematic investigation which should utilize
the facts of all biological disciplines that throw light upon the relation of organisms and their conditions of life (Dubinin, 1951).

Systematics has as its goal the illumination of the system of organisms, i.e.,
the true existing relationship between their separate groupings. Such a goal
should determine the position of this science among the other branches of
biological knowledge, namely as one of the most important general biological
disciplines, which utilizes comparative data arising from the remaining detailed
branches of biology. Such an appraisal of the significance of systematics is the
only right one methodologically.

As for the qualities of the data which would permit us to illuminate the historical development of organisms, their value in demonstrating the modifica-

tions of organisms in time is significant. Such, first of all, is the diversity of the organic world which directly and accurately reflects the changes of organisms in their history. Especially important is the illumination of the nature of the process of development--the clarification of the internal problems and of their solutions which influence the process of development itself. Many-sided knowledge of the features of organisms allows us to recognise the chief problem or the main link in their development. The possibility of making a discovery of this nature for every development is established by a determination of the functional importance and, in that way, the role in the activity of the chief or 'governing' features of organisms. The explanation of the role in the activity of these main traits which distinguish different organisms, is the most significant condition in the discovery of the nature of the conflicts which have taken place in their development. There can be no doubt that the so-called 'governing' features which characterize taxa are nothing else than the reflection of the solution of the problem which brought about the divergence of the given group of organisms. There cannot be any other conception of the nature of the governing features (sometimes they are called 'guiding' or 'predominant' traits or 'features'): the negation of the connection of these traits with the conflicts of development is equivalent simply to a negation of the dialectic understanding of the process of development. Only systematics - a science which examines and compares an organism as a complete unity and not as a metaphysical complex of separate systems of organs - can produce reliable opinions concerning the nature of the problems that are processes of evolution. In this way the most important significance of the study of the diversity of organisms is the task of systematics.

Far more complex and difficult is the illumination, from material on the diversity of organisms, of the processes of formation of new forms, of the change from quantitative, gradual modifications to qualitative, rapid transformation. The physiological-biochemical side of the investigation undoubtedly has paramount significance in the solution of this problem; but the interpretation will be reliable only if the nature of the conflicts is understood.

The chief, although indirect, demonstration of the presence of rapid qualitative changes in the historical development of the organic world is really the existing system of interdependent species, genera, families and so forth. Only with the dialectical path of development could such a system, consisting of sharply-separated, but really interdependent groups, be brought into being. Natural selection, though very important in the evolution of organisms in general, could not in itself create the system of organisms known to us. In fact, if uniform development should have a place in nature, then it would be impossible to establish clearly the delimited taxa. The number of 'transitional' forms would be very large, the reality of taxa would simply be absent and any kind of classification would be in great measure subjective. Such a state of affairs to some extent had a place in the early period of the development of systematics, when it was a purely formal and superficial classification. Very many especially clear classifications were correctly explained by systematics in spite of the unsound methods of investigation. Such, for example, are many of Linné's genera, which

indicate real, existing groupings, usually of a higher level than generic. Similar
is the relative ease in determining the distinctions between species of the majority of groups of organisms. All this is convincing and, however indirectly, bears
witness to the dialectical character of the process of morphogenesis and species-
building in particular.

The general conclusions concerning the value of the evidence of systematics
are clear. This area of biological knowledge which utilizes comparative data
from other divisions of biology, permits us thoroughly and closely to outline
the interconnection of organisms and other phenomena of nature; to demonstrate
the changes of organisms in time and finally to throw light upon the nature of
conflicts which influence the derivation of concrete groups of organisms. However, the illumination of the features of development as gradual, quantitative
changes into rapid qualitative variations on the basis of the study of diversity,
i.e. by means of comparative-descriptive ('systematic') methods, can be carried
out only indirectly and insufficiently.

Evidence from other divisions of biology.

It remains to examine data obtained during studies of individuals, i.e. of on-
togenetics in the wide sense; structure (morphological data), processes of activity
(physiological-biochemical features) and so forth. All data obtained in this way
are inevitably unilateral, illuminating only one system of organs, one physiological process or one distinct function or feature.

The accurate association of morphological and physiological investigation
although giving a precise illumination of the nature of the change of the given
function and organs, cannot completely illuminate the evolution of organisms,
since it cannot be applied to the variation of organisms in time, and is not applied
to their diversities, the chief evidence of the historical development of a given
group of organisms.

Individual biological disciplines can thoroughly illuminate the relations of
organisms with the phenomena of nature surrounding them. They may explain
the features of the process of individual development and show how quantitative
modifications merge into qualitative ones; in fact cover all those sides of the
organism, the knowledge of which does not require full comparison of individual
different organisms one with another. In this last case the data of any separate
biological discipline are insufficient.

*More thorough study of the systematics of organisms — the basis of a solution
to the problem.* — Reviewing the importance of the evidence of historical development which we get from different branches of biological knowledge, the following definite conclusions can be drawn. The ineffectiveness of unilateral utilization of the data of any one branch of biology is clear. The necessity of investigating the systematics of organisms and their real connections with each other,
which alone may allow us to throw light upon the nature of the problems of
whole living organisms, is obvious. By utilizing completely all the knowledge of
organisms which we get from different divisions of biology, their differences and

similarities, systematics also throws light upon an extremely important group of
facts, accessible only to it. This is especially valuable in the absence of direct
paleontological data, which are often unique historical documents.

A consideration of the possibilities of creating a separate branch of biology for
the illumination of the historical development of organisms and, in particular the
establishment among them of such an important place for systematics, is quite
opportune. Until recently the area of use of the data of systematics was problematical: its role has usually been limited to narrow, purely secondary questions of the
determination of organisms. (It is necessary to note that the incomplete value the
limited data of the separate disciplines has for the illumination of the historical
development of organisms is not something new and previously unknown.) Appraising with open eyes the limitedness the unilateral data of independent biological disciplines has, it is possible to obtain scientific solutions to the questions posed
by the historical development of the organic world; solutions which would not only
show how the process was going on, but also why it went this way rather than that,
and indicating the moving force of this development, namely the presence of concrete conflicts.

Such are the characteristics of the investigative methods towards the historical development of organisms. The inherent complexity in the use of knowledge obtained
from different disciplines and the necessity of a thoroughgoing study of diversity
as the foundation of these works, directly corresponds to the problems of systematics
in its true sense.

Internal problems in the historical
development of organisms

An examination of the features of organisms and an explanation of the character of
the problems which have influenced their development play the most important part
in solving our problem. Only an examination of the real determining features ('chief
link in development,' according to V. E. Ruzhencev, 1953, p. 24) can allow us to discover the nature of the conflicts. A simple investigation of 'diagnostic features' and
an attempt to establish on the basis of these, the conflicts in development, definitely
cannot answer the question; the establishment of governing features or of a chief
link in development is an obligatory preliminary condition during the investigation
of conflicts in the historical development of a group. It is only possible to carry out
an analysis and to clear up the features of the conflicts when the traits of the animal's
organization which influenced their derivation have been determined. The principal
consequence of this compulsory condition is the impossibility on the basis of a
'purely' paleontological study of revealing the real governing features of the development of extinct groups of organisms (Rohdendorf, 1950a).

Consequently the illumination of the functional characteristics of the observed
features of fossils is particularly important. The ordinary, purely morphological
description of the structure of an animal is necessary for the recognition of diversity. It is definitely sufficient for the discovery of the real governing features
and for an understanding of the very nature of historical development. In spite of

great difficulty the functional understanding of the observed structural
traits of organisms (both now living and especially extinct) is very important and turns out to be the compulsory condition for the success of the
work (Rohdendorf, 1950a).

Nature of internal conflicts

In what do the internal conflicts in the evolution of organisms consist? Examining phylogenesis as a historical process of modifications of the successive on-
togenies of organisms indivisibly associated with them by conditions of life, it
is natural to see this process in modifications of the features of activity. The
living animal undergoes processes of exchange, individual development, reproduction and settlement, all passing under the control of the nervous apparatus -
this is the basis, the obligatory condition for the existence of the living thing.
Only with the presence of these processes is possible the long existence of organic forms, their historical development. Therefore, the conclusion is certain that
it is the internal conflict among the needs of the vital processes which moves the
process of development. This is the real process by which modification of the
conditions of life mold the derivation of the new.

My problem is only the general determination of the character of the conflicts
among the main processes of activity, and the concrete consideration of separate
cases of historical development. As it is easy to see, each vital process, every
phenomenon, has its own inherent features and in fact the characteristics of
conflicts in each separate case are quite peculiar and unrepeatable. This shows,
with great clearness, the importance of a general determination of the features
of the internal conflicts of vital processes; any detailed and exhaustive investigation inevitably compels us to go far into a consideration of a few detailed processes and makes it impossible to encompass the whole problem with the necessary
latitude.

A general determination of the properties of internal conflicts in the activity
of an organism, outlined above, results in the clarification of those requirements
which are 'produced' by the organism for the conditions of living and for the
accomplishment of the necessary processes.

The internal conflict in an organism is a very complex system of interrelations
of processes. These most distinctive conflicts nevertheless fall into two large
groups — those which concern the organism as a whole and those which are confined to one part of its activity: respiration, excretion, etc. The first group are
internal problems from the standpoint of the organism, but external from the
standpoint of the respiratory or excretory system itself.

Examples of these conflicts are indicated in Table 4.

These examples may easily be multiplied. It is sufficient to look at any concrete examples of conditions of life of an animal, or at the examples in different
forms by the material of different animals. There were conflicts between processes connected with the different environmental requirements; for example,
between respiration, nutrition, growth and reproduction. To determine these is

Table 4

Aspects of conflicts

Need for respiration (in water or terrestrial medium)	Need for nourishment (on dry land or in water)
Need for nourishment (in water or on dry land)	Need for individual development (in water or on dry land)
Need for reproduction	Need for individual development and growth
Need for nourishment in definite conditions	Need for reproduction in different conditions

not difficult. It is somewhat more difficult to clear up the conflicts between reproduction and individual development but, in this case, the depth of the differences of these interdependent processes directly denotes their conflict. Examining the conflicts, for example, between the features of the processes of exchange (nutrition, respriation) and of the processes of individual development, we observe them easily in differences of the same processes. Actually the realization of exchange in the time of development continually changes as a result of increase (growth) and as a result of modification (rebuilding) of the organism. The fact that there are these differences demonstrates the presence of conflicts.

In conclusion it is necessary to emphasize that these conflicts only are the true basis of development. A consideration of any kinds of so-called 'conflicts' between predator and prey, between the organism and the conditions of its habitat, between these or other competing organisms, shows that such relations are not the immediate cause of the modification of an organism. These are only conditions which influenced that or another solution of the internal conflicts. A review of the accurate conception of internal conflicts and of the action of external factors upon the organisms is an important and vast methodical problem, the development of which appears independent, and which I do not have the opportunity to touch upon.

It is very important to have in view that the negation of the importance of internal conflicts to the organism and the addition of factors of the external environment as significant, direct moving forces of organic development, is a well-known form of mechanism which, under the appearance of the 'struggle for materialism', in fact replaces the dialectic.

Table 5

Process of activity	Aspects of conflict	
Exchange as a whole	Anabolism - assimilation of required substances and the building up of living material	Catabolism - excretion of waste substances and the decomposition of living material
Nutrition	Need for nourishment by the most caloric substances	Need for obtaining water by the organism
Nutrition	Need for rapid and abundant absorption of nutritive substances	Need for complete assimilation of food
Nutrition	Need for feeding on easily accessible substances	Need to secure chemical full-value of nutritive material
Respiration	Need for the greatest intensification of the respiratory process by means of extraction of the maximum mass of oxygen	Need for the most rapid delivery to the tissues of the body of oxygen and elimination of carbon dioxide
Individual development	Need for most rapid attainment of a sexually mature condition	Need for growth and rebuilding of the organism in development
Individual development	Need for building up of larval organs and other temporary structures	Need for reduction of temporary structures and building up of organs of the definitive state
Individual development	Need for quick growth and development of the earliest minute stages in ontogenesis	Need for intensive feeding and protection of these stages
Reproduction	Need for the formation of sex products in the greatest quantity	Need for the formation of the most viable sex products
Reproduction	Need for the guarantee of impregnation of the greatest possible number of oocytes	Need for the guarantee of the development of the greatest number of oocytes

Concrete examples of internal conflicts in the
evolution of some animal groups

The evolutionary paths of dipterous insects examined in the factual part of this
book showed the exceptional importance of a dialectical analysis of the processes
of development as an uninterrupted chain of contradictory processes, their solutions and change to new conflicts. The last stage in the final part of the investigation, is a detailed description of concrete conflicts in the historical development of
dipterous insects which are known best of all to the author, and that of other
groups of animals. In fact, the examples of concrete conflicts cited below are
independent outlines of separate episodes in the history of the given stems of
animals. In these examples, which are described according to a plan, is indicated
first of all the nature of the problem. The various aspects of the conflict are then
unravelled - those or other requirements of activity between which arose contradictory relations. In the conclusion of each example is revealed the solution
to the conflict which is realized in those or other conditions of life. In addition,
examples are given of concrete groups of animals which originated as a result
of a given solution to the conflict.

Example 1. – From the history of the derivation of insects of the order Diptera.
The internal conflict of the organism is between locomotion and nutrition in the
larval stage.

Conflict

A. Need for a solid substrate for locomotion of the larva, provided by three
pairs of walking legs.

B. Need for nutrition in a
moist, semiliquid substrate,
forming comparatively large
masses.

Solution

When surface layers of nutritive substrate were in short supply, or desiccated, or
of poor nutritional value, the larvae developed the ability to live submerged in the
deep mass of substrate--in the semiliquid layers. The need for nutrition, 'B', overcame
the need for a solid surface to walk on, 'A'. The larvae lost their walking legs and
acquired a wormlike body. This turned out to be the chief prerequisite to the
formation of the first representatives of the order of dipterous insects.

In different conditions, where there was a more moist surface medium, the larvae
lived on the surface layers of substrate. Here 'A' proved to be predominant. The
larvae retained walking legs and developed the capacity to feed on relatively solid,
dry substrates. This solution did not lead to the derivation of apodous larvae. As
a result, there arose a 'sister' group to the Diptera, closer to the extinct representatives of the mecopteroids, but which remains undiscovered.

Example 2. – From the history of the derivation of insects of the order Diptera. The internal conflict is between locomotion and development.

Conflict

A. Need for movement for accommodation of descendants in locally disposed substrates in which the larvae live.

B. The possibility of performing slow flight by means of two pairs of wings possessing insignificant lifting powers.

Solution

The conditions led to a single solution--the development of strong flight by means of an anterior pair of wings: in short, the development of the dipterous condition. 'A' was the chief determinant, and 'B', the lesser, became the subject for change. As a result dipterous insects came into being. Had the conflict been resolved in the favour of 'B' and the insects had retained feeble flight by means of the ancestral four wings, then dispersal to new larval feeding grounds would have been difficult and such insects would have died out. (That may possibly have happened in the history of the group, but we do not have any paleontological evidence.)

Example 3. – From the history of the derivation of ancient original groups of the order Diptera--of the infraorders Bibionomorpha and Tipulomorpha. The internal conflict is between development and exchange.

Conflict

A. Need for nutriment by means of attaining the most moistened layers of the nourishing substrate of the larvae.

B. Need for respiration by means of contact with atmospheric air for all spiracles of the segments.

Solution

In conditions of increased moisture, free water for example, 'A' was again predominant and the larvae developed the ability to dwell in the deepest, dampest parts of the substrate. This influenced the modification of the non-predominant side of the conflict, the means by which the tracheal system made contact with atmospheric air: there developed a metapneustic respiration by means of the most posterior pair of spiracles becoming enlarged and complicated while other spiracles atrophied. As a result of this solution to the conflict there arose new representatives of the ancient tipulomorphs.

However, where conditions were of only moderate moistness (for instance in terrestrial biotopes) 'B' proved to be the main consideration, and the larvae retained respiration by all segmental spiracles. This influences the solution of the lesser problem, the need for nourishment only attainable in the dampest parts of the substrate. There developed a change of nutriment by passing over into substrates which were considerably more rare and locally disposed, but which were also very rich in nourishment, such as rotting wood, which is infested by mycelia, the fibrous inner bark of trees and fruiting bodies of the higher fungi. From this solution there sprang up new representatives of the ancient bibionomorphs.

Example 4. — From the history of the derivation of different groups of tipulomorphs. The internal conflict is between development and exchange.

Conflict

A. The need for preservation of contact with atmospheric air during respiration of the larvae.	*B.* Need for nutriment for the larvae in muddy or detritus masses on the bottom of a reservoir

Solution

In slowly flowing, shallow reservoirs the chief aspect of the problem turned out to be 'A'. The larvae brought about contact with the surface film for respiration. Nutrition was then modified by the development of predation, the ability to feed on microorganisms on the surface film of water and on the surface of plants and stones. This resulted, through a chain of numerous secondary conflicts, in the appearance of different tipulomorphs--crane flies (Tipulidea), blood-sucking mosquitoes (Culicidea), dixideans (Dixidea).

On the other hand, where the water had a rapid current, the larvae had difficulty in preserving contact with the surface film for respiration, and they solved the conflict in favour of 'B'. They fed on the muddy masses at the bottom of the reservoir and developed the ability of osmotic respiration of oxygen dissolved in the water. The tracheal system of these larvae became closed. Thus developed the vast group of tipulomorph-chironomideans, to which belong the non-biting midges (Chironomidea), the blood-sucking midges (Ceratopogonidae), and blackflies (Simuliidae).

Example 5. — From the history of the derivation of fungoid and ground midges (Fungivoridae and Sciaridae). The internal conflict is between locomotion and development.

Conflict

A. The need for nourishment of the larvae in locally disposed substrates: fruit bodies of higher fungi and rotting wood which is infested by mycelia.	*B.* Weak flight ability and imperfect running powers.

Solution

In the habitat of coniferous woods in a cold climate with sudden changes of season, the major aspect proved to be 'A', the need for nourishment from local temporary substrates. This led to minor modifications in locomotive ability; there developed powerful running organs, 'muscular' legs, which permitted the insect to dig in the litter of woods. There was also an improvement in the organs of flight; a powerful muscular system and perfect wings developed. This solution led to the evolution of the family of fungus gnats (Fungivoridae).

However, in the deciduous woods of a moderate climate, the need to develop locomotive ability gave way to the easier solution of widening the range of nutrients, and here the insects acquired the ability to feed on dead leafy plants owing to an improvement in the digestive system of the larvae, which became capable of digesting cellulose. This turned out to be an entirely useful acquisition which greatly extended the nutritive resources of the insect. It brought about the evolution of ground midges (Sciaridae), the most important of the co-makers of humus.

Example 6. — From the history of the derivation of gall midges, the larvae of which live in the tissues of living plants (Cecidomyiidae). The internal conflict is between exchange and development.

Conflict

A. The need for nourishment of the larvae on the tissues of rotting leaves of vascular plants on the surface of the ground.

B. The need for substantial dampness for the egg, larval, and pupal stages of development.

Solution

In moderate or sub-tropical conditions where there might have been less moisture in plant societies or the rapid drying of plant residues, the major problem was how to guarantee stable conditions for development. The minor aspect, 'A', the need to feed on rotting plants, fell before the greater need for damp conditions of development, 'B', and the ability to live in the tissues of living plants came into being. This most important development was brought about by changes in the digestive system, the change to extra-intestinal digestion (see Rohdendorf, 1946, p. 92-94). Changes in the methods of oviposition also occurred and here was produced the prerequisite for the formation of a vast group of insects–gall midges.

Example 7. — From the history of the derivation of some groups of culicid Culicidea. The internal conflict is between feeding and respiration during individual development.

Conflict

A. The need for food at all stages of development—larvae and adults.

B. The need for respiration in atmospheric air during development.

Solution

Where the larvae inhabited small stagnant reservoirs which were poor in oxygen but rich in microorganisms, such as might occur in a moderate or warm climate, 'B' proved to be the major necessity. The minor problem of feeding, 'A', was changed. There developed microphagia: feeding on microorganisms. The winged insects developed blood-sucking on vertebrates, and on this basis, blood-sucking mosquitoes (Culicidae) developed.

In the conditions posed by large reservoirs, lakes, or cold flowing waters which were rich in oxygen and plankton, 'A' became the determining factor and the respiration of the larvae changed. The larvae became active predators, acquiring the capacity of osmotic respiration. They broke away from the surface film and populated the whole mass of water. The winged insects (mosquitoes) later on reduced the capacity of feeding down to complete aphagia. Through a chain of secondary changes, this led to the evolution of the group of culicid-chaoborids (Chaoboridae).

Example 8. — From the history of the derivation of the chief groups of winged insects, Pterygota of infraclasses of Neoptera and Palaeoptera. The internal conflict is between shelter and locomotion (Rohdendorf, 1949, p. 120).

Conflict

A. The need for rapid movement: flight or running for the gaining of food, dispersal, and reproduction

B. The need for shelter from predators and abiotic factors (rain).

Solution

The synthesis was made whereby the need for shelter from enemies, 'B', was made unnecessary by the acquisition of greater locomotive powers, 'A'. Long powerful wings were developed which allowed the insect to perform rapid flight. No adaptation was made towards living in more sheltered places. Thus the group Palaeoptera evolved.

In conditions of dense vegetation, however, the development of rapid movement, 'A', was not such a satisfactory answer, and minor modifications led to the complete folding of the wing organs. This solution influenced the chief prerequisites of the derivation of the group Neoptera, of neopterous insects as a whole.

Example 9. – From the history of different groups of fishlike vertebrates (Pisces). The internal conflict is between individual development and reproduction.

Conflict

A. The need for the quickest individual development towards the attainment of a sexually mature state

B. The need for the greatest reproduction, the maximal increase of descendants

Solution

In warm reservoirs with a moderate quantity of plankton and a large number of predators, the major necessity was to be 'A', the rapid attainment of sexual maturity. The maximal increase in the number of descendants, 'B', was incompatible and this led to changes. The animals developed viviparity: that is, the number of descendants from one individual was much reduced but, on the other hand, the most exposed period of individual development was greatly shortened. This solution was reached by many groups; for example, by sharks (Elasmobranchii).

In conditions of moderate or cold reservoirs which were rich in plankton but relatively poor in predators, a maximal increase in the number of descendants, 'B', was the answer to the problem. This led to changes in the rate of individual development. The duration of development was increased, accompanied by extreme fertility in the adult form. Such a solution occurred in the history of very many groups of fishlike forms: for example, of herrings (Clupeiformes).

Example 10. – From the history of the derivation of some groups of Mesozoic reptiles–Archosauria. The internal conflict is between exchange and locomotion (according to the data of Efremov, 1953).

Conflict

A. The need for rapid movement for the recovery of food and protection from enemies

B. The need for protection of the body from overheating

Solution

In conditions of rich plant feeding the need for rapid movement, 'A', won over the need for protection from overheating. Changes led to the animal being able to rise on the posterior limbs to look around and survey the region. This production of a bipedal condition turned out to be a most important development in many groups of reptiles.

In an exposed terrain, 'B' proved to be the most important, the need for protection from overheating. This was brought about by the development of thick bone or skin armour. 'A' was of secondary value and changes included a reduction of the powers of locomotion. Such a solution of the conflict influenced the derivation of the group of armoured archosaurs, the ankylosaurs.

Example 11. – From the history of the derivation of some groups of Permian four-footed vertebrates (Theromorpha). The internal conflict is between the processes of exchange (according to the data of Biushkov, 1951).

Conflict

A. Need for continuous respiratory movements for exchange of air in the lungs	*B.* Need for continuous presence of food in the mouth cavity for its grinding and swallowing

Solution

A high temperature makes a necessity of an intensive exchange of air by continuous respiratory movements. This proved to be the most important function of the animal. A special adaptation was made to allow for the secondary process, 'B', so that there was no interruption in respiration while food was in the mouth – the secondary palate was produced. This solution proved to be the most important prerequisite for the derivation of the first mammals.

Example 12 – From the history of the derivation of some groups of hoofed mammals (Ungulata). The internal conflict is between the processes of feeding.

Conflict

A. The need for devouring a large quantity of plant mass – grassy plants or leaves of woody plants	*B.* The need for a greater grinding of plant mass for complete digestion

Solution

Where there was a shortage of pasture and an abundance of predators and, consequently, the frequent necessity for escape, 'A' proved predominant. The answer was achieved by the animal developing the ability to swallow a large volume of plant material unchewed and to regurgitate it and chew it later on when a safe resting place had been reached ('chewing the cud'). This solution led to the emergence of the group of ruminants (Ruminantia).

Where pastures were more plentiful, predators fewer, and running away easier, the need to grind the food completely at the time of eating, 'B', proved to be the strongest factor. These animals developed particularly strong lophodont teeth which permitted them to break up plants during their feeding, simultaneously developing a special symbiont fauna of flagellate protozoa in the intestinal tract to assist digestion. These traits were prerequisite for the formation of some groups of the odd-toed ungulates (Perissodactyla).

I have confined myself to a consideration of these 12 examples of concrete conflicts in the historical development of animals. The selection is quite at random: a great many more could be given.

Conclusions

1. The chief condition for success in the solution of questions of historical development of organisms is the methodical review of the phenomena. The unilateral development of the features of the organism has special importance; it illustrates the characteristic of development, the moving power of which is internal conflicts, their solution and replacement by new ones. A determination of the functional features of observed morphological traits turns out to be most important. On this basis it is obligatory to examine the phylogenetic changes of the organisms as a dialectic process of development, from all sides and consecutively.

2. The investigation of the historical development of the Diptera, a vast group of contemporary insects, having a wide distribution and an important significance in all biocenoses of dry land and of fresh water, served as an example. Rich and completely new materials on the most ancient Diptera of the Mesozoic turned out to be particularly useful paleontological evidence which served as a base for phylogenetic constructions.

3. The consideration of mecopteroids, related to the dipterous fossils, and the appraisal of the chief features of development and way of life of contemporary Diptera allowed us to determine the chief conflict in development which influenced the derivation of this order of insects. This was the necessity for nourishment of the larva within separated semiliquid nutrient strata. The solution to this conflict was the development of a wormlike apodous larva, and the development of the two-winged condition which permitted adults a high-frequency wing beat which increased the lifting power in flight.

4. The study of the features of Diptera of the contemporary fauna is revealed by statistics on the number of species of the families of the order. These data have significance as one of the characteristic traits of every group of animals, reflecting their history and interconnection with the environment. These data are still insufficiently appraised and investigated.

5. A consideration of the structure of the system of the order Diptera, namely of taxa of suborder rank, led to the conclusion that the widely known groupings – Nematocera, Brachycera and Cyclorrhapha – lack naturalness and phylogenetic unity. The last two groups illustrate only the peculiar phenomenon of oligomerization. The third 'suborder', although a nearly natural phylogenetic

grouping in regard to the level of categories, is unacceptable as a taxon of the
rank of suborder. With the exception of the most peculiar eastern Asiatic genus
Nymphomyia all present-day Diptera constitute one taxon of the rank of suborder. *Nymphomyia* shows clear traits of similarity with Triassic representatives
of the group Dictyodipteridae: a forward-directed free head in the pupa, large
prothorax, enormous simple eyes and other features denoting attachment to a special
suborder of Diptera, the Archidiptera, a relict of early Mesozoic fauna.

6. Present Diptera (suborder Eudiptera) are characterized by the mouth parts of
the pupa, directed backwards along the body, small simple eyes, and moderately developed prothorax. This suborder is divided into a series of associations of superfamilies, namely infraorders, 12 of which are represented in the contemporary fauna.

7. A consideration of the features of the activity and organization of the present-
day representatives of all the infraorders of Diptera and their systematics allowed
us to arrive at the origins of the derivation of the given groups. A close appraisal
of the features provided the possibility of indicating the nature of the conditions
and conflicts in history which influenced the derivation of 'governing' features
of concrete groups. The chief thing was the acknowledgement of the obligatory
adaptive character of all new traits in the organization of the animal and the negation of the role in historical development of 'accidentally' arising deviations.

8. The determination of the relative time of derivation of the features proved
to be the most important. Ultimately, the establishment of the original governing
traits became possible as a result of a functional analysis of the features; a determination of their role in activity made any kind of assumption concerning the
'equivalence' of different 'combinations of features' of formal systematists unnecessary. The establishment of the time of derivation of the given features themselves governed the time of derivation of other traits. Such selection of concrete
systematic unities is very fruitful and permits us to reveal the sides of historical
development which up to then had been indistinct.

9. A consideration of all known paleontological evidence in dipterous insects
has been developed. Until recently the Diptera were known only from the Lower
Jurassic fauna of western Europe; in the present investigation a relatively rich
faunistic complex of the Upper Triassic of Issyk-kul is described; these most
ancient remnants of Diptera have great significance for an explanation of the paths
of historical development of the order.

10. The Upper Triassic Diptera of Issyk-kul are reported to the number of
54 Triassic species distributed among 33 genera (only two of them known earlier
as Jurassic species), 18 families (four of them known earlier from Jurassic or
contemporary genera), 13 superfamilies (four known earlier on the basis of recent
families) and four infraorders (two known earlier on the basis of recent representatives). These numbers bear witness to the extreme unusualness of Upper Triassic
Diptera, sharply distinguished even from the Lower Jurassic, to say nothing of
the more recent Diptera. Especially important is the presence in the Upper Triassic fauna of two peculiar infraorders not represented in the most recent faunas.
Two other infraorders, the tipulomorphs and bibionomorphs, although represented
in the fauna of the Jurassic and Cenozoic, are sharply distinguished according to

their superfamily composition: out of five superfamilies of tipulomorphs, three
are purely Upper Triassic; the same may be said concerning the five superfamilies
of bibionomorphs.

11. The Lower Jurassic Diptera known on the basis of literature data according to remnants of the deposits of Lower Jurassic age of western Europe and
eastern Siberia belong to 53 species of 17 genera (only 14 discovered in Lower
Jurassic fauna), 11 families (only four in the Lower Jurassic fauna), seven superfamilies (only one Lower Jurassic), and three infraorders. The general composition
of the Lower Jurassic fauna is still insufficiently studied; the data which we have,
show the nearness of it to the Upper Jurassic and a sharp distinction from the
Triassic. Significant is the appearance in this faunistic complex of the infraorder
Asilomorpha, richly represented in recent times. The tipulomorphs are represented
by the well-developed superfamily Tipulidea and by the extinct, Lower Jurassic
Eoptychopteridea. The bibionomorphs are characterized by the first appearance
of the superfamily Bibionidea, represented by two Jurassic families, and the superfamily Fungivoridea also with two Mesozoic families. The superfamily Rhyphidea
is represented already by two families, widespread also in the Cenozoic.

12. The Middle Jurassic Diptera are the most investigated of all the Mesozoic
forms: the total number of described species exceeds 60. Such a fauna as Karatau
(Kazachstan), from which is reported 64 species, belonging to 55 genera (two of
them discovered in the Lower Jurassic or in the Upper Triassic), 25 families (six
of them distributed in the faunas of another age, Triassic, Lower Jurassic or
Cenozoic), 12 superfamilies (only two extinct, Middle Jurassic) and four infraorders.
The total composition of the Karatau Middle Jurassic fauna is comparatively well
known. The most characteristic traits are: the great variety of bibionomorphs
forming a clear majority of the represented Diptera of all faunas, the diversity of
asilomorphs, and finally the presence of two most peculiar superfamilies of tipulomorphs.

13. The Upper Jurassic Diptera of western Europe are still insufficiently studied
and the composition of this fauna cannot be judged. Altogether about 20 species,
the attachment of which to families may not be indicated exactly as a result of
the imperfect account of their remnants, have been described.

14. The Diptera of the Cretaceous period are up to now almost unstudied,
if one does not consider the descriptions of individual species of chironomids of
the Lower Cretaceous of eastern Asia and remnants of a few forms of chironomids
from deposits of the Laramide formation of Canada.

15. The Diptera of the Tertiary period began to be studied very long ago; remnants of Diptera from Baltic amber were described at the middle of the Eighteenth
Century. At present more than 2,000 descriptions of Tertiary species of Diptera
are known in the literature. The composition of the Tertiary fauna of Diptera
demonstrates its great similarity to the contemporary; it is sufficient to say that
the number of peculiar purely Tertiary families is very small and it is possible
that such are absent. The number of peculiar extinct genera is also small. So great
a closeness of the Tertiary Diptera to the contemporary forces us to investigate
the fossil remnants with special attentiveness and in close connection with an in-

vestigation into the system of independent families or narrower groups of the present-day fauna on the scale of the whole planet; knowledge of tropical faunas proves to be a compulsory condition for an accurate determination of Tertiary species and often this cannot be made on the proper level. Simultaneously, this circumstance does not permit us, without a detailed study, to use literature data on the Tertiary insects which was obtained without a monographic treatment of the material and which is always incomplete or even simply inaccurate.

16. The survey made of the features of contemporary and fossil Diptera allowed us to ascertain the phylogenetic relations of these insects. The schemes of the phylogenesis of the Diptera show that the most ancient were the infraorders Diplopolyneuromorpha and the Dictyodipteromorpha, which were already very scarce at the end of the Triassic and which bore the character of relics. These two infraorders form the special suborder Archidiptera, one represented in the present-day fauna by the scarce relics, the Nymphomyiidae. The dictyodipteromorphs turned out to be the original group for two infraorders - the tipulomorphs and the bibionomorphs which, in their turn, served as the sources of derivation of all the recent groups of Diptera. It is especially important to note the splitting off from one group of Bibionomorpha, namely the rhyphideans, of the important infraorder Asilomorpha in the middle of the Triassic period. Towards the end of the Mesozoic the asilomorphs reached great diversity and gave the beginning to the largest Cenozoic infraorder of Diptera, the Myiomorpha which, in the contemporary geological epoch, contains the greatest number of species. The general picture of the phylogenetic relations of the chief groups of the order Diptera differs substantially from earlier schemes by other authors: these last schemes have been investigated critically and their proper positions serve as a basis for the proposed scheme.

17. A review of the phylogenetic relations is not confined to the establishment of the interrelations of large groupings, infraorders, but also touches on the connections of superfamilies and the majority of families. Particular interest is represented by many fossils, groups now extinct - the Triassic Diplopolyneuromorpha, Eopolyneuridea; the Jurassic Eoptychopteridea, Tanyderophryneidea, Dixamimidae and Mesophantasmatidea. These highly specialized Diptera bear witness to the extreme unusualness of the faunas of the past, in the composition of which lived groups clearly distinct from contemporary forms.

18. The final division of the whole investigation is devoted to a consideration of the evidence for the historical development of organisms. The incompleteness and unilateral nature of the data, which are received from different branches of biological knowledge, compel us to aim at a wide as possible range of facts from different disciplines to illuminate the path and causes of historical development. Consideration of the value of the evidence of the historical development of organisms, paleontology, ontogeny, comparative morphology, physiology, and systematics, makes clear the scantiness and unilateral nature of knowledge. The greatest role in the illumination of historical development is played by systematics, a science which investigates the system of organisms, the chief evidence of their historical development.

19. The last division of the final part is devoted to a consideration of the nature of conflicts in the historical development of organisms. A solution to this problem is directly linked with the detection of the nature of the vital process as a whole, i.e., with the fundamental problem of biology and all natural science. The present investigation has the limited objective of throwing light upon only the process of historical development; therefore a close consideration of the problem of the nature of a vital process and its natural laws is not developed. The accomplishment of the processes of exchange, growth, reproduction, development, settlement, and defence, their inter-connections and harmony with the conditions of life, is an obligatory condition in the existence of the organism. It is natural to see the conflict in the process of development just in these phenomena. The needs of the different aspects of activity prove to be real internal conflicts of the living organism, by which also are accomplished changes in the organism as a result of alterations of the conditions of its life, namely its evolution.

20. The internal conflicts of the organism consist of contradictory needs by different aspects of activity. They can be divided into strictly internal conflicts and the internal conflicts of the separate aspects of activity (for examples of these and others see p. 317-318). One must emphasize that these conflicts only are the real basis in development; any kind of so-called 'conflicts' between predator and prey, between the organism and conditions of existence, between these and other competing organisms, are not the immediate cause of changes in organisms, being only conditions which caused one or another solution to the internal conflicts.

21. A review of the correct understanding of internal conflicts and of the influence of external factors on the organism is an important independent methodological problem, which it is impossible to consider in the present work. The negation of the importance of the internal conflicts of the organism, the addition of factors of the external environment of the immediate moving forces of organic development, is a well-known form of mechanism which, under the condition of the 'struggle for materialism', practically abolishes the dialectic.

22. Everything said up to now about the nature of conflicts in the historical development of an organism requires a possibly more complete illumination by concrete examples. The necessary examples show different stages of the historical development of some groups of animals - dipterous insects (Examples 1-7), the chief groups of winged insects (Example 8), fishlike vertebrates (Example 9), Mesozoic archosaurs (Example 10), Permian four-footed vertebrates (Example 11) and ungulate mammals (Example 12). In these examples is indicated the two sides of the conflict and the path of its solution. Each example terminates with an indication of the group which appeared as a result of the described solution.

23. A consideration of the conflicts in the historical development of organisms shows the importance of a detailed selection of the nature of each contradictory process, and how imperative it is that the sides of the conflict should be clarified and their principal and non-principal sides established. Only with such a concrete selection of the phenomena is it possible to judge precisely the nature of the conflict, the paths of its solution and its replacement by another conflict.

24. The determination of the sides of conflict in the historical development of

organisms and from them the chief and non-chief is very important. The chief side turns out to be the one which remains unchanged. The non-chief side is the one that is changed and, therefore, is especially important as providing material for modifications, when arise the new conflicts of the organism.

25. Reorganization of the non-principal side of the conflict thus forms those new characteristics which it has not been the custom to name as 'distinguishing traits' originating in new organisms. While a solution of the question seems at first glance to be surprising, it is clearly accurate.

26. The investigation made into the historical development of dipterous insects and the attempt at general conclusions concerning the characteristics of this process have as their primary objective the illumination of the phenomena of development in the evolution of organisms. The principal finding turned out to be the determination of the correct method of investigation of the phenomena of historical development. There is no doubt that the determination of progress in development, on the one hand is combined with following through investigations in the field of fundamental questions in biology and the problems of the production of life and, on the other hand, as the very restricted form with concrete investigation into the history of a group of organisms.

In fact, the actual results of the present investigation touched upon the far more important and serious problem of the nature of the processes in the development of organisms--the internal conflicts of the processes of development--than the originally furnished goal of the work - questions concerning progressive development.

BIBLIOGRAPHY

BEKLEMISHEV, V. N. 1944. Ecology of the malarial mosquito. Med: State Publ., M., 1-299. (in Russian).

BEKLEMISHEV, V. N. 1948. Concerning the relations between the systematic position of causative organisms and carriers of infectious diseases of terrestrial vertebrates and Man. – Med. Parasitol., 5: 385-400. (in Russian).

BEKLEMISHEV, V. N. 1951. Parasitism of arthropods on terrestrial vertebrates; the paths of its derivation. Ibid., 2: 151-160; 3: 233-241. (in Russian).

BEKLEMISHEV, V. N. 1954. Parasitism of arthropods on terrestrial vertebrates. II. The chief trends of its development. Ibid., 1: 3-20. (in Russian).

BEKLEMISHEV, V. N. 1957. Some general questions of the biology of blood-sucking lower Diptera. Ibid., 5: 562-566. (in Russian).

BERG, C. O. 1960. Biology of snail-killing Sciomyzidae (Diptera) of North America and Europe, Verhandl. XI. Internat. Kongr. f. Entomol., Bd. 1, S. 197-202.

BRANDT, E. 1879. Vergleichend-anatomische Untersuchungen über das Nervensystem der Zweiflügler (Diptera). – Horae Soc. Entomol. Ross., Bd. 15.

BRAUER, F. M. 1869. Kurze Charakteristik der Dipterenlarven. – Verhandl. k. k. Zool., bot. Ges. Wien, S. 846.

BRAUER, F. M. 1883. Die Zweiflügler des Kaiserlichen Museums 31, Wien, Bd. 111. System. Studien auf Grundlage der Diptera Larven nebst einer Zusammenstellung von Beispielen aus der Literatur über dieselben und Beschreibungen neuen Formen. Denkschr. k. k. Akad. Wiss., Wien, Math. – naturwiss, Cl., Bd. XL, VII, S. 1-100.

BRAUER, Fr., BERGENSTAMM, J. E. v., 1889-1894. Die Zweiflügler des Kaiserlichen Museums zu Wien. Vorarb. zu einer Monographie der *Muscaria schizometopa*. Teil. 4. Denkschr. k. k. Akad. Wiss., Bd. 56, S. 69-180; Bd. 58, S. 305-447; Bd. 60, S. 89-240; Bd. 61, S. 537-624.

BREEV, K. A. 1950. Concerning the behavior of bloodsucking Diptera and of gadflies during an attack of them on northern deer and the resulting reactions of deer. – 1. – Parasitol. Coll. ZIN* Acad. Sci. USSR, XII: 167-198. (in Russian).

BRODIE, P. B. 1845. A history of the fossil insects in the secondary rocks of England. London, p. 33-34, 121-122.

BURAKOVA, L. V. 1931. On the biology of *Scatopse fuscipes* Meig. a possible pest of mosquitoes (*Phlebotomus*). – Parasitol. Coll. ZIN Acad. Sci. USSR, II: 113-118. (in Russian).

CARPENTER, F. M. 1935. Fossil insects in Canadian amber. – Univ. Toronto, Studies, Geol. Ser., vol. 38, 69.

CARPENTER, F. M. 1937. Insects and Arachnids from Canadian amber. – Univ. Toronto, Studies, Geol. Ser., vol. 40, p. 7-62.

CHERESHNEV, N. A. 1951. Biological features of the gadfly *Gastrophilus pecorum* Fabr. (Diptera, Gastrophilidae). Reports Acad. Sci. USSR, 77, 4: 765-768. (in Russian).

*ZIN   Zoological Institute of the Academy of Sciences, U.S.S.R.

CHERESHNEV, N. A. 1953. New data on the biology and morphology of the gastric gadfly *Gastrophilus pecorum* F. (Diptera, Gastrophilidae). – Trans. Inst. Zool. Acad. Sci. Kazach. SSR, I: 84-101. (in Russian).

CHERNOVA, O. A. 1961. Concerning the systematic position and geological age of mayflies of the genus *Ephemeropsis* Eichwald (Ephemeroptera, Hexagenitidae). Entomol. Rev., 4: 858-869. (in Russian).

COCKERELL, T. D. A. 1927. Fossil insects in the British Museum. – Ann. and Mag. Natur. Hist., Bd. IX, 20, 585-594.

COLLESS, D. H. 1962. A new Australian genus and family of Diptera (*Nematocera*: Perissomatidae). – Austral. J. Zool., vol. 10, III, 519-535.

CURRAN, C. H. 1934. The families and genera of North American Diptera. N. Y., p. 1-512.

CZERNY, L. 1934. Musidoridae (Lonchopteridae). In: E. LINDNER. Die Fliegen, Bd. V, S. 1-16.

DAVITASHVILI, L. Sh. 1948. History of evolutionary paleontology from Darwin to the present day. Pub. by Acad. Sci. USSR: 1-576. (in Russian).

DUBININ, V. B. 1951. Concerning the species criteria in parasitic animals. – Parasitol. Coll. ZIN Acad. Sci. USSR, XIII: 5-28. (in Russian).

EDWARDS, F. M. 1926. Preliminary note on the new genus of Scatopsid flies from New Zealand. – Ann. and Mag. Natur. Hist., Ser. 9, vol. 9, 267-269.

EDWARDS, F. M. 1926. The phylogeny of Nematocerous Diptera. A critical review of some suggestions. Verhandl. III. Internat. Kongr. f. Entomol., Zürich, S. 111-130.

EFREMOV, I. A. 1940. Taphonomy – a new branch of paleontology – Publishing House of Acad. Sci. USSR, Dept. of Biol. Science: 405-413. (in Russian).

EFREMOV, I. A. 1950. Taphonomy and geological annals. Vol. 1. – Trans. Paleontol. Institute of the Acad. Sci. USSR, XXIV: 1-178. (in Russian).

EFREMOV, I. A. 1953. Questions of the study of dinosaurs. – Nature, June: 26-37. (in Russian).

EICHWALD, E. 1865. *Lethaea rossica* ou paléontologie de la Russie. Stuttgart, Bd. 11, S. 1192-1195.

ENGEL, E. O. 1930-1938. Asilidae. In: E. LINDNER. Die Fliegen. Bd. IV, 2, S. 1-491.

GILYAROV, M. S. 1949. Features of the soil as a habitat and its significance in the evolution of insects. Pub. by Acad. Sci. USSR, 1-279. (in Russian).

GIRSCHNER, E. 1896. Ein neues Musciden-System auf Grund der Thoracalbeborstung und der Segmentierung des Hinterleibs. – Illustr. Wochenschr. Entomol., Neudamm., Bd. 1. S. 12-16, 30-32, 61-64, 105-112.

GLUMAĆ, S. 1960. Phylogenetical system of the Syrphid-flies (Syrphidae, Diptera) based upon the male genital structure and the type of the larvae. Verhandl, XI, Internat. Kongr. f. Entomol., Bd. I, S. 202-206.

GRABAU, A. W. 1923. Cretaceous fossils from Shantung. – Bull. Geol. Surv. China, vol. V, 2, 164-181.

GRAUVOGEL, L. 1947. Note preliminaire sur la faune du Grès à Voltzia. – C. r. Soc. gèol. France, 3 mars, p. 19-92.

GRUNIN, K. J. 1947. The solution of the puzzle of the biology of the gadfly *Portschinskia magnifica* Pl. – Entomol. Rev. XXIX, 3-4: 214-223. (in Russian).

GRUNIN, K. J. 1949. Concerning one biological feature of subcutaneous gadflies of the genus *Oestromyia* Br. – Reports, Acad. Sci. USSR, 67: 193-195. (in Russian).

GRUNIN, K. J. 1950a. Larvae of stage 1 of gadflies of the family Oestridae and Hypodermatidae and their significance for determining phylogenesis. – Parasitol. Coll. ZIN Acad. Sci. USSR, XII: 225-271. (in Russian).

GRUNIN, K. J. 1950b. On the question of the transition of gadflies to a new host. – Entomol. Rev., XXXI, 1-2: 85-89. (in Russian).

GRUNIN, K. J. 1955. Gastric gadflies (Gastrophilidae). Fauna of the USSR, new series, No. 60, Insecta Diptera, XVII, 1: 1-96. (in Russian).

GRUNIN, K. J. 1957. Naso-pharyngeal gadflies (Oestridae). Fauna of the USSR, new series, No. 68, Insecta Diptera, XIX, 3: 1-147. (in Russian).

GRUNIN, K. J. 1962. Subcutaneous gadflies (Hypodermatidae). Fauna of the USSR, new series, No. 82, Insecta Diptera, XIX, 4: 1-238. (in Russian).

HANDLIRSCH, A. 1906-1908. Die Fossilen Insekten. Leipzig, S. 1-1430.

HANDLIRSCH, A. 1937-1939. Neue Untersuchungen über die fossilen Insekten. – Ann. Naturhist. Mus. Wien. Bd. 48, S. 1-140; Bd. 49, S. 1-240.

HARDY, D. E. 1960. The Diptera of Hawaii, Verhandl. XI. Internat. Kongr. f. Entomol., Bd. I, S. 167-168.

HENDEL, Fr. 1928. Zweiflügler oder Diptera. II. Allgemeiner Teil. Die Tierwelt Deutschlands. Bd. II, S. 1-135.

HENDEL, Fr. 1937. Ordnung der Pterygogenea: Diptera-Fliegen. In: W. KÜKENTHAL. Handbuch der Zoologie, Bd. IV. Berlin und Leipzig, IV, 3, 9-11, S. 1729-1998.

HENNIG, W. 1948-1952. Die Larvenformen der Dipteren, Bd. I-III. Berlin, S. 1-185; 1-458; 1-628.

HENNIG, W. 1953. Kritische Bemerkungen zum phylogenetischen System der Insekten. – Beitr. Entomol., Bd. 3, Sonderh., S. 1-85.

HENNIG, W. 1954. Flügelgeäder und System der Dipteren. – Beitr. Entomol., Bd. 4, H. 3/4, S. 245-388.

HENNIG, W. 1955. Das Flügelgeäder der Gattung *Allactoneura.* – Beitr. Entomol., Bd. 5, H. 1/2, S. 127-128.

HENNIG, W. 1955-1962. Muscidae. In: E. LINDNER. Die Fliegen, Bd. VII, 638, S. 1-768.

HENNIG, W. 1958. Die Familien der Diptera Schizophora und ihre phylogenetischen Verwandschaftsbeziehungen. – Beitr. Entomol., Bd. 8, H. 5/6, S. 505-688.

HERTING, B. 1957. Das weibliche Postabdomen der Calyptraten Fliegen (Diptera) und sein Merkmalswert für die Systematik der Gruppe. – Z. Morph. u. Ökol. Tiere, 45, S. 429-461.

JAMES, M. T. 1937. A preliminary review of certain families of Diptera from the Florissant Miocene beds. – J. Paleontol., vol. II, 2, p. 241-247.

JAMES, M. T. 1939. A preliminary review of certain families of Diptera from the Florissant Miocene beds. II, vol. 13, 1, p. 42-48.

JOBLING, B. 1936. A revision of the subfamilies of the Streblidae and the genera of the subfamily Streblinae (Diptera, Acalypterae), including a redescription of *Metelasmus pseudopterus* Coquillet and a description of two new species from Africa. – Parasitology, vol. 3, p. 355.

JOBLING, B. 1939. A redescription of *Pseudostrebla ribeiroi* Costa Lima and the description of a new genus and species of the Streblidae from Brasil. – In: Arbeiten über Morph. und taxon. Entomologie, Bd. 6, N 3, S. 268-275.

JOHANNSEN, O. A. 1909. Diptera, fam. Mycetophilidae. – Genera insectorum, Bruxelles, fasc. 93, p. 1-70.

KOMÁREK, J. 1914. Über die Blepharoceriden aus dem Kaukasus und Armenien. – Sitzungsber. Königl. Böhm. Ges. Wiss., Prag, S. 1-19.

KRÖBER, O. 1924-1925. Therevidae. In: E. LINDNER. Die Fliegen, Bd. IV$_3$, 26, S. 1-60.

KRÖBER, O. 1925. Tabanidae. Ibid., Bd. IV$_2$, 19, S. 1-146.

KRÖBER, O. 1926. Omphralidae. Ibid., Bd. IV$_3$, 27, S. 1-8.

LINDNER, E. 1924-1925. Rhagionidae. In: Die Fliegen, Bd. IV$_1$, 20, S. 1-49.

LINDNER, E. 1925-1928. Die Fliegen der palaarktischen Region (Handbuch), Bd. 1, S. 1-280.

LINDNER, E. 1930a. Phryneidae (Anisopodidae, Rhyphidae) und Petauristidae. In: Die Fliegen, Bd. II$_1$, 1, S. 1-22.

LINDNER, E. 1930b. Blephariceridae und Deuterophlebiidae. In: Die Fliegen II$_1$, 2, S. 1-37.

LINDNER, E. 1930c. Thaumaleidae (Orphnephilidae). In: Die Fliegen, Bd. II$_1$, 3, S. 1-16.

LINDNER, E. 1936-1938. Stratiomyiidae. In: Die Fliegen, IV$_1$, 18, S. 1-218.

LIPINA, N. N. 1949. Larvae of the Tendipedids of Teletsk lake, of its tributaries and river Bii. – Trans. ZIN Acad. Sci. USSR, VII, 4: 193-212. (in Russian).

LOPES, de SOUZA H., MONTEIRO, L. 1959. Sôbre algunas espécies brasileiras de *Stylogaster* Macq., com a descrição de quatro espécies novas (Dipt., Conopidae). – Studia entomol. Rio de Janeiro., vol. 2, 1-4, 1-24.

MARTINI, E. 1929-1931. Dixidae und Culicidae. In: E. LINDNER. Die Fliegen. Bd. III, 12, 11, S. 1-398.

MARTYNOV, A. V. 1923. On the two main types of wings of insects and their significance for a general classification of insects. – Trans. I All-Russian congress of zoologists, anatomists and histologists in Petrograd, 15-21 December, 1922: 88-89. (in Russian).

MARTYNOV, A. V. 1925a. On some results of the investigation of insects of the Jurassic shales of Turkestan. – Reports, Russ. Acad. Sci.: 105-108. (in Russian).

MARTYNOV, A. V. 1925b. Preliminary note concerning the fossil insects of the Jurassic shales of Karatau. – Trans. Mid-asiatic committee of antiquity, I: 193-199.(in Russian).

MARTYNOV, A. V. 1927. Jurassic fossil Mecoptera and Paratrichoptera from

Turkestan and Ust-Balei (Siberia). Publishing House of Acad. Sci. USSR: 651-666.

MARTYNOV, A. V. 1937. Liassic insects of Shuraba and Kizil-Kii. – Trans. PIN* Acad. Sci. USSR, VII, 1: 1-232. (in Russian).

MARTYNOV, A. V. 1938. Outlines of the geological history and phylogenesis of the orders of insects. Part 1. – Trans. PIN Acad. Sci., VII, 4: 1-149. (in Russian).

MARTYNOVA, O. M. 1948. Materials on the evolution of the Mecoptera. – Trans. PIN Acad. Sci. USSR, XIV, I: 1-76. (in Russian).

MESNIL, L. P. 1939. Essai sur les Tachinaires (Larvaevoridae). Monogr. Stat. et Labor. de recherches agronom. Paris, p. 1-67.

MESNIL, L. P. 1944-1956. Larvaevoridae (Tachinidae). In E. LINDNER. Die Fliegen, Bd. VIII, 64, S. 1-512.

MONCHADSKI, A. S. 1936. Larvae of mosquitoes (Fam. Culicidae) of the USSR and adjacent countries. Findings on the fauna of the USSR. – Publishing House of Acad. Sci. USSR, 24: 1-383. (in Russian).

MONCHADSKI, A. S. 1937. Evolution of the larva and its connection with the evolution of adult mosquitoes within the limits of the family Culicidae. Publishing House of Acad. Sci. USSR, OMEN: 1329-1351. (in Russian).

MONCHADSKI, A. S. 1940. Diptera. In the book: "Life of fresh waters", I: 233-263. (in Russian).

OLSUFEV, N. G. 1929. Studies on parasites of the Asiatic locust (*Locusta migratoria* L.) of the order of Diptera and their superparasites. Publishing House of Applied Entomology, IV, I: 61-120. (in Russian).

OLSUFEV, N. G. 1936. On the microscopic anatomy of the head and digestive tract of larvae of *Tabanus*. – Parasitol. Collection ZIN Acad. Sci. USSR, II: 247-278. (in Russian).

OLSUFEV, N. G. 1937. The horseflies (Tabanidae). Fauna of the USSR, new series, No. 9, Insecta Diptera, VII, 2: I-XIII, 1-434. (in Russian).

PERFILEV, P. P. 1937. Mosquitoes. Fauna of the USSR, new series, No. 10, Insect Diptera, III, 2: I-VIII, 1-145. (in Russian).

PING, C. 1928. Cretaceous fossil insects of China. Paleontol. sinica, XIII, ser. B, 1. p. 1-42.

ROBACK, S. S. 1951. A classification of the muscoid calyptrate Diptera. – Ann. entomol. Soc. Amer., vol. 44, p. 327-361.

ROBACK, S. S. 1954. The evolution and taxonomy of the Sarcophaginae. Illinois biol. monographs, vol. XXIII, 3-4, p. 1-181.

ROHDENDORF, B. B. 1937. Family Sarcophagidae. I. Fauna of the USSR, new series, No. 12, Insecta Diptera, XIX, 1: I-XV, 1-501. (in Russian).

ROHDENDORF, B. B. 1938. Dipterous insects of the Mesozoic of Karatau. I. Brachycera and part of the Nematocera. – Trans. PIN Acad. Sci. USSR, VII, 3: 29-67. (in Russian).

ROHDENDORF, B. B. 1939. Concerning the Miocene fauna of insects in the vicinity of the city of Voroshilovska. – Nature, 12: 85-88. (in Russian).

ROHDENDORF, B. B. 1940. The section 'Excavations, finds, reports', concerning the finding of the Stavropol Miocene fauna. − Paleontol. Rev., 2: 78. (in Russian).

ROHDENDORF, B. B. 1943. On the evolution of the flight of insects − Reports. Acad. Sci. USSR, 40, 4: 187-189. (in Russian).

ROHDENDORF, B. B. 1946. Evolution of the wing and phylogenesis of the nematocerous Diptera Oligoneura (Diptera, Nematocera) − Trans. PIN Acad. Sci. USSR, XIII: 1-108. (in Russian).

ROHDENDORF, B. B. 1947a. The fauna of dipterous insects of the Jurassic of Karatau and the significance of it for an understanding of the paths of evolution of the order. − Reports, Acad. Sci. USSR, 55, 8: 757-760. (in Russian).

ROHDENDORF, B. B. 1947b. On the occurrence of the Liassic insects of Sogut. − Reports Acad. Sci. USSR, 57, 6: 607-608. (in Russian).

ROHDENDORF, B. B. 1948a. Venation of the wings of insects and its significance for systematics. − Publishing House of Acad. Sci. USSR, series of biol., 2: 199-203. (in Russian).

ROHDENDORF, B. B. 1948b. A new family of parasitic muscoid Diptera from the sands of the region beyond the Volga. Reports, Acad. Sci. USSR, 63, 4: 455-458. (in Russian).

ROHDENDORF, B. B. 1949. The evolution and classification of the flying apparatus of insects. − Trans. PIN Acad. Sci. USSR, XVI: 1-176. (in Russian).

ROHDENDORF, B. B. 1950a. Concerning the investigation of concrete paths of the evolution of insects. − Publishing House of Acad. Sci. USSR, series of biol., I: 78-96. (in Russian).

ROHDENDORF, B. B. 1950b. Progressive development in evolution. − Publishing House of Acad. Sci. USSR, 6: 53-58. (in Russian).

ROHDENDORF, B. B. 1951. The organs of movement of dipterous insects and their origin. − Trans. PIN Acad. Sci. USSR, XXXV: 1-180. (in Russian).

ROHDENDORF, B. B. 1957. Paleoentomological investigations in the USSR − Trans. PIN Acad. Sci. USSR, LXVI: 1-102. (in Russian).

ROHDENDORF, B. B. 1958. (Rohdendorf, B. B. The main features of phylogenetic relics. XVth Intern. Congr. of Zool. London 1958, Sect. 1, Paper 28: 1-2).

ROHDENDORF, B. B. 1958-1959. (Rohdendorf, B. B. Die Bewegungsorgane der Zweiflügler − Insekten und ihre Entwicklung. Wiss. Zeitschr. d. Humboldt − Univer. zu Berlin, math-naturwiss. Reihe VIII, 1: 73-119, 2: 269-308; 3: 435-454).

ROHDENDORF, B. B. 1959a. Phylogenetic relicts. − Trans. of Institute of the morphology of animals. Acad. Sci. USSR, 27: 41-51. (in Russian).

ROHDENDORF, B. B. 1959b. Questions of paleozoological systematics. − Jour. Paleontol., 3: 15-26. (in Russian).

ROHDENDORF, B. B. 1960. Features of ontogenesis and their significance in the evolution of insects. The Ontogeny of Insects. Acta symposii de evolutione insectorum, Praha, 1959: 56-60. (in Russian).

ROHDENDORF, B. B. 1961a. The most ancient infraorders of Diptera from the Triassic of central Asia. − Jour. Paleontol. 3: 90-100. (in Russian).

ROHDENDORF, B. B. 1961b. (Rohdendorf, B. B. Neue Angaben über das System der Dipteren. XI. Internationaler Kongress für Entomologie, Verhandlungen, I: 151-158).

ROHDENDORF, B. B. 1962. The order Diptera. In coll: 'Foundations of paleontology'. Tracheata and chelicerata: 307-344. (in Russian).

RUBTSOV, I. A. 1936. A new species of midge (*Simulium oligocenicum* sp. n.) from amber. — Reports, Acad. Sci. USSR, II (XI), 8 (97): 347-349. (in Russian).

RUBTSOV, I. A. 1937. On the evolution of bloodsucking midges (Simuliidae, Diptera). — Publishing House of Acad. Sci. USSR, OMEN: 1289-1327. (in Russian).

RUBTSOV, I. A. 1939. On the evolution of gastric gadflies (Gastrophilidae) in connection with the history of their hosts — Zool. Jour., XVIII, 4: 669-684. (in Russian).

RUBTSOV, I. A. 1940. Geographic distribution and evolution of gadflies in connection with the history of their hosts — Nature, 6: 48-60. (in Russian).

RUBTSOV, I. A. 1951. On the morphology and evolution of the abdomen and genital appendages of fly-phasids (Diptera, Phasiidae s.l.). Reports of the All Union Entomol. Society 43: 171-249. (in Russian).

RUBTSOV, I. A. 1956. Gnats (fam. Simuliidae): Fauna of the USSR, new series, No. 64, Insect Diptera, VI, 6, 2nd Ed: 1-860. (in Russian).

RUKAVISHNIKOV, B. I. 1930. Materials for the study of flies which are parasites in the larval and adult phases of the locust (*Locusta migratoria* L.). — Trans. of Pres. of Growth, series of entomol., I, I: 191-261. (in Russian).

RUZHENCEV, V. E. 1953. The chief questions of paleozoological systematus in the light of Mitchurin's biology. — Materials of the paleontol. council on the paleozoic. Pub. by Acad. Sci. USSR: 5-36. (in Russian).

SABROSKY, C. W. 1960. The present status of the systematics of Diptera. Verhandl. XI. Internat. Kongr. f. Entomol., Bd. 1, S. 159-166.

SACK, P. 1933. Nemestrinidae. In: E. LINDNER. Die Fliegen, Bd. IV$_1$, 22, S. 1-42.

SACK, P. 1934. Mydiadae. Ibid., Bd. IV$_1$, 23, S. 1-29.

SACK, P. 1936. Cyrtidae (Acroceridae). — Ibid., IV$_1$, 21, S. 1-36.

SAVCHENKO, E. N. 1961. Flies — long legged ones (fam. Tipulidae). Subfam. Tipulinae: genus *Tipula* L.. (part I). Fauna of the USSR, new series, No. 79, Insecta Diptera, II, 3: 1-487. (in Russian).

SCUDDER, S. H. 1890. The Tertiary insects of North America. — U. S. Geol. Surv. Terr. vol. 13, p. 1-734.

SHTAKELBERG, A. A. 1925. A new fossil representative of the genus *Tubifera* Mgn. (Diptera, Syrphidae). — Russ. Entomol. Rev., XIX: 89-90. (in Russian).

SHTAKELBERG, A. A. 1928. On the occurrence of *Nycteribosca kollari* Fefld. (Diptera, Streblidae) in Turkestan. — Russ. Entomol. Rev. XXII, 1-2: 133. (in Russian).

SHTAKELBERG, A. A. 1937a. Fam. Culicidae — bloodsucking mosquitoes (subfam. Culicinae). Fauna of the USSR, new series, No. 11. Insecta Diptera, III, No. 4: 1-258. (in Russian).

SHTAKELBERG, A. A. 1937b. On the dipterofauna of the homes of vertebrates in the vicinity of Ashkabad. – Trans. Committee for the study of the production of forces, series Turkm., 9: 283-288. (in Russian).

SHTAKELBERG, A. A. 1948. Diptera. In the coll.: 'Animal world of the USSR', II, Desert: 162-179. (in Russian).

SHTAKELBERG, A. A. 1950. Diptera. In the coll.: 'Animal world of the USSR', III, Steppes: 162-213. (in Russian).

SHTAKELBERG, A. A. 1953. Diptera. In the coll.: 'Animal world of the USSR', IV, Forest: 228-317. (in Russian).

SHVANVICH, B. N. 1949. The course of general entomology. Pub. by 'Sov. Science': 1-900. (in Russian).

SIMPSON, G. C. 1945. The principles of classification and a classification of mammals. – Bull. Amer. Mus. Natur. Hist., vol. 85, p. 1-350.

SMIRNOV, E. S. 1923. On the structure of the male genital system of dipterous insects. Biol. News, I, M.: 39-56. (in Russian).

SMIRNOV, E. S. 1936. On the system of the family Syrphidae. – Moskovoskoe obschcestvo ispytateli prirody (Bulletin) div. biol. XXV, (4): 249-255. (in Russian).

SPEISER, P. 1908. Die geographische Verbreitung der Diptera pupipara und ihre Phylogenie. – Z. wiss. Insektbiol., Bd. 4, S. 241-246, 301-305, 420-427, 437-447.

STATZ, G. 1940. Neue Dipteren (Brachycera et Cyclorrhapha) aus dem Oberoligozän von Rott. – Palaeontographica, A, Bd. 91, S. 120-174.

TILLYARD, R. J. 1926. The insects of Australia and New Zealand. Sydney, vol. I-XVI, p. 1-560.

TILLYARD, R. J. 1929. Permian Diptera from Warner's Bay, N.S.W. – Nature, vol. 123, p. 778-779.

TILLYARD, R. J. 1933. The Panorpoid complex in the British Rhaetic and Lias. British Museum (N.H.) Fossil Insects, N3, p. 7-79.

TILLYARD, R. J. 1937. The ancestors of Diptera. – Nature, vol. 139. p. 66-67.

TOKUNAGA, M. 1932. A remarkable dipterous insect from Japan, *Nymphomyia alba, gen. et sp. nov.* – Annot. zool. japon., vol. 13, 5, p. 559-566.

TOKUNAGA, M. 1935. A morphological study of a Nymphomyiid fly. – Philippine J. Sci., vol. 56, 2, p. 127-214.

TOKUNAGA, M. 1936. The central nervous, tracheal and digestive systems of a Nymphomyiid fly. – Philippine J. Sci., vol. 59, 2, p. 189-216.

VIUSHKOV, B. P. 1951. Some questions of the evolution of toothed reptiles. – Advances of Contemp. Biol., Vol. XXXII, No. 2 (5): 271-280. (in Russian).

ZERNY, N., BEIER, M. 1936. Lepidoptera-Schmetterlinge. In: W. KÜKENTHAL. Handbuch der Zoologie, Bd. IV, 2, Insecta 2, 7-8. Berlin und Leipzig, S. 1554-1726.

ZIMIN, L. S. 1948. Finding of larvae of the synanthropic flies of Tadjikistan. Findings on the basis of the fauna of USSR. Pub. Acad. Sci. USSR, 28: 1-115. (in Russian).

ZIMIN, L. S. 1951. Present day flies (Muscidae). Fauna USSR, new series, No.

45, Insect Diptera, XVIII, 4: 1-286. (in Russian).
ZIMINA, L. V. 1960. New data on the system of the Conopidae based on the material of fauna of the USSR. — Zool. Jour., XXXIX, 5: 723-733. (in Russian).